Birkhäuser

Trends in the History of Science

Trends in the History of Science is a series devoted to the publication of volumes arising from workshops and conferences in all areas of current research in the history of science, primarily with a focus on the history of mathematics, physics, and their applications. Its aim is to make current developments available to the community as rapidly as possible without compromising quality, and to archive those developments for reference purposes. Proposals for volumes can be submitted using the online book project submission form at our website www.birkhauser-science.com.

More information about this series at http://www.springer.com/series/11668

Maria Teresa Borgato •
Erwin Neuenschwander • Irène Passeron
Editors

Mathematical Correspondences and Critical Editions

Birkhäuser

Editors
Maria Teresa Borgato
Department of Mathematics and Computer Science
University of Ferrara
Ferrara, Italy

Erwin Neuenschwander
Institute of Mathematics
University of Zurich
Zurich, Switzerland

Irène Passeron
Institut de Mathématiques de Jussieu-Paris Rive Gauche
CNRS, Sorbonne Université
Paris, France

ISSN 2297-2951 ISSN 2297-296X (electronic)
Trends in the History of Science
ISBN 978-3-319-73575-7 ISBN 978-3-319-73577-1 (eBook)
https://doi.org/10.1007/978-3-319-73577-1

Library of Congress Control Number: 2018955311

This book is published under the imprint Birkhäuser, www.birkhauser-science.com by the registered company Springer Nature Switzerland AG part of Springer Nature.
The registered company address is: Gewerbestrasse 11, 6330 Cham, Switzerland

Preface

Collected Mathematical Works and Correspondences: A Short History and an Introductory Summary

The publication of collected mathematical works and correspondences has a long and distinguished tradition. Apart from the editions of classical Greek and Roman works in the Renaissance, it started in the sixteenth and seventeenth centuries with the publication of, among others, the works of François Viète (*Opera mathematica*, London 1589, or Leiden 1646), Niccolò Tartaglia (*Opere*, Venice 1592/93 and 1606), Christoph Clavius (*Opera mathematica*, 5 vols, 1612), Simon Stevin (*Œuvres mathématiques*, 6 vols, Leiden 1634), and Galileo Galilei (*Opere*, 2 vols, Bologna 1655/56). Around the same time, the first editions of mathematical correspondences (*Commercium epistolicum* of John Wallis, 1658; and John Collins, 1712, etc.) appeared, and in the eighteenth century, there followed some twenty further editions of collected mathematical papers (Jacob and Johann Bernoulli, Pierre de Fermat, Bernard Lamy, Pierre Louis Moreau de Maupertuis, Jacopo Riccati, Gilles Personne de Roberval, Robert Simson, etc.).

The nineteenth and twentieth centuries saw a highpoint in the publication of multivolume and often nationalistically driven collected mathematical works. The most voluminous of these were the editions of the works of Leonhard Euler (ca. 100 volumes, 1911–), Gottfried Wilhelm Leibniz (ca. 50 volumes, 1923–), Augustin-Louis Cauchy (27 vols, 1882–1974), Johannes Kepler (26 vols, 1937–2017), Christiaan Huygens (22 vols, 1888–1950), Galileo Galilei (20 vols, 1890–1909), Paul Tannery (17 vols, 1912–50), Joseph-Louis Lagrange (14 vols, 1867–92), Pierre-Simon de Laplace (14 vols, 1878–1912), Arthur Cayley (13 vols, 1889–97), René Descartes (12 vols, 1897–1910), Nicolas de Condorcet (12 vols, 1847–49), Carl Friedrich Gauss (12 vols, 1863–1933), etc. The best of these editions include, besides the collected works, also unpublished papers, letters, commentaries, translations, biographies, bibliographies, etc., that are often unavailable anywhere else.

These developments led in the twenty-first century, the age of "Digital Humanities," to large-scale projects such as the D'Alembert edition with its new interface *D'Alembert en toutes lettres*, or platforms like *Circulation of Knowledge and Learned Practices in the 17th-century Dutch Republic*, which currently contains about 20,000 letters that were written by and sent to various scholars who were active in the Netherlands in the seventeenth century. Similar projects are *Early Modern Letters Online,* or even the Stanford project *Mapping the Republic of Letters*, which includes also other media besides printed works and correspondences. For a compilation and description of the approximately one thousand printed editions of collected mathematical works and correspondences published to date, we refer the reader to the "Collected works in Mathematics and Statistics Collection" of the Stanford University Libraries or to the comprehensive bibliography of Steven W. Rockey. The latter appeared in a print version in 1991 and is now available online in an updated version on the website of the Mathematics Library of Cornell University.

The present volume originates from a symposium held at the sixth ESHS Conference in Lisbon in 2014. It presents 16 mostly ongoing projects on editions of collected works and correspondences. The first group of papers deals with the edition of large-scale collected works and correspondences in the past, present, and future. Luigi Pepe gives a short summary of the publications in this field and describes in detail the history and scope of Lagrange's collected works. Eberhard Knobloch deals with the history of Series VII and VIII of the Leibniz edition, whereas Karin Reich and Elena Roussanova offer a critical survey and inventory of the edited works of Carl Friedrich Gauss. Irène Passeron and Alexandre Guilbaud describe the mathematical correspondence of D'Alembert and its digital edition, Sulamith Gehr presents the online edition of the Bernoulli letters, and Philip Beeley contributes a survey of the history of the nine-volume print edition of the correspondence of John Wallis.

A further group of papers considers the renewal of mathematical research in Italy at the time of the Risorgimento, the Italian unification around 1870. Maria Teresa Borgato and Iolanda Nagliati provide an overview of the creation and consolidation of a network of personal relations among Italian mathematicians and leading European scholars in this period. Other papers of this group contain detailed descriptions and evaluations of the correspondences of Giusto Bellavitis, Enrico Betti, Francesco Brioschi, Luigi Cremona, Placido Tardy, etc. The third and last group of papers presents a variety of other projects on European mathematical correspondences from different centuries. Nicolas Rieucau discusses the scientific correspondence of Condorcet, Erwin Neuenschwander describes the major correspondences of B.L. van der Waerden, and Catherine Goldstein and Scott A. Walter present different aspects of the correspondences of Hermite-Lipschitz and Hilbert-Poincaré. For further information about the sixteen contributions, we refer the reader to the following introduction and the abstracts of the authors.

The present volume of course cannot give a comprehensive overview of the vast topic of the publication history of collected works and correspondences over the last five centuries. Nevertheless, we hope that it will prove useful to future editors in accomplishing their task and that it will promote further historical work in important fields of study such as knowledge transfer and communication networks, where scientific, societal, and economic interests all come into play.

Zurich, Switzerland Erwin Neuenschwander

Introduction

Correspondences and editions of collected works: problems, situations, perspectives

Letter writing has always been very important for the spreading of scientific ideas, even in times of a great number of specialized journals.

The correspondences on mathematical issues or those of interest in the history of mathematics involve a vast field of topics, not only those of a scientific nature. They include letters between mathematicians and from mathematicians to politicians, publishers, and men and women of culture. Leibniz, Euler, D'Alembert, Lambert, Lagrange, Laplace, Gauss, Hermite, and Cremona are undoubtedly authors of great interest and their letters are precious documents, but the correspondence of less well-known authors can also make an important contribution to the history of science.

All of these kinds of correspondence constitute an essential component in the reconstruction of biographies, as well as the genesis of scientific ideas, in analyzing relations and debates and, ultimately, in the correct dating and interpretation of various memoirs. Their publication is, therefore, important for the success of critical editions of the works of great mathematicians (Galileo, Newton, Wallis, Huygens, Euler, the Bernoulli family, etc.).

In dealing with our subject, one must also take into account the varying editorial standards and formats for editions carried out in the past, especially in the nineteenth century, the most prolific period for collected works (Galileo, D'Alembert, Lagrange, Laplace, Huygens, Cauchy, Fourier, Weber, Gauss, Riemann, Kronecker, Dirichlet, etc.). They vary greatly in their presentation and structure; generally, they contain only printed works. At times, they are ordered chronologically, or according to discipline or type of publication. Only rarely are the correspondences, whether complete or partial, included in the edition.

Variety in editorial criteria is also to be found in twentieth-century editorial projects, some of which are still ongoing (Galileo *Edizione Nazionale*, Leibniz, Bernoulli, Brioschi, Betti, D'Alembert, ...) and are gradually being supported by digitalization processes. In fact, the digital editions make mathematical works of the past increasingly available to a wider public and facilitate the research process of scholars by allowing them to easily access and browse rare texts. This poses new problems in addition to those of the traditional printed editions, particularly in the choice of the target audience and corresponding suitable technical tools.

The editors of the present volume invited scholars to reflect on these topics in a symposium within the frame of the 6th International Conference of the European Society of History of Science entitled "Communicating Science and Technology," held in Lisbon, September 4–6, 2014. The topic generated considerable interest and the symposium on "Mathematical Correspondences and Critical Editions" was a great success in terms of participation and debate. Subsequently, a project aimed at collecting these contributions came into being, and other scholars were invited to intervene on the theme, since the publication of collected works and correspondences is of major interest in the field of the history of mathematics.

This volume contains sixteen contributions by various researchers from five different European countries. It offers a fairly broad spectrum, albeit partial, of the research being carried out, as well as the arguments under debate, such as the complementary role of printed and digital editions, integral and partial editions of correspondences, reproduction techniques of manuscripts, pictures and formulas, and tools for identifying dates and correspondents. These problems may involve different approaches according to the period and the subject, in this wide-ranging volume that focuses on correspondences and collected works from the seventeenth to the twentieth century with reference to all mathematical sciences.

Our intention was not to present a simple collection of various projects of editions, but rather to relate correspondences and works and compare the various types of edition, the problems encountered, and the solutions found to solve them. Of particular interest was the way in which the editions of correspondences and works should be linked and prioritized. For example, in the edition of Huygens, letters precede the works; for Lagrange, letters follow the works; and for Favaro's edition of Galileo, letters follow the works in the final volumes of the series, but were collected before and organized within a unique editorial plan. Important editions, however, like those of Laplace and Cauchy, do not contain the correspondence.

All of the contributions are related to editorial projects of correspondences or collected works. In some cases, the papers deal with projects of print editions (Leibniz, Wallis, Lagrange, Gauss) or online or mixed editions (Bernoulli, D'Alembert, Poincaré). In other cases, they refer to a correspondence between two mathematicians, relevant for specific mathematical contents (Germaine–Gauss, Betti–Brioschi, Hermite–Lipschitz), or are aimed at reconstructing a particular period for the history of mathematics (Cremona, Tardy), or a network of relations (D'Alembert, van der Waerden). Other articles discuss policy and methods for dating letters and discovering unknown correspondents (D'Alembert, Condorcet, ...), or critically examine previous non-satisfactory editions (Lagrange, Gauss).

It is not our aim to create an exhaustive discussion of the best method for producing an edition, which depends on many variables, such as the historical period and range of correspondences, the multiplicity of correspondents and overlapping with other editions, as well as the contents and target audience. We believe, however, that a volume that allows us to compare various situations by presenting a reasonably wide picture may be a publication that arouses considerable interest for many scholars of the history of mathematics.

The very first article of the volume, for instance, poses fundamental questions regarding editions of correspondences: should they be complete or partial? Should they feature only unedited material? Should previously published material be included? In the case of a partial publication of selected letters, on what basis should the criteria be chosen: subject (scientific, political, private etc.), importance, or correspondents? Whatever the choice, a census of all existing documents must be as thorough as possible.

The author of the first article, Philip Beeley, describes the stages that led to the edition of the correspondence of John Wallis, beginning with Christoph J. Scriba's research carried out at Oxford in the early 1960s. Scriba's cultural and methodological background originated from the Leibniz edition carried on by the German Academy of Sciences at Berlin, according to the guidelines of the Leibniz scholar Joseph Ehrenfried Hofmann. Beeley's paper outlines Scriba's profound and systematic investigation into Wallis's manuscripts, letters, and other materials, at Oxford, Cambridge, London, and Vienna, which produced a whole series of card catalogues and a list of Wallis's correspondence of 800 or so letters. From the initial idea of publishing only a significant selection of Wallis's letters, of interest for the history of science in general or the history of mathematics in particular, the project went in a new direction with the discovery of up to over 2000 new letters. After 30 years, this led to a collection of up to ten volumes, the first four of which were published from 2003 to 2014, with the fifth currently being printed. Philip Beeley entered the project when he was a doctoral student at the Technische Universität of Berlin, taking over as the successor of the previous collaborator, Sigmund Probst, at the beginning of November 1996. Together with Scriba, and after Scriba's death, he acted as editor of John Wallis's entire correspondence in chronological order, complete with a critical analysis and introductory essays on the themes discussed in the letters. Beeley explains, in his contribution, the various choices that had to be made during the course of the project, due to the development that took place within the methodology of the historiography of science in the last decades of the previous century, shifting from an internal historiographical approach to a more general survey of the history of ideas.

Eberhard Knobloch's essay introduces us to one of the biggest edition projects ever planned: the Leibniz edition, which cannot possibly be described in a few pages. It includes more than 50 published volumes of the expected 130 and has been a reference point for other edition projects. Eberhard Knobloch provides a detailed examination of its VII series, modified in 1975, and exclusively devoted to the manuscripts concerning mathematics (30 volumes), whereas the scientific, medical, and technical writings were to be published in a new series, the VIII, which was to follow. In 1976, Knobloch was assigned the editorial work on the first two volumes of Series VII, the first to be completed within 10 years, and the second within the following 5 or 7 years, with the help of a research assistant, Walter S. Contro.

Knobloch describes the difficulties and events involved in that edition, which accompanied the increasingly professional and academic growth of the young but already experienced researcher, and which were affected by the period of unification of Germany. One of the first and foremost tasks was the identification and dating of Leibniz's writings during his stay in Paris, which were to fill eight volumes of Series VII. Owing to the fact that most of the handwritten manuscripts are not dated and that dating the manuscripts to a period of exactly half a year is generally not possible, it became clear that a strict chronological ordering of the handwritten manuscripts was impracticable. The solution came in the form of defining thematic groups for the entire Parisian period and developing a chronological order within these groups. In the case of volume VII, 1, the following groups were identified: geometry, number theory, and algebra.

With the assistance of various colleagues, Knobloch was able to publish the first four volumes of Series VII between 1990 and 2008 (the series was then completed in two other volumes). Knobloch was mainly involved in Leibniz's writings on actuarial and financial mathematics, also published in Series IV, Political Writings (IV, 4 (1680–1692), section VII Statistics, Life Insurance, Pensions). Knobloch's subsequent involvement in Series VIII further demonstrated how important it is for a historian of sciences and editor of old scientific texts to have not only the required scientific knowledge but also philological skills, since a lack of either one of these could lead to serious misinterpretations. A project of such dimension also revealed the need for international cooperation among scholars and institutional agreements with other countries (France and Russia in this case), as well as adequate funding. Digitalization of Leibniz's manuscripts was carried out first. Series VIII, leaving aside previously attempted editions of some manuscripts containing serious mistakes, includes up to ninety percent of unedited material. Knobloch points out these errors and presents significant examples (various apparatuses and instruments linked to pneumatics and mechanics) of Leibniz's procedure in the field of Natural Sciences and Technology. The first two volumes of Series VIII appeared in 2009 and 2016.

Sulamith Gehr describes the history of another celebrated edition project, namely, the publication of the letters of the mathematicians of the Bernoulli family. The project started in the 1930s in the form of a classical book edition and continued as an online edition in the past decade.

The Bernoulli family's correspondence includes a vast network of over 400 correspondents, among whom can be found the foremost scientists of the seventeenth and eighteenth centuries, and, as such, presents various problems of classification and organization. Gehr's work reminds us of the important role played by epistolary commerce in the transmission of knowledge and scientific debate in past centuries. In the specific case of the Bernoulli family, there is an important historical precedent: the publication edited by Gabriel Cramer of the *Commercium philosophicum et mathematicum*, which appeared in 1745 and featured the correspondence between Johann I Bernoulli and Gottfried Wilhelm Leibniz, following, as it did, just a few years after the publication of the *Opera omnia* by Johann I Bernoulli in 1742 may be considered as a completion of it. Further plans of epistolary publication followed, namely, the idea of publishing the correspondence between Johann I Bernoulli and the Marquis de L'Hôpital, begun by Johann III Bernoulli but never concluded. In the nineteenth century, the letter

Asian Beekeeping in the 21st Century

Panuwan Chantawannakul
Geoffrey Williams · Peter Neumann
Editors

Asian Beekeeping in the 21st Century

Springer

Editors
Panuwan Chantawannakul
Faculty of Science
Chiang Mai University
Chiang Mai, Thailand

Geoffrey Williams
Department of Entomology & Plant Pathology
Auburn University
Auburn, Alabama, USA

Peter Neumann
Institute of Bee Health
University of Bern
Bern, Switzerland

ISBN 978-981-10-8221-4 ISBN 978-981-10-8222-1 (eBook)
https://doi.org/10.1007/978-981-10-8222-1

Library of Congress Control Number: 2018940007

Printed on acid-free paper

This Springer imprint is published by the registered company Springer Nature Singapore Pte Ltd.
part of Springer Nature
The registered company address is: 152 Beach Road, #21-01/04 Gateway East, Singapore 189721, Singapore

Preface

Asia is regarded as the center of honey bee diversity. Both European and Asian honey bees not only produce honey and other bee products (royal jelly, bee pollen, bee wax, and propolis), but also play a vital role in maintaining the local fauna and sustaining agricultural crops. Species distributions of most honey bees overlap in southeast Asia. This likely increases the potential for interspecies transmission of pests and parasites, and their spread to other parts of the world by human translocation. The decline of honey bee populations is of great concern. Global colony losses of the European honey bee are believed to be caused, in part, by pests and parasites originating from Asia, such as the mite *Varroa destructor*, the microsporidian *Nosema ceranae*, and some bee viruses. This book provides insights to readers from local researchers concerning the history of beekeeping, current bee diversity, bee and flora, development of modern beekeeping, as well as setbacks caused by bee diseases and parasites across different geographical areas in Asia.

We thank Dr. Alvin Yoshinaga who provided comments on several chapters in this book, as well as the students in the Bee Protection Laboratory, Faculty of Science, Chiang Mai University (Ms. Sasiprapa Krongdang, Mr. Chainarong Sinpoo, Ms. Patcharin Phokasem, Ms. Wannapha Mookhploy, and Dr. Jakkrawut Maitip) who assisted us during the compilation of this book. Panuwan Chantawannakul would like to thank Prof. Dr. Siriwat Wongsiri who inspired her to do bee research in Thailand; she also thanks Prof. Dr. Tawee Tunkasiri, Prof. Dr. Saisamorn Lumyoung, and Assistant Prof. Dr. Sakunee Bovonsombut, and Assistant Prof. Dr. Nopasit Chakpitak for providing strategic guidance throughout her career. P.C acknowledges Thailand Research Fund (RSA6080028), National Research Council of Thailand, Chiang Mai University fund, and Thai government fund. We also thank the International College of Digital Innovation for the "Smart Insect" project. Geoffrey Williams would like to acknowledge support by the Swiss National Science Foundation Project 31003A_169751. We also are thankful to all our research collaborators, COLOSS friends (prevention of honeybee COlony

LOSSes Honey bee research association, http://www.coloss.org), and the University of Bern, Switzerland, which provided facilities during early versions of the book.

We also owe our appreciation to our families and friends for supporting us during our work on bee research, and this book in particular.

Chiang Mai, Thailand — Panuwan Chantawannakul
Auburn, USA — Geoffrey Williams
Bern, Switzerland — Peter Neumann

Contents

1 **The Overview of Honey Bee Diversity and Health Status in Asia** 1
Panuwan Chantawannakul and Samuel Ramsey

2 **Beekeeping in Turkey: Bridging Asia and Europe** 41
Aslı Özkırım

3 **Beekeeping in Parts of the Levant Region** 71
Nizar Haddad and Lisa Horth

4 **Apiculture in Israel** 95
Victoria Soroker, Slabezki Yossi, and Nor Chejanovsky

5 **Beekeeping and Honey Hunting in Nepal: Current Status and Future Perspectives** 111
Ratna Thapa, Sunil Aryal, and Chuleui Jung

6 **Current Status of the Beekeeping Industry in China** 129
Huoqing Zheng, Lianfei Cao, Shaokang Huang, Peter Neumann, and Fuliang Hu

7 **Beekeeping on Taiwan Island** 159
Mei-Chun Lu

8 **Beekeeping in Korea: Past, Present, and Future Challenges** 175
Chuleui Jung and Myeong-lyeol Lee

9 **Beekeeping in Mongolia** 199
Khaliunaa Tsevegmid, Selenge Dooshin, Samuel Ramsey, and Panuwan Chantawannakul

10 **Bee Diversity and Current Status of Beekeeping in Japan** 223
Mikio Yoshiyama and Kiyoshi Kimura

11 **Beekeeping in Vietnam** 247
Pham Hong Thai and Tran Van Toan

12 Bee Diversity and Current Status of Beekeeping in Thailand 269
Panuwan Chantawannakul

13 Social Bees and the Current Status of Beekeeping in Indonesia 287
Sih Kahono, Panuwan Chantawannakul, and Michael S. Engel

14 Management and Conservation of Philippine Bees 307
Cleofas R. Cervancia

15 Conclusions and Future Perspectives 323
Panuwan Chantawannakul, Samuel Ramsey, Geoffrey Williams, and Peter Neumann

Chapter 1
The Overview of Honey Bee Diversity and Health Status in Asia

Panuwan Chantawannakul and Samuel Ramsey

Abstract Traditional honey bee hunting and beekeeping are vital to the economic and spiritual lives of Asians. Bee products are known as not only food/food supplement but also traditional medicine for aiming to promote good health, especially in eastern regions. Honey bees also play crucial roles in pollination. Asia is regarded as the homeland of honey bees as it hosts at least nine honey bee species. The European honey bee was introduced from Europe, North America, and Oceania to Russia, Japan, India and other countries in Asia. The growth of global human population size, globalized trade economic wealth, and technological developments in transportation efficacy has promoted the transmission of bee diseases, parasites and pests. A great concern over honeybee population decline has accelerated research in bee diseases, parasites, and pests. This chapter provides an up-to-date information on bee diseases, parasites, and pests in Asia.

Keywords Asian honey bee • Thai bee • *Apis dorsata* • Thai honey • *Apis cerana*

1.1 Introduction

Honey bees provide invaluable ecosystem services throughout Asia. Agricultural ecosystems, in addition to tropical and mountainous regions, depend on their ability to pollinate a large variety of crops and wild plants. Furthermore, their production

P. Chantawannakul (✉)
Department of Biology, Faculty of Science, Chiang Mai University, Chiang Mai, Thailand

International College of Digital Innovation, Chiang Mai University, Chiang Mai, Thailand

Environmental Science Research Center, Faculty of Science, Chiang Mai University, Chiang Mai, Thailand

S. Ramsey
Department of Entomology, University of Maryland, College Park, MD, USA

P. Chantawannakul et al. (eds.), *Asian Beekeeping in the 21st Century*,
https://doi.org/10.1007/978-981-10-8222-1_1

of honey, brood, propolis, wax, bee pollen, royal jelly, and bee venom provision people with food, nutritional supplements, and traditional medicinal treatments. Dramatic losses in European honey bee populations in recent years have prompted interest in Asian honey bee research. These losses are attributed, in large part, to parasites, pests, and pathogens that Asian honey bees have survived and thrived with for millennia.

At least nine species of honey bees are native to Asia making it the homeland of honey bee diversity. The European honey bee (*Apis mellifera*) is also found in Asia as an introduced species. Inhabitants of Asia have had a very close association with these insects for generations with records of native beekeeping going back more than 2000 years (Chantawannakul et al. 2016). Any quest to better understand the intricacies of honey bee biology should begin here.

1.2 *Apis* Species Diversity

The genus *Apis* is divided into three sub-genera: *Apis* (the cavity-nesting bees), *Micrapis* (the dwarf honey bees), and *Megapis* (the giant honey bees). Cavity-nesting honey bees (*Apis cerana*, *Apis nigrocincta*, *Apis koschevnikovi*, and *Apis nuluensis*) form the most widespread, diverse group of the three assemblages. *A. cerana* appears to be composed of at least two species and researchers have proposed dividing it into *A. cerana* and *Apis indica* (Arias and Sheppard 2005; Lo et al. 2010). The cavity-nesting species are characterized by their behavior of building multiple combs within protected structures like tree hollows or boxes provided by humans. Honey bee diversity generally is highest in the tropical areas of Asia. The evolutionary shift towards cavity-nesting allowed for these species to expand their geographic range to several environments other than the tropics.

By contrast, the dwarf honey bees (*Apis andreniformis* and *Apis florea*) build single comb nests in open air as do the giant honey bees. The inherent exposure to the elements limits their range to tropical and subtropical regions. Dwarf honey bees tend to build comb around the branch of a tree and will abscond leaving it and its contents behind if threatened. Land development and accompanying deforestation have reduced the distribution of these species in recent years.

Giant honey bees (*Apis dorsata* and *Apis laboriosa*) are known for building their large, single frame colonies at high elevations (especially *Apis laboriosa*). Typically, these colonies are constructed on cliff faces, under tree branches, or on large man-made structures such as apartment buildings or water towers. The comb can be more than a meter across (Oldroyd and Wongsiri 2006). Their stores of honey and brood attract opportunistic human honey-hunters as well as a host of specialist pests both vertebrate and invertebrate. A third species was proposed for this sub-genus in *Apis breviligula* of the Philippines (Lo et al. 2010).

1.3 Interactions with Humans

Humans in Asia have been associated with honey bees for millennia. For much of this history, humans were opportunistic honey-hunters. Honey was the only sweetener available through most of human history, and as such was highly sought after. Ancient depictions of honey hunting have been discovered on rock faces in India dating back perhaps 8000 years (Mathpal 1984). Ancient people are depicted scaling tall trees with ladders or rope, sheering comb with long sharpened branches, and collecting it in baskets. There is also evidence that people would annually stake claim to *A. dorsata* nests, the earliest of which comes from ancient China (between 265 and 290 AD).

Management of honey bees in hives was first developed using *A. cerana* and occurred as far back as 300 BC (Chantawannakul et al. 2016). Western Asians in countries like Afghanistan and Pakistan kept these colonies in hollow logs, clay pots, or straw baskets. More recently, the preferred species for beekeeping, both in Asia and abroad, has shifted to the more commercially profitable *A. mellifera.* This species produces, on average, 15 times more honey per year than *A. cerana* (Chantawannakul et al. 2016). In addition, its disinclination to abscond and general ease of management further precipitated this shift. However, keeping *A. cerana* is not without its merits. This species requires less overall investment than *A. mellifera.* It is also more resistant to parasites and disease than the European honey bee. This has not escaped the notice of enterprising beekeepers in Asia. Large-scale beekeeping operations utilizing *A. cerana* can be found in temperate areas of China and India (Chantawannakul et al. 2016). The subspecies of *A. cerana* found in these areas has a decreased likelihood to abscond than other subspecies in Asia. Adoption of native beekeeping has been met with resistance in other areas of Asia with Malaysia being one of the last countries to adopt native beekeeping in 1936 (Crane 1999).

Though not as well adapted to the biotic or abiotic factors common to Asia, *A. mellifera* is kept to a far greater extent than *A. cerana.* No other area of the world accommodates more managed *A. mellifera* colonies than Asia (FAO 2014). Colonies were widely imported first to Russia from North America, Europe, and Oceania in the late 1700s. India, Indonesia, and Japan followed in the 1800s. Almost every Asian country accommodated *A. mellifera* by the 1980s (Arai et al. 2012; Sanpa and Chantawannakul 2009; Crane 1999; Wu et al. 2006).

While honey bees are revered by people groups across Asia, anthropogenic factors are proposed as the primary driver in their potential decline (Oldroyd and Wongsiri 2006). Little study has focused on the health of populations of native honey bees, likely due to their migratory nature, but researchers and honey-hunters alike have anecdotally reported sharp declines in native honey bees. Honey hunting, competition with *A. mellifera,* pesticide exposure, interspecific transfer of novel pathogens, deforestation, and anthropogenic climate change have all been cited as stress factors. There is evidence to suggest that these declines are less pronounced than those of *A. mellifera* in recent years but the implications are no less concerning (Li et al. 2012a; van der Zee et al. 2012; Yang and Cox-Foster 2005).

1.4 Health Stressors

Honey bee colonies comprise a dense aggregation of highly related individuals making a single colony a perfect breeding ground for parasites and disease. This has likely contributed to the emergence of social grooming and other hygienic behaviors common to eusocial organisms. Absconding and seasonal migration are also important in reducing the intensity and duration of infection. These behaviors help to maintain manageable or asymptomatic levels of parasites and pathogens within colonies. It is usually when colonies are stressed (because of inadequate nutrition, pesticide exposure, environmental issues, etc.) or exposed to invasive organisms that they lose their ability to properly defend against these issues. Subsequently, disease or parasitosis can then develop.

1.4.1 Mites

Mites (Subclass: Acari) have evolved to efficiently exploit nearly every ecological niche on this planet owing in no small part to their rapid reproductive rate and staggering adaptability (Walter and Proctor 1999). Over millions of years of evolutionary history, four genera *Varroa* (Varroidae, Mesostigmata), *Euvarroa* (Varroidae, Mesostigmata), *Tropilaelaps* (Laelapidae, Mesostigmata), and *Acarapis* (Tarsonemidae, Prostigmata) have developed intimate association with honey bees to the extent that they cannot exist apart from them. Asian honey bees are the original host to the two major groups, the *Varroa* and *Tropilaelaps,* as well as the less economically important *Euvarroa. Varroa* are primarily parasites of the cavity-nesting bees. The *Euvarroa* specifically parasitize dwarf honey bees and the *Tropilaelaps* originally specialized in parasitizing the giant honey bee species. The most pathogenic of the genus *Acarapis* appear to primarily parasitize *A. mellifera* but have also been found in association with Asian honey bee species.

All of these parasitic mites likely started as opportunistic predators of pollen scavenging mites within cavity-nesting honey bee colonies or as phoretic mites feeding opportunistically on host secretions, later shifting to an obligate parasite lifestyle (Walter and Proctor 1999). The cavity-nesting bees still play host to the greatest diversity of mite species (both pathogenic and nonpathogenic). The existence of these parasites in sympatry with multiple honey bee species potentially promotes interspecific exchange of parasites between them.

1.4.1.1 *Varroa*

Four mite species have been described within the genus *Varroa*: *Varroa jacobsoni* (Oudemans 1904), *Varroa underwoodi* (Delfinado-Baker and Aggarwal 1987), *Varroa rindereri* (de Guzman and Delfinado-Baker 1996), *Varroa destructor*

(Anderson and Trueman 2000). All four species are obligate parasites of honey bees. *V. underwoodi* and *V. rindereri* were found parasitizing *A. cerana* in Nepal and *A. koschevnikovi* in Borneo, respectively (Delfinado-Baker and Aggarwal 1987; de Guzman and Delfinado-Baker 1996). The emergence of a *Varroa* species, presumably *V. jacobsoni,* destroying colonies worldwide prompted more study of the genus. With further evaluation, it became apparent that the species being called *V. jacobsoni* was actually composed of two distinct species and the one threatening bees around the world was not *V. jacobsoni* but a new species aptly named *"V. destructor"*. This new parasite had achieved a nearly cosmopolitan distribution and is presently found in virtually every *A. mellifera* colony around the world excluding Australia and small outlying island territories (Rosenkranz et al. 2010).

All species of *Varroa* successfully parasitize cavity-nesting honey bees. *Varroa jacobsoni* has the widest host range parasitizing all five cavity-nesting honey bee species (de Guzman and Delfinado-Baker 1996; Delfinado-Baker et al. 1989; Koeniger et al. 2002; Woyke 1987; Otis and Kralj 2001). By contrast, *V. destructor* has only been recorded in *A. cerana* and *A. mellifera* colonies (Anderson and Trueman 2000). *Varroa underwoodi* is restricted to *A. cerana, A. nigrocincta,* and *A. nuluensis.* It is found throughout most of the range of *A. cerana* where it frequently co-infests colonies with *V. jacobsoni or V. destructor* (Oldroyd and Wongsiri 2006). *Varroa rindereri* distinguishes itself as the most species specific, found thus far only infesting *A. koschevnikovi* though specimens have been collected from hive debris of *A. dorsata in* Borneo (Koeniger et al. 2002).

Varroa are relatively large for mites. They are visible with the unaided eye though analysis of detail requires magnification. The adult females are dark red in color. The idiosoma is covered entirely by a convex dorsal shield that is ellipsoid in shape being wider than long. The attachment points of their appendages are obscured by this protective shield when viewed from above.

Distinguishing between *Varroa* species can be difficult, owing to their small size and morphological similarity. There are, however, several characters such as body size, shape, peritreme dimensions, and, to some extent, chaetotaxy (arrangement, size, and number of hairs on an organism) that can be of use in determining species. Size appears to be the most helpful characteristic across species. *Varroa underwoodi* is arguably the easiest to recognize. This species is the smallest on average across haplotypes (average length = 0.745; average width = 1.146 mm) and has a more triangular shape than its congeners when viewed ventrally. It also has several setae along the margin of both sides of the dorsal shield that are noticeably longer relative to its size than those of other *Varroa* species (de Guzman and Rinderer 1999; Delfinado-Baker and Aggarwal 1987). *Varroa rindereri* is the largest of the four species, though only slightly larger than *V. destructor,* with average measurements of length = 1.180 and width = 1.698 mm. Morphologically, it appears very similar to *V. destructor* aside from having four more marginal setae on the edge of the dorsal shield and one fewer seta on the sternal shield (de Guzman and Rinderer 1999).

As of yet, no consistent morphological characters have been determined to distinguish between *V. jacobsoni* and *V. destructor* other than size. *Varroa destructor* are visibly larger (average length = 1.167; average width = 1.709 mm) than *V. jacobsoni* (average length = 1.063; average width = 1.552 mm) (Anderson and Trueman 2000). *Varroa destructor* usually appear more ovular than the more rounded *V. jacobsoni*. Wide variation exists within and between *V. destructor* and *V. jacobsoni* populations in Asia and appears to play an important role in the pathogenicity of these parasites. Presently, 24 haplotypes are recognized, 15 for *V. jacobsoni* and 9 for *V. destructor* (Anderson and Trueman 2000; de Guzman and Rinderer 1998, 1999; Warrit et al. 2006; Navajas et al. 2010; Solignac et al. 2005; Fuchs et al. 2000; Zhou et al. 2004). The Korean (K) haplotype is the most successful and virulent being distributed throughout the geographic range of *A. mellifera*. The second haplotype, Japan/Thailand (J), is competitively inferior to the K haplotype and has a limited distribution encompassing Japan, Thailand, and the Americas (Anderson and Trueman 2000; de Guzman 1998; Garrido et al. 2003; Munoz et al. 2008). Recent genetic analysis of *Varroa* in Asia revealed hybridization between haplotypes and immigration into the population. More work is needed to determine the virulence of these hybrid mites.

The increased virulence and competitive success of *V. destructor* is due in large part to the ability of two of its haplotypes to exploit both worker and drone brood for reproduction. For reasons that are not yet fully understood, *V. jacobsoni* appears unable to reproduce in worker brood of either *A. cerana* or *A. mellifera* (Anderson 1984; Anderson and Fuchs 1998; Anderson and Sukarsih 1996). Drone brood is seasonally produced and relatively less significant to the healthy functioning of the colony thus limiting the reproductive success of this parasite and its impact on the colony. In *V. destructor,* the ability of the parasite to exploit worker brood is likely dependent primarily on its haplotype (De Jong 1988). In the Philippines, Papua New Guinea, and Java, three unresolved haplotypes of *V. destructor* reproduce only in *A. cerana* drone brood (Anderson 2004).

Variable reproductive and infestation rates are proposed to be a result of interaction between bee behavior and the haplotype of the mite. *A. cerana* are well known for their ability to effectively remove mites from adult bees (Peng et al. 1987a, b). This prevents species like *V. destructor* which have been shown to feed on adult bees during their phoretic phase (Ramsey 2018) from further damaging worker populations. It also removes individual mites from the population before they are able to reproduce (either for the first time or again). Hygienic behavior, also observed widely in *A. cerana*, decreases the reproductive output of the mite population when infested developing bees are removed from brood cells (Boecking and Spivak 1999; Boot et al. 1997; Fries et al. 1996a; Peng et al. 1987b; Büchler et al. 1992). However, grooming alone does not account for why *A. mellifera*, which does not exhibit these behaviors as consistently, does not have its worker brood exploited by *V. jacobsoni*. Further, *V. destructor* have been observed periodically entering a worker cell with a viable larva yet producing no offspring which suggests heretofore undiscovered mechanisms of host resistance (Rosenkranz et al. 2010; Locke 2016; Strauss et al. 2015).

Life Cycle

The success of *Varroa* as honey bee parasites is a direct result of the high level of synchrony between the mite's life cycle and that of the host. It is noteworthy, however, that most of what we know of *Varroa's* life cycle is based on studies of *V. destructor* on *A. mellifera* not on its original host *A. cerana*. Further study is necessary to determine other intricacies in and deviations from the standard *Apis/Varroa* relationship.

Varroa's life cycle consists of two phases: the reproductive and the phoretic phase. In the reproductive phase, a mated female called the "foundress mite" invades a fifth instar brood cell (worker or drone brood are acceptable hosts though drone brood is preferred). She invades this cell between 1 and 20 h of capping (with 1–3 h before capping being the period of peak attractiveness in worker brood (Beetsma et al. 1999)). She quickly climbs across the larva and embeds herself in any brood food remaining at the bottom of the cell. About 10 h after the capping of the cell, the larva consumes the remaining brood food and begins to spin a cocoon. The foundress mite then climbs onto the larva and begins to feed on it.

Researchers have long believed that *Varroa* feed exclusively on the hemolymph of immature and potentially adult bees but recent work has shown the target food source to be the abundant fat body tissue present in both adult and immature bees (Ramsey 2018). Feeding on this nutrient-dense host tissue, the foundress mite produces a large egg representing a sizeable portion of her body weight every 30 h. Each egg develops directly into the nymph stage (skipping the larval stadium entirely). The first egg always develops into a male while subsequent eggs all develop into females. The nymphs grow at a remarkable rate ostensibly consuming the same host tissue, through the same wound as the mother. They reach sexual maturity after about 5 days with males reportedly developing slightly faster. Assuming one foundress mite enters the cell, the male produced will mate with his sisters. Any female mite having reached maturity by the time the host honey bee reaches maturity will emerge with it as it chews its way out of the capped cell. Any mites that have not reached maturity before the adult bees emerge die within the empty cell.

What follows for the mated females is the phoretic phase. *Varroa* attach themselves to worker bees and can potentially be ferried to other colonies. Mites prefer to attach to middle-aged nurse bees and can be found on foragers as well but prefer not to attach to newly emerged bees. They are difficult to observe in this stage resulting from their tendency to wedge themselves between the overlapping plates on the underside of the bee's abodmen (or metasoma). Their cryptic nature during this stage and the erroneous expectation that they should be conspicuous can lead beekeepers, when evaluated via visual inspection, to conclude that they have little to no *Varroa* in their colonies until populations grow out of control. The duration of this stage is variable, shown to last anywhere between 1 and 13 days (usually around 4–5 days but can be much longer when brood is not present to initiate the reproductive phase as in winter periods (Beetsma et al. 1999; Martin 1998)). Most *Varroa* females go through reproductive cycles more than once. Two to three times appears to be the average though as many as eight have been observed (Fries et al. 1996b; Martin and Kemp 1997; Martin et al. 1997).

Damaging Effects

While the genus as a whole is very well specialized to develop on honey bees, the tendency of *V. destructor* to lead to the mortality of its host implies an incipient evolutionary relationship. The (K) haplotype of *V. destructor* more closely resembles a parasitoid than a parasite as its success almost always results in the death of its host. Much of the damage attributed to *Varroa destructor* infestation is a result of the viruses vectored by these parasites (i.e., deformed wings, paralysis, shortened life spans); however, a number of deleterious effects associated with *Varroa* feeding may be associated with the process of feeding itself. *Varroa* are known to weaken the immune system of their host, promote precocious foraging, decrease flight performance, decrease vitellogenin titers, increase overwintering mortality, reduce metabolic rate, increase water loss, and to reduce the ability of the host to store protein. This diverse array of pathologies may result from this parasite feeding on fat body tissue as it is involved with all of these functions.

Advanced *Varroa* infestations can be identified by conspicuous symptoms. Capped brood cells are often opened prematurely (a condition called "bald brood") and the immature bee inside chewed down. Spotty brood patterns and high proportions of bees with deformed wings are also typical in heavily infested colonies. This pathology is often referred to as Parasitic Mite Syndrome (PMS). If left untreated, infested colonies usually die within 1–3 years of infestation. Fries et al. (2003) and Rosenkranz et al. (2006) both reported that a colony with 30 mites per hundred bees or greater during the summer has no chance of surviving the winter.

Detection and Management

Early detection of *Varroa* by sight can be difficult because of their cryptic nature. For this reason, monitoring methods have been developed to assess the size of mite populations. One such method is the "sugar roll". A half cup of approximately 300 bees is coated with powdered sugar. The powdered sugar dislodges the mites by disrupting purchase between the mite's ambulacrum (or foot pad) and the host's exoskeleton. The mites fall off of the host or can be easily groomed off. From there they can then be shaken from the cup of bees through a fine mesh screen onto a lightly colored surface. Lightly misting the surface with water will dissolve the powdered sugar allowing for the mites to be easily counted. More than nine mites, or three per hundred bees, is the threshold at which neglecting to treat would likely lead to irreversible damage to the colony. The same count can be accomplished using alcohol or soapy water in lieu of powdered sugar; however, the powdered sugar method is preferred because the bees are relatively unharmed by the experience and can be returned to the colony. Mite populations can also be monitored using screened bottom boards in tandem with sticky boards beneath the colony to quantify the levels of mites that naturally fall from bees within the colony. As the population in the colony grows, the number of mites falling on to the sticky board should increase proportionally. It is recommended that beekeepers monitor their mite levels regularly especially before and after any treatment to verify the efficacy of the treatment method.

Since *A. mellifera* colonies are so negatively impacted by the presence of *Varroa*, treatment is required to lower mite populations to manageable levels. The organophosphate coumaphos (Checkmite+), pyrethroids tau-fluvalinate (Apistan) and flumethrin (Bayvarol), and the amidine (Amitraz) are commercially available synthetic acaricides (Rosenkranz et al. 2010). These chemical measures are simple and cost-effective but can taint wax and contaminate honey. It is recommended that they be used well before or after the nectar flow in that region. It is also advisable for the beekeeper to remove honey supers from the colony prior to applying treatment.

Naturally occurring compounds such as organic acids and essential oils (formic acid, oxalic acid, lactic acid, and thymol) are applied as alternatives to the synthetic acaricides because of consumer concern about the contamination of hive products. However, these products are volatile compounds and as such, their effectiveness is variable depending on temperature and humidity. A successful attempt to control evaporation rate was made in the tropics with lemon grass oil and porous ceramics used to limit evaporation. The porous ceramics were able to control the level of the lemon grass oil volatiles in the hive environment for a defined period of time; in this case, 1 week for one dose (Chantawannakul et al. 2016). This method was not only effective at controlling *Varroa,* but reduced populations of *Tropilaelaps* and inhibited growth of Chalkbrood disease in *A. mellifera* colonies. This method has only been tested in the field in Thailand, however, and still requires further evaluation in other climates and temperatures.

Several other methods of mite remediation have shown promise and would help to further diversify integrated pest management (IPM) efforts to control this parasite. Attempts at biological control employing entomopathogenic fungi (*Beauveria, Metarhizium, or Verticillium)* appear to have potential as acaricides but have seen little success as control measures for *Varroa* (Chandler et al. 2000; Shaw et al. 2002; Meikle et al. 2012). Neem extract and other herbal plant products have also been proposed for use as acaricides but have never been tested widely for efficacy. Breeding for mite resistance and the usage of RNAi both appear to be promising prospects and are still being evaluated for efficacy. A *Varroa* sex pheromone incorporated into the wax foundation reportedly caused a 20% reduction in the reproductive rate of mites within the colony. In addition, 10–25 mM of lithium chloride, a known miticide, can kill mites within 2–3 days of feeding.

1.4.1.2 *Tropilaelaps*

The genus *Tropilaelaps* comprises four species. *Tropilaelaps clareae* was the first to be described. Surprisingly, the original specimens were collected from field rats and dead *A. mellifera* near *A. mellifera* colonies in the Philippines (Delfinado and Baker 1961). It took nearly a decade for the actual host of this species to be correctly identified as *A. dorsata* (Bharadwaj 1968; Laigo and Morse 1968). Twenty years later, *Tropilaelaps koenigerum* was described from specimens collected in Sri Lanka, India, Thailand, and Borneo (Delfinado-Baker and Baker 1982). In 2007, two more species were added to the genus in *Tropilaelaps mercedesae* and *Tropilaelaps thaii*. Molecular analysis revealed, similar to *V. jacobsoni*, that *T. clareae* was actually two distinct species; the second being named *T. mercedesae*. All four species are parasites of giant honey bees in varying regions of Asia but several strains of *T. mercedesae* and *T. clareae* have shifted to parasitizing *A. mellifera* to devastating effect (de Guzman et al. 2017).

There are 26 known haplotypes of *T. mercedesae* with 20 parasitizing *A. dorsata* and 6 on *A. mellifera*. Sixteen haplotypes are currently recognized for *T. clareae* with four on *A. mellifera* and 12 on *A. dorsata* (Anderson and Morgan 2007). Before the recognition of *T. mercedesae* as a separate species, *T. clareae* was considered to be a widespread parasite of *A. mellifera* far beyond its original host's geographic range (Indonesia, Philippines, Malaysia, Laos, Vietnam, Thailand, Burma, Bhutan, Nepal, Sri Lanka, India, and Pakistan). They could be found in South Korea, Taiwan, China, Hong Kong, Iran, Afghanistan, Kenya, and the western Pacific island of New Guinea (Anderson and Morgan 2007). Researchers have shown cautious optimism that the spread of *Tropilaelaps* beyond tropical and subtropical regions will be limited by its need for constant brood production (Woyke 1985). However, the establishment of *Tropilaelaps* in South Korea and some areas of China calls this perspective into question as honey bee brood is limited or absent altogether during the winter common to this region (de Guzman et al. 2017). This suggests that *Tropilaelaps* may have similar capacity to survive in temperate climates to that of *V. destructor*. Further, *Tropilaelaps* found in Kenya show the ability of this parasite to survive lengthy intercontinental transit (Kumar et al. 1993).

Tropilaelaps mercedesae appears to be the least species specific of any of the bee mites which should be considered further cause for concern in countries that have recently detected this species within their borders. This species has been shown to parasitize representatives of all three sub-genera of the genus *Apis* (de Guzman et al. 2017). Its ever-widening host range has even been expanded to encompass a bee outside the genus *Apis* having been found in association with *Xylocopa*, or carpenter bees (Abrol and Putatunda 1996). When present in *A. florea*, *T. mercedesae* outcompete that host's usual parasitic mite (*E. sinhai)*. Based on the plasticity already shown in this organism, it seems likely that *Tropilaelaps* will soon be introduced into several other countries if strict inspection systems and quarantine measures are not implemented expeditiously.

Tropilaelaps appear to be unable to feed on adult bees though this has yet to be confirmed experimentally. They starve in the absence of brood as they appear to lack the requisite mouthpart structuring to feed on adult bees. Despite this apparent inability to exploit adult bees, *T. mercedesae* is competitively superior to *V. destructor* due in large part to its higher reproductive rate. It is now considered to be the most successful parasite of *A. mellifera* throughout mainland Asia (Buawangpong et al. 2015; Burgett et al. 1983; Chantawannakul et al. 2016).

Thus far, it appears *Tropilaelaps* have not heavily impacted Asian honey bee populations. This is likely because of their coevolutionary history with these mites and similar species. When *Tropilaelaps* populations grow in the colonies of their original host (the giant honey bees), the occupants abscond leaving the majority of the mites trapped in the immobile brood. Seasonal migration and the broodless conditions that follow also help to naturally lower populations. Social grooming also appears to play a role in substantially reducing mite populations. In cage studies, *A. cerana, A. mellifera, and A. dorsata* artificially inoculated with *T. mercedesae* showed substantial differences in their respective grooming rates. Within 6 h of inoculation, *A. cerana* had already removed 2/3 of the mites and *within* 24 h had fully eliminated the population of parasites. The next highest rate of grooming efficiency was observed in *A. dorsata* with *A. mellifera* showing the lowest rate (Khongphinitbunjong et al. 2012). These findings correspond to the rates of infestation in these three species with *A. cerana* rarely infested by *Tropilaelaps*, relatively low levels in *A. dorsata*, and the high infestation levels reported in *A. mellifera.*

Tropilaelaps can generally be described as small, reddish-brown, pill-shaped mites. A dorsal shield fully covers the idiosoma which is longer than wide; likely the trait to which they owe their characteristic speed and maneuverability (Delfinado-Baker and Baker 1982). The base of each appendage is also obscured by the dorsal shield in this genus. The first pair of legs is noticeably longer and thinner than the following three pairs. These thin legs are sensory organs used like antenna rather than as locomotory appendages (a common trait in much of the Acari).

Similar to *Varroa*, distinguishing between *Tropilaelaps* species can be difficult. Light microscopy is often necessary to see key morphological characters and even then, positive identification can be challenging. Though there is some overlap in the size range of these mites, size can still be a helpful character to use for quick identification. *T. mercedesae* is the largest of the four species (average length = 0.979 mm; average width = 0.542 mm). The similarly sized *T. thaii* (average length = 0.890 mm; average width = 0.492) and *T. clareae* (average length = 0.882 mm; average width = 0.484) are just slightly smaller than *T. mercedesae. Tropilaelaps koenigerum* is noticeably the smallest of the four with length ranging between 0.684 and 0.713 mm and width ranging between 0.428 and 0.456 mm (Anderson and Morgan 2007).

Life Cycle

Varroa provide the foundational information for our understanding of honey bee brood parasites thus, much of our description of *Tropilaelaps* (and *Euvarroa*) will focus on how these other life cycles compare to what we tend to think of as the standard parasitic mite life cycle. Similar to *Varroa,* the life cycle of *Tropilaelaps* can be separated into a reproductive phase in which the brood is exploited and a phase that, at this time, appears to be a phoretic phase. Adults appear to be used exclusively as a mode of transit and not as food sources. In the reproductive phase, a mated *Tropilaelaps* female enters the partially capped cell of a late-stage bee larva. (Drone brood is preferred when parasitizing *A. mellifera* but no such preference has been observed with this parasite on its native host). *Tropilaelaps* females become gravid and are able to lay an egg just 2 days after mating. They do not need to feed on host larvae to stimulate ovogenesis like *Varroa* which require about 60 h with their host in order to start laying eggs. Oviposition in *Tropilaelaps* has been observed as early as the cocoon spinning phase which typically takes place about 10 h post-capping. Foundress *Tropilaelaps* also lay eggs at a faster rate than *Varroa*, potentially laying once every 24 h as opposed to every 30 h.

The egg develops into a six-legged larval stage progressing quickly to protonymph, deutonymph, and adult in about 6–9 days (Woyke 1987). Males appear to require about 24 h less time to mature than females. This observation makes sense in light of the seemingly reversed gender order in *Tropilaelaps* oviposition by comparison to *Varroa*. Foundress mites normally produce a female as their first and sometimes their second offspring with the male egg being more common in successive layings. When the mature host bee finally chews its way through the capping, both male and female *Tropilaelaps* emerge. *Tropilaelaps* may also be able to mate outside of the cell which likely reduces the chances of inbreeding depression. Research conducted in Thailand has shown that a single foundress is likely to produce 1–2 offspring on *A. mellifera* worker brood. However, that number increases if multiple foundress mites share the same cell, increasing to 2–3.

The relatively rapid reproduction reported in *Tropilaelaps* may be an adaptation to offset their high rates of reproductive failure. These rates vary depending on the region but can be quite substantial. A 7–18% rate of reproductive failure was observed in studies conducted in Afghanistan and Vietnam while around 30% of females observed in a study in Thailand did not reproduce. More recently, rates of 50–93% were observed in *A. mellifera* and about 65% in *A. dorsata*. Reproductive rates would have to be high to maintain populations in the face of such high reproductive failure. The reasons for the observed reproductive failure in these ostensibly mated females likely involve an interplay between the parasite's haplotype, the species of the host and biological/behavioral differences inherent therein. Though *T. mercedesae* has been observed parasitizing a wide variety of hosts, the entire population in a given colony may not be able to successfully exploit the host. The persistence of the population may be the result of a small number of individuals with an important mutation that allows for reproduction in a new host system.

The phoretic phase for *Tropilaelaps* has been estimated to be about 1.3 days, substantially shorter than that of *Varroa*. This rapid turn-around between reproductive cycles further boosts the reproductive rate of the mite and shortens the most vulnerable stage of their life cycle. *Tropilaelaps* mites collected from *A. mellifera* hive debris show high rates of mutilation suggesting that when the mites are groomed they are aggressively removed from the bodies of adult bees. Brood infestation in *A. mellifera* colonies can range from 2 to 54% while infestations on adult bees usually range from 1 to 3% (Woyke 1984, 1987). The apparent inability of *Tropilaelaps* to adequately feed on adult bees is believed to play a role in the shortening of this phase as well. These observations introduce the question of how these parasites are able to spread to new colonies efficiently having truncated the stage associated with relocation.

It has been suggested that *Tropilaelaps* may spread by scurrying between close aggregations of their host during their brief phoretic phase. Their original host *A.*

dorsata is known to form dense aggregations (sometimes more than 200 colonies observed on the same tree or cliff face) (Oldroyd and Wongsiri 2006). *Tropilaelaps* appear to be maladapted morphologically to hold on well to flying hosts thus they may rely on the aggregation behavior of their native host to facilitate spread.

Our understanding of the life cycle of *Tropilaelaps*, much like that of *Varroa,* is based primarily on the most damaging representative of the genus (*T. mercedesae*) on its new adapted host (*A. mellifera*) rather than its original host. A life cycle based entirely on such a specific case runs the risk of presenting exceptions as the rules themselves. Conclusions about the natural behavior of this parasite should only be drawn from these observations with caution, if at all.

Damaging Effects

Under laboratory conditions, parasitism by *T. mercedesae* on *A. mellifera* worker brood led to heightened titers of Deformed Wing Virus (DWV), symptomatic DWV infection (i.e., deformed wings and shortened abdomens), reduction in weight, and shortened life span of the host (Khongphinitbunjong et al. 2016). *Tropilaelaps mercedesae* is a confirmed vector of DWV (Dainat et al. 2009; Forsgren et al. 2009). Black queen cell virus (BQCV) has been detected in *Tropilaelaps* populations as well (though there exists no positive evidence that they actively transmit this virus to their host). Similar to the effects of *Varroa* feeding, *Tropilaelaps* parasitism reduces total protein titers of the host pupa and apparently alters host immune response (Khongphinitbunjong et al. 2016).

The feeding dynamic in *Tropilaelaps* has not been well studied. Similar assumptions made about the host tissue *Varroa* feed on have also been made about *Tropilaelaps.* Verification of whether they feed specifically on hemolymph or fat body would be helpful in better understanding exactly how this parasite impacts its host. It has been observed that *Tropilaelaps* do not create a single feeding site on their host but feed from multiple locations on brood causing more widespread damage than attributed to *Varroa* (de Guzman et al. 2017).

At the colony level, *Tropilaelaps* infestation substantially reduces the worker population. Several of the workers present are observed with physiological deformities that impair their ability to tend to the needs of the colony. As a result, secondary parasites such as wax moth and small hive beetle are frequently seen in association with late-stage parasitism. Eventually, these factors lead to the outright collapse of the colony.

Detection and Management

Bald brood and irregular brood patterns can be signs that a *Tropilaelaps* infestation is progressing. In addition, the presence of mites running rapidly along the comb can be a conspicuous sign as well. Detection methods such as sugar rolls/alcohol washes may be of limited utility with this parasite considering the brief duration of its phoretic phase. A survey of 300 adult bees would likely yield low numbers of mites even in a highly infested colony leading a beekeeper to conclude a lower infestation rate than is accurate.

Analyzing infestation rates in brood would likely yield a more accurate picture. Opening 100–200 brood cells to determine the presence of *Tropilaelaps* has been recommended to achieve the best assessment of the population but this method is prohibitively time consuming under most circumstances. However, the so-called bump method appears to be a quick and accurate method of determining approximate infestation levels even when populations are small. To test via the "bump method," a brood frame containing purple-eyed or older pupae should be struck against a paper covered table. It is helpful if this table is coated in powdered sugar to hamper escape of the mites. The powdered sugar can be dissolved later by spraying a small amount of water to facilitate counting the mites. The success of this method can be increased by scratching the cell cappings with a capping scratcher before bumping the frame as it will allow for release of mites within capped brood cells (Pettis et al. 2013; de Guzman et al. 2017).

Without chemical intervention, *A. mellifera* colonies rarely survive more than a few months of heavy *T. mercedesae* infestation (Burgett et al. 1983; Crane 1990). Beekeepers across Asia have been adapting *Varroa* control methods to combat this pest, as there are currently no products labelled to specifically address this pest. Apistan is a potential option being used in several countries. The hive should be treated for two cycles of 28 days at a recommended dosage of 2–3 Apistan strips per colony (Burgett and Kitprasert 1990). Strips of filter paper can be soaked in a solution of 15% potassium nitrate and 12.5% amitraz to then be used as a hive fumigant (Anderson and Roberts 2013). Amitraz can be sprayed on the walls of the hive and the comb three to four times at 4-day intervals to reduce mite levels but this method is not recommended when an abundance of eggs and young brood are present. Coumaphos and flumethrin are also options and as with the aforementioned treatments should not be used within 8 weeks of the nectar flow in a given area.

Sulfur and camphor have been used as low-cost alternatives to the synthetic acaricides. Smoke from tobacco leaves in the smoker has also been shown to be an effective method of inducing mite drop (Anderson and Roberts 2013; Burgett and Kitprasert 1990; Camphor et al. 2005; Kongpitak et al. 2008). In the Philippines and Thailand, Alagaw leaves (*Premna odorata*) and lemon grass (*Andropogon citratus*) are used to manage *Tropilaelaps* and *Varroa* with varying levels of success (Booppha et al. 2010; Cervancia 1993). Thymol, formic acid, and a combination of thymol and oxalic acid are among the most effective chemical treatment measures for *Tropilaelaps* (Garg et al. 1984; Hoppe et al. 1989; Mahmood et al. 2011, 2012; Raffique et al. 2012; de Guzman et al. 2017). The same lemon grass oil/ceramic system described for *Varroa* control earlier in this chapter has also shown effectiveness in controlling *Tropilaelaps* (Booppha et al. 2010). Inducing broodless conditions in a colony has also been proposed as a control method for *Tropilaelaps* to exploit its inability to survive for more than a few days in the absence of brood (Woyke 1984). Breeding for mite resistance has not gained popularity among beekeepers in Asia. A combination of multiple methods, chemical, and otherwise is recommended to maximize efficacy and avoid promoting resistance in mite populations.

1.4.1.3 *Euvarroa*

Thus far, only two species of *Euvarroa* have been described, both parasites of dwarf bees. *Euvarroa sinhai* was first observed on *A. florea* in India (Delfinado and Baker 1974) and *Euvarroa wongsirii* was described from specimens collected from *A. andreniformis* in Thailand. Both species appear to have minimal impact on their respective hosts. Neither has been observed reproducing on worker brood. They are apparently confined exclusively to drone brood which may explain the remarkable observation that these mites can persist in their phoretic phase for 4–10.5 months (Mossadegh 1990). This is likely an adaptation to allow for their persistence on a seasonally produced resource.

Though not observed in worker brood on their original hosts, *E. sinhai* has shown the ability to reproduce on worker brood of *A. mellifera* and the ability to survive for an extended period on *A. cerana* (Koeniger et al. 1993; Mossadegh 1990). Hive debris containing *Euvarroa* has been collected from *A. mellifera* and *A. dorsata* colonies. While parasitism of other *Apis* is rare, these findings show potential for a future host shift in this species and warrants further study.

Euvarroa resemble *Varroa* but are distinguished from this related genus and from each other by shape and chaetotaxy. *Euvarroa wongsirii* is wider posteriorly with a somewhat triangular shape by comparison to the pear-shaped *E. sinhai.* Both species have marginal setae that are distinctly longer than those of *Varroa* species (39–40 in *E. sinhai* and 47–54 in *E. wongsirii*).

Life Cycle

Little is known about most aspects of the *Euvarroa* life cycle. The gaps in our knowledge of this parasite are typically filled with speculation from the life cycle of the closely related *Varroa* genus or studies conducted on *A. mellifera* rather than on the original hosts of these parasites. Laboratory study of *E. sinhai* using *A. mellifera* worker brood determined that the development from egg to adult takes about 6–7 days for females and 5 for males. The females were shown to produce 4.3 progeny per worker brood cell (Mossadegh 1990). On its natural host, *E. sinhai* has been observed with up to 15 mites developing in a single drone cell.

Detection and Management

All honey bee colonies have parasitic mites associated with them but only *A. mellifera* colonies are treated to reduce mite levels thus there are no treatment strategies currently in place for parasites of dwarf honey bees.

1.4.1.4 *Acarapis*

Mites of the genus *Acarapis* are by far the smallest of the acarine obligate bee parasites, measuring less than a tenth of a millimeter in width on average. They are also the only obligate parasitic mite with no association with brood, a fact that surely contributes to their relatively subdued impact on colony health by comparison to the *Tropilaelaps* and *Varroa*. They appear to be primarily parasites of *A. mellifera*. The genus comprises three species (*Acarapis externus, Acarapis dorsalis,* and *Acarapis woodi)* of which only *A. woodi* (the tracheal mite) is known to cause any significant level of damage to its host. First discovered in Europe nearly a century ago, *A. woodi* is now present in some areas of Asia.

India was the first Asian country to report the presence of *A. woodi* (Michael 1957; Milne 1957). Since then, this species has been observed parasitizing *A. mellifera* in Egypt, Iran, Jordan, Kuwait, Lebanon, Palestine, Syria, and Israel (Amr et al. 1998; Gerson et al. 1994; Matheson 1993; Mossadegh and Bahreini 1994; Rashad et al. 1985; OIE 2004). In the 1980s, *A. woodi* was found infesting *A. indica* in India, Pakistan, Bangladesh, and China (Delfinado-Baker and Baker 1982). More recently, *A. woodi* was detected in failing *A. cerana japonica* colonies in Japan (Kojima et al. 2011). *Acarapis dorsalis* and *A. externus* have been collected on multiple continents but in Asia have only been reported in Iran (Mossadegh and Bahreini 1994). Unfortunately, details of the extent of these infestations are sparse.

Life Cycle

All mites of the genus *Acarapis* are reported to feed on hemolymph (Mossadegh and Bahreini 1994). In *A. woodi,* females invade the tracheae of newly emerged bees (apparently preferring drones to workers or queens). They then feed through the tracheal wall. After a few days of feeding, the female will begin ovipositing 1 egg each day for about 6 days (Royce et al. 1988). Eggs hatch into larvae that also feed on bees. The male mites develop more quickly and wait to mate with the developing females. The mated females later emerge from the tracheae to seek out new hosts. Development from egg to adult takes about 11 days on *A. mellifera*. Further research is necessary to provide a detailed life history of *A. dorsalis* and *A. externus*. Of what is known of them, it is clear that they are external parasites of honey bees and do not enter the tracheae to complete their life cycle. They develop from egg to adult in about 8–9 days and tend to lay between 2 and 5 eggs (de Guzman et al. 2001; Ibay 1989; Royce et al. 1988; Eckert 1961).

Damaging Effects

Despite their small size, the symptoms of *A. woodi* infection can be very conspicuous. Bees from colonies with advanced infestations are often seen walking around lethargically in front of the colony. Because the causative agent of this ailment is virtually invisible to the unaided eye, this is often the first sign of infestation witnessed by beekeepers. Infested individuals also exhibit distension of the abdomen, uncoupled hind and forewing, and damaged tracheae (Sammataro and Yoder 2011). Heavily infested colonies may fail to adequately regulate the temperature of the colony leading to chilling of the brood and reduced bee longevity. Virus-like particles have been observed in *A. woodi* (Liu 1991) but there is not enough evidence to determine that tracheal mites are able to vector viruses like *Varroa* and *Tropilaelaps*. More study in this area is warranted.

Detection and Management

Several of the symptoms associated with tracheal mite infection overlap with those of other ailments thus positive confirmation of the presence of *Acarapis* is necessary. The presence of *A. woodi* in a colony can be reliably determined only by dissection of the thorax of bees to verify the presence of mites or damage caused by feeding. The mites appear as miniscule white spheroids which can be very difficult to resolve against the lightly colored inner tissue of the bees thorax. To more easily view the mites or the damage they cause, the honey bee can be immersed in a 5% potassium hydroxide solution for about 23 h at 43 °C and then stained for 10 min with 1% methylene blue solution. Scarring from mite feeding can readily be seen as a darkening of the tracheae which is obvious when compared with uninfested bees. For identification of the mites themselves, a small section of trachea is removed and mounted on a glass slide. Under 200× magnification, *A. woodi* can be readily identified.

Management of this mite is distinct from that of *Varroa* and *Tropilaelaps*. There is often no need for use of chemical pesticides. One of the most effective methods is to place a grease patty (225 g/colony), made of hydrogenated vegetable oil and granulated sugar in a 1:2 ratio on top of the frames. Oil from the patty covers the hairs of the bees when they come to feed on the patty. These oils impede the movement of the mites between hosts and can inhibit mites from infesting newly emerged bees. If chemical control is necessary, food grade menthol crystals (25 g for one story hives, 50 g for two story) can be placed in a porous chamber to facilitate vaporization. The crystals can also be mixed with oil in a 1:1 ratio and heated at 65 °C. A sheet of cardboard can then be immersed in the menthol-oil mixture and placed at the bottom of the hive. Two applications are recommended separated by 1 week. An absorbent material immersed in food grade formic acid (65%) can also be placed in hives to reduce or eliminate mite populations.

1.4.1.5 *Neocypholaelaps*

The genus *Neocypholaelaps* (Family: Ameroseiidae, Mesostigmata) comprises roughly two dozen nonpathogenic mites typically referred to as pollen mites. These mites are opportunistic, phoretic parasites, and typically use honey bees and other pollinators as a means of transport to spread out their populations between flowers (Oldroyd and Wongsiri 2006). Periodically, these mites can be inadvertently transported back to honey bee colonies where they take up residence in stored pollen. These mites can also climb on to bees in such high numbers that they alter the behavior of the bees. It is these circumstances that make them noteworthy to beekeepers and researchers alike, especially in Asia where diversity of these mites is especially high. Though they have not seen as much research as the other acarines discussed above, researchers have questioned whether these mites can influence the amount of pollen foragers bring back to the hive (Eickwort 1990).

Neocypholaelaps have been observed covering the thorax, abdomen, and legs of workers potentially reducing the capacity of the pollen baskets. It is also possible that heavily infested honey bees may return to the colony prematurely. *Neocypholaelaps* infested *A. cerana* have been observed returning to the colony and performing a shaking dance that elicits other members of the colony to remove the parasites (Oldroyd and Wongsiri 2006). Further study is necessary to determine how these parasites affect their hosts both directly and indirectly.

1.4.2 Viruses

Most studied honey bee viruses are small icosahedral particle containing a positive-sense single-stranded RNA genome. Their effects have been felt far back in human history, even prior to germ theory with Aristotle describing symptoms of what was likely chronic bee paralysis virus (CBPV) nearly 2000 years ago. Of the 26 or so described viruses (Chen et al. 2004; Allen and Ball 1996; Chen and Siede 2007; de Miranda et al. 2013), most have been isolated from *A. mellifera* (with notable exceptions being Kashmir bee virus (KBV), Apis iridescent virus (AIV), and Thai/Chinese sacbrood virus (TSBV/CSBV) all of which were isolated from *A. cerana*). The status of *A. mellifera* as a cosmopolitan species promotes the emergence of novel viral interactions in its wide range of geographic locations. The host and geographic range for these infections has expanded rapidly in recent history because of apicultural shipments between countries (Tsevegmid et al. 2016). This observation is of particular concern in Asia where all *Apis* species are present and further host shift is likely. Black queen cell virus (BQCV) and DWV may have already shifted to exploiting Asian honey bees (Mookhploy et al. 2015; Zhang et al. 2012).

Transmission of bee viruses can occur horizontally (between individuals of the same generation) or vertically (between generations). Pathogens transmitted horizontally tend to show greater virulence. Horizontal transmission can occur via a number of means (e.g., close physical contact, trophallaxis, consuming contaminated food, oral-fecal transmission). Vertical transmission occurs when viruses are transmitted in sperm or to eggs. Another of the most common routes of transmission is via a parasitic vector (e.g., *Tropilaelaps* or *Varroa*). Israeli acute bee paralysis virus (IAPV), KBV, BQCV, acute bee paralysis virus (APBV), *Varroa destructor* virus (VdV-1), and DWV can be vectored horizontally, vertically, or via vector. Several viruses have been detected in stored pollen, honey, and royal jelly (ABPV, BQCV, CBPV, KBV, SBV, DWV, and IAPV) (Chen et al. 2006b).

The emergence of unexplained colony collapses throughout the world in 2006 prompted renewed interest in honey bee viruses leading to the availability of information in much greater depth. The most common bee viruses found in different species of *Apis* in Asia can be classified into two groups: Iflaviridae (SBV, DWV, and VDV-1) and Dicistroviridae (BQCV, KBV, ABPV, IAPV).

1.4.2.1 Iflaviridae

SBV was the first virus discovered in *A. mellifera* (White 1913). SBV infection of *A. mellifera* has been reported in Thailand, Vietnam, South Korea, Japan, and China (Forsgren et al. 2015; Kojima et al. 2011; Sanpa and Chantawannakul 2009; Yoo et al. 2009; Ai et al. 2012). A related but physiochemically distinct sacbrood virus was detected in *A. cerana in* Thailand circa 1976 (Bailey et al. 1982). This virus was named Thai sacbrood virus (and is also known as Chinese sacbrood virus). TSBV showed greater virulence than SBV reportedly causing the death of more than 90% of managed *A. cerana* populations in Kashmir (Abrol and Bhat 1990) and has been detected in populations of *A. dorsata* and *A. florea* in India (Allen and Ball 1996).

The most prevalent of the Iflaviridae is DWV likely due in large part to the ubiquity of its vectors *V. destructor* and *T. mercedesae.* Their ability to vector viruses in tandem with their concomitant capacity to depress their the immune system of their host leads to markedly high virus titers and rapid disease progression. Virulence is dependent on transmission route as well. Infections transmitted directly to the host hemocoel (e.g., via parasitic feeding) are known to cause rapid mortality. DWV has been detected in populations of *A. florea, A. cerana, A. mellifera,* and *A. dorsata* and is believed to be able to infect all honey bee species (Chantawannakul et al. 2016). Multiple variants of DWV have been discovered and it has been recently proposed to be comprised of three types. DWV (type A) consists of DWV and Kakugo virus (KV), type B includes VDV-1, which was first isolated from *Varroa*, and type C which was recently discovered in asymptomatic bees in Devon, UK (Lanzi et al. 2006; Zioni et al. 2011; Martin et al. 2012). DWV type A (formerly known as DWV) is found to be more dominant in bees expressing clinical symptoms of DWV (i.e., deformed wings).

1.4.2.2 Dicistroviridae

Like DWV, BQCV and IAPV likely have a host range encompassing all honey bee species. BQCV is widely prevalent in managed colonies of both *A. mellifera* and *A. cerana* in (China, South Korea, Thailand, Japan, and Vietnam) (Choe et al. 2012; Forsgren et al. 2015; Li et al. 2012a; Mookhploy et al. 2015; Yang et al. 2013). BQCV samples have been recovered from *A. florea* and *A. dorsata* in Thailand and China (Mookhploy et al. 2015; Zhang et al. 2012). IAPV has been detected in both *A. mellifera* and *A. cerana* as well in China and Japan. In these countries, they have been associated with colony losses especially when co-infected with *V. destructor.* ABPV has been detected in *A. cerana* (Choe et al. 2012) and in *A. mellifera* in China (Ai et al. 2012; Forsgren et al. 2015) and Thailand (Sanpa and Chantawannakul 2009).

Damaging Effects

Much of our understanding of the pathological effects of these viruses is derived from studies conducted under laboratory conditions. Virus is injected directly into bees or they are artificially exposed to viruses in their feed. Even when inoculated with levels that are biologically relevant, the artificial nature of these studies has the potential to exaggerate or otherwise present an inaccurate picture of the natural effect that these viruses have on their host. There is ample evidence to suggest that most infected bees are primarily or wholly asymptomatic until stress factors such as starvation, poor nutrition, or exposure to parasites weakens them enough for clinical symptoms to arise. Natural tissue tropism of honey bee viruses is still poorly documented, and this fact should be considered when reading detailed descriptions of bee virus-linked pathologies.

DWV (DWV type A) and VDV-1 (DWV type B) are closely related viruses and are believed to have similar pathologies associated with them. DWV leads to malformed wings, shortened abdomens, stunted growth in general, and a substantially reduced life span. When symptomatic, they emerge from capped cells with useless wings creating an added burden for the colony as they still consume food but are unable to help the colony with vital functions such as foraging and the like. DWV usually persists asymptomatically in honey bee populations but exhibits far greater virulence when vectored by *Varroa* or *Tropilaelaps* mites.

SBV infection in a colony is characterized by infected larvae that degenerate into fluid filled sacs. Infection occurs after larvae ingest contaminated food and death occurs soon after. Heavily infected colonies are characterized by an irregular brood pattern with sunken or perforated cappings.

Dicistroviruses are prevalent among and within honey bee populations; however, their symptoms and mortality are inconsistent. CBPV affects adult bees and causes abnormal trembling of the body and wings followed by eventual loss of motor coordination. Infected bees are typically hairless, shiny, and darker in appearance. Guard bees have been observed attacking and otherwise impeding the return of these individuals to the colony. This is potentially a form of social immunity as it would reduce the likelihood of the infection being passed to healthy bees. CBPV appears to be less virulent than other Dicistrovirids. ABPV and IAPV cause paralysis and sudden death 1 day after infection, while CBPV needs much higher viral titers to cause paralysis in honey bees and even then, requires several days. The most common cause of death in developing queen larvae is BQCV. It causes a conspicuous blackening of the larvae that is difficult to mistake for other pathologies.

There are several described viruses for which there is no recognized pathology (e.g., bee virus X (BVX), bee virus Y (BVY), and *Apis mellifera* filamentous virus (AmFV). In most cases, it is expected that bee colonies go on with covert infections of these viruses, not showing any distinct symptoms despite virus replication (de Miranda et al. 2013; Sanpa and Chantawannakul 2009).

Co-infections with multiple viruses are common potentially amplifying the effects of each virus. DWV and ABPV were found co-occurring in 17% of samples collected. Three viruses were detected in 13% of samples and 4% of samples evidenced concurrent infection with 4 different viruses (DWV, ABPV, SBV, and KBV).

Detection and Management

There is currently no direct method to manage honey bee viruses but that does not mean that the beekeeper is left without options for prophylaxis or remediation. Effective management of honey bee viruses is best achieved by managing their vectors (*Tropilaelaps and Varroa*) and contributors to colony stress. The observation of individual *Varroa* harboring up to five different viruses stands as an important reminder that allowing mite populations to persist untreated can be a costly mistake (Chantawannakul et al. 2006). In areas where *Varroa* is still absent (e.g., Uganda), no evidence of several known bee viruses can be detected including CBPV, ABPV, and IAPV.

In addition, the utility of good beekeeping practices in limiting virus transmission is difficult to overstate. Practices such as promoting proper bee nutrition, maintaining clean equipment, managing parasitic mites, and using disease-resistant stock when available can often be all that is needed to avoid viral outbreak (Chen et al. 2006a). Vaccination might be an alternative control as well, as RNA interference (RNAi) has been used successfully as a preventative treatment for IAPV (Hunter et al. 2010). This method does, however, face challenges when considering the high cost of production and the high genetic diversity of honey bee viruses.

1.4.3 Fungi

The warm, humid, dark nature of the honey bee colony has the potential to serve as a climate controlled incubator for fungal spores if not adequately managed. This is especially the case for the cavity-nesting species who maintain the most precisely climate controlled nests of all the *Apis* sub-genera. Furthermore, the ubiquity of trophallaxis in honey bee colonies all but ensures the spread of internal fungal pathogens that make it into the colony.

1.4.3.1 *Nosema* spp.

Nosema (Nosematidae, Microsporidia) are among the most successful pathogens of honey bees worldwide. This genus comprises spore-forming fungi called Microsporidia that parasitize several insect species. Two species are known to specialize in infesting the digestive system of honey bees: *Nosema apis and Nosema ceranae*. The first to be described was *N. apis* in the early 1900s (Zander 1909). Its original host is thought to be *A. mellifera* (Fries 1993). Survey data has identified *N. apis* in *A. mellifera* and *A. cerana* but no other honey bee species (Klee et al. 2007; Rice 2001). *Nosema ceranae*, however, appears to have originated in Asia as it was first described as a parasite of *A. cerana* in China (Fries et al. 1996a). Furthermore, Asian honey bees appear to be less susceptible to infection than *A. mellifera* potentially reflecting a coevolutionary history. A survey of infection among honey bee species showed infection levels of 77.5%, 22.2%, 45.4%, and 37.5% for *A.*

mellifera, A. cerana, A. florea, and A. dorsata, respectively. Thus far, *N. ceranae* has been detected throughout the region in Taiwan, Vietnam, China, Thailand, Turkey, Indonesia, Solomon Islands, Japan, and Jordan (Botías et al. 2012; Chaimanee et al. 2010; Haddad 2014; Huang et al. 2007; Klee et al. 2007; Liu et al. 2008; Whitaker et al. 2011; Yoshiyama and Kimura 2011). *N. apis* has been detected in Israel, Indonesia, and Turkey (Rice 2001; Whitaker et al. 2011; Gatehouse and Malone 1999).

Since 2003, *N. ceranae* appears to have replaced *N. apis* as the dominant microsporidium parasitizing honey bees worldwide. The ecological implications of this shift warrant further study. Though both parasites are virtually indistinguishable in appearance, there is evidence to conclude that they differ in several key aspects of how they affect their host. *Nosema apis* only infects midgut epithelial cells; however, *N. ceranae* is apparently able to infect other tissues throughout the digestive and endocrine system being found in the malpighian tubules, hypopharyngeal glands, salivary glands, and fat body (Chen and Huang 2010).

Life Cycle

The *Nosema* life cycle begins when a spore is ingested by a honey bee. Microsporidia can only survive outside the host cell as a spore and to ensure reproduction they shed tens of millions of spores into the environment via their host's excrement. These spores can remain viable in the environment for up to 4.5 years (Steche 1985). In the digestive system of a host bee, the spore ruptures at one end firing a specialized organelle called a polar tube into a nearby gut epithelial cell. This organelle injects the infectious spore content into one of the host cells. *Nosema* have a reduced, compact genome that is specialized for parasitism. The spore itself is lacking in mitochondria to produce its own energy and is wholly reliant on the host for metabolic function. Once safely inside the host's gut epithelial cell, the vegetative state of the parasite begins to divide. Within 6–10 days the epithelial cell is filled with new spores. When this swollen epithelial cell inevitably ruptures, these new spores are shed into the host's digestive system and pass into the environment with the excrement of the host (Fries 1988).

Damaging Effects

Nosema spp. infection leads to the shortened life span of individual workers. Infection has also been linked to the early onset of foraging behavior in honey bees due to elevated levels of juvenile hormone (Lin et al. 2001). By comparison to *N. apis, N. ceranae* infections appear to be heavier, potentially a result of their expanded host tissue range. The production of millions more spores per bee may increase the infected individual's nutritional requirements leading to terminal morbidity of the host (Martín-Hernández et al. 2011). Infection levels tend to be higher when bees are confined to the colony for extended periods (such as the winter or extended rainy periods as are common in the tropics) likely because the bees are unable to leave the colony to evacuate their waste. In temperate regions, *N. ceranae* infection is often associated with a slow buildup of honey bee populations following the winter.

Infection of *A. cerana* workers with *N. ceranae* appeared to be linked to lower levels of key microflora, particularly *Bifidobacterium* and Pasteurellaceae, both of which produce several antibiotic compounds important to host defense against parasites (Li et al. 2012b). Further, research suggests that *N. ceranae* may also promote the outbreak of other bee diseases, as this infection has been observed in association with heightened rates of Chalkbrood infection (Hedtke et al. 2011).

Despite a host of evident pathological effects on the individual level, researchers have yet to establish a consistent causal link between high *N. ceranae* levels and colony mortality. The answer likely lies within a complex interplay of climate conditions, nutritional status, infective dose, co-infection with other pathogens, and host genetics.

Detection and Management

Symptoms of *Nosema* infection can be indistinct and frequently overlap with those of other pathologies. Observation of these symptoms should prompt attempts at verification via more reliable means. Trembling workers, dead bees around the hive, and workers producing brown excrement are commonly reported as symptoms of *N. apis* infection. *N. ceranae* infections are often covert producing few if any distinct symptoms. Verification of these infections requires the viewing of honey bee gut contents under light microscopy. Standard practice is to take a sample of 100 honey bees from a colony, crush them to expose the gut contents, and add 100 ml of water to dilute the sample. When added to a glass slide or hemocytometer, spores can readily be viewed in a sample at 400×. Species specific identification is prohibitively difficult with light microscopy alone and can be performed using PCR-based techniques (Fries et al. 2013).

Both honey bee-associated *Nosema* species can be managed using the fungicide fumagillin but varying rates of success imply that further research into control methods is necessary. The quantity of product to be used and the timing of usage may require adjustment to reach consistent efficacy (Huang et al. 2013; Williams et al. 2008; Williams et al. 2011; Akrantanakul 1987).

1.4.3.2 Chalkbrood

The fungal disease Chalkbrood results from the infection of *Ascosphaera apis* (Eurotiomycetidae, Ascomycota). A disease of *A. mellifera,* the etiological agent of Chalkbrood was first described in Europe in the early twentieth century. The highly infectious nature of this parasite has allowed for it to steadily expand its geographic range. It was detected in the USA in 1965 and soon after in Canada (Hitchcock and

Christensen 1972). Since then it has achieved a cosmopolitan distribution. In Asia, incidence of Chalkbrood is widespread having been reported in China, Russia, Korea, Japan, Philippines, Thailand, Turkey, and Israel (Aronstein and Murray 2010; Cervancia 1993; Heath 1985; Oldroyd et al. 2006; Reynaldi et al. 2003). Incidence of Chalkbrood is reportedly higher during the rainy season in Thailand (June–July) likely due to the constant high ambient humidity (Chantawannakul et al. 2005; Theantana and Chantawannakul 2008). In recent years, the genome of *A. apis* has been sequenced aiding in efforts to better understand biology of this infectious organism (Qin et al. 2006).

Life Cycle

As with *Nosema*, the Chalkbrood life cycle begins with the ingestion of food infested with spores. However, unlike *Nosema*, *A. apis* spores need to be consumed by honey bee larvae to complete their life cycle. Young larvae (between 1 and 4 days old) are most susceptible to infection (Aronstein and Murray 2010). The spores germinate in the lumen of the larval gut and produce hyphae which penetrate the gut wall invading the peritrophic membrane, epithelial cells, and the basal membrane usually within 3 days of infection. This invasion of tissues results in substantial internal damage both mechanical and enzymatic.

Soon after, the larva begins to display clinical symptoms culminating in fungus eventually breaking out of the hind end of the body of the larva spreading a thick coat of mycelium over its host. At this point, the larva begins to take on the eponymous "chalky" appearance and texture as spores are produced on the surface of the host. The larva swells to about the size of the cell and usually dies in an upright position. Larvae at this stage are frequently called "mummies" and are conspicuously black or white. Each black "mummy" is covered in about ten billion infectious spores (Aronstein and Murray 2010). The purpose of the white mummies has yet to be determined and warrants further study.

Chalkbrood spores are long-lived and can persist on bees themselves, in their food, wax, or on improperly sanitized hive equipment. These spores can remain viable for 15 years or more (Anderson et al. 1997; Flores et al. 2005). This ensures that hive boxes and frames from previously infected colonies will remain infective for years to come. The introduction of Chalkbrood to Turkey is believed to be the result of contaminated beeswax imported into the country (Tutkun et al. 1993).

Damaging Effects

Chalkbrood infection rarely results in the death of a honey bee colony but infections of *A. apis* reduce colony strength and productivity decreasing honey production by up to 50% (Heath 1982; Aronstein and Murray 2010). Chalkbrood thus far has not made a host shift to Asian honey bees likely because the tropical conditions that these species prefer are not optimal for the growth of this pathogen. It is not considered to be an infection of great concern in most of Asia.

Detection and Management

Chalkbrood disease is prevalent in cool, humid conditions and often establishes in poorly ventilated colonies. Reducing excess moisture within the colony, maintaining adequate colony strength, and hygienic practices are key strategies in controlling Chalkbrood disease in apiaries. Chalkbrood-resistant lines of honey bees have been developed and are another nonchemical approach to manage this parasite. It is imperative when faced with Chalkbrood infection within an apiary that the beekeeper avoid moving or using equipment that has not been adequately sanitized. Exchanging frames or hive boxes between colonies or the usage of the same hive tool between colonies will only facilitate the spread of *A. apis.* Usage of lemon grass oil with a porous ceramic delivery system has proved an effective emerging technique for inhibiting the growth of *A. apis* (Booppha et al. 2010). Other medicinal plant extracts and volatiles (*Cinnamomum cassia*, *Piper betel* extract; oils containing citral and azadirachtin extracted from Neem) have been employed as well to varying levels of success (Liu and Ho 1999; Davis and Ward 2003; Chantawannakul et al. 2005).

1.4.4 Bacteria

It should not be surprising that one of the most infamous honey bee infections known to beekeeping is the result of a bacterial infection. Nevertheless, many bacteria species exist in commensal or mutually beneficial associations with honey bees supplementing host immune function, detoxifying xenobiotics, or providing key nutrients that they otherwise could not obtain from their exclusive diet of pollen and sugar. However, it is the two pathogenic species that are the topic of this section: *Paenibacillus larvae* and *Melissococcus plutonius*, the causative agents of American foulbrood (AFB) and European foulbrood (EFB), respectively.

1.4.4.1 American Foulbrood (AFB)

American foulbrood is so devastating a disease that several countries impose strict inspection, reporting, and treatment standards for dealing with it. The disease is recognized by the World Organization for Animal Health (OIE) because of its high virulence and highly infectious nature. The causative agent, *Paenibacillus larvae* has now achieved a nearly cosmopolitan distribution, including Asia (Genersch 2010). This disease has been detected in *A. cerana* in India (Singh 1961) and *A. mellifera* in Taiwan (Yen and Chyn 1971). However, a survey conducted in Thailand in 2003 showed no evidence of *P. larvae* infection in Thai *A. mellifera* (National Bureau of Agricultural Commodity and Food Standards 2008).

Asian honey bees are rarely reported to show clinical signs of AFB infection. When artificially inoculated with *P. larvae* spores, *A. cerana* brood did contract the infection but consistent with the persistently hygienic nature of this insect, infected larvae were removed before reaching the infectious stage (Chen et al. 2000).

Life Cycle

P. larvae is a food-borne pathogen. When ingested by honey bee larvae, the spores germinate in the midgut, multiply, cross the gut barrier, and eventually invade the hemocoel (Garcia-Gonzalez and Genersch 2013). Young larvae are far more susceptible to infection than older larvae. One-day old larvae can reportedly be infected with as few as ten spores while infection of two-day old larvae requires millions of spores. The infected larva progresses to a pre-pupa, spins a cocoon, and dies soon after (Chantawannakul and Dancer 2001). Highly bioactive proteases produced by the bacterium digest the larva into a semi-liquid form. The larva then dries to a hard scale on the floor of the cell (Genersch 2010). The scales are highly infective containing millions of spores that can retain viability for several decades (Hasemann 1961).

Detection and Management

AFB is easily recognizable by the markedly pungent odor that emanates from infected colonies. The foul odor is one of the first signs of infection. The colony can then be inspected for sunken, discolored cell cappings which likely contain infected larva or the remains thereof. Infection is typically confirmed by sticking a matchstick into the semi-liquid mass beneath the capping. The lytic enzymes produced by *P. larvae* transform honey bee brood into a gel that forms a strand or "ropes" when probed. This character is diagnostic and unfortunately signals the end of a colony under most circumstances.

Because of the extreme durability and longevity of *P. larvae* spores, burning infected colonies is the only recommended treatment in many countries (Chantawannakul and Dancer 2001). Burning infected colonies can pose a severe financial burden to beekeepers so some countries have allowed the usage of chemotherapeutic antibiotics such as oxytetracycline hydrochloride (OTC) or sulfathiazole. These antibiotics only affect the vegetative stage of this bacterium and have been shown to promote antibiotic resistance. Furthermore, they can contaminate bee products and move into the human food chain (Murray et al. 2007). Several attempts have been made to develop alternatives to burning colonies (e.g., biological control via antagonistic bacteria, treatment using propolis, essential oils, or royal jelly) with limited success. Further research into alternative control measures is necessary.

1.4.4.2 European Foulbrood

M. plutonius, a non-spore forming bacterium is the etiological agent of European foulbrood (EFB). *M. plutonius* has been detected throughout Asia in *A. mellifera* colonies in Vietnam, Japan, Thailand, and recently in Saudi Arabia (Akratanakul 1987; Aronstein and Murray 2010; Ansari et al. 2016). EFB has not, however, been detected in New Zealand. Infection has been reported in *A. cerana, A. mellifera, A.*

florea, and A. laboriosa (Chantawannakul et al. 2016). *Melissococcus plutonius* isolated from *A. mellifera* in different regions were found to be markedly similar genetically and serologically. However, they were distinct from a highly prevalent strain found in *A. cerana*. This strain was discovered to be highly virulent (Takamatsu et al. 2014). Due to the highly infective nature of this pathogen, it too is listed by the OIE and is subject to several laws and regulations varying by region.

Life Cycle

Larvae are infected by *M. plutonius* after consuming contaminated food. The bacterium then invades the gut tissue, multiplying rapidly. Younger larvae are more easily infected by this parasite than older larvae and are usually killed by day 4 or 5. Unlike *P. larvae*, *M. plutonius* do not produce spores and as such their survival outside of a living host is estimated in months rather than years. The exact mechanisms for pathogenesis in *M. plutonius* are still unknown as are many of the details of its life cycle. It is known, however, that larvae do periodically survive infection by *M. plutonius* developing, albeit more slowly, into undersized adults. These surviving bees help to perpetuate the disease with the high volume of bacteria shed in their feces immediately prior to pupation.

Detection and Management

EFB shares some conspicuous pathologies with AFB. High larval mortality leading to a spotty brood pattern and occasional sunken brood cappings can make the two difficult to distinguish. EFB is also characterized by a distinctly putrid odor and as a result was considered, for quite some time, to be the same disease as AFB. In the field, the two can be differentiated by the diagnostic character of roping larvae. Larvae infected by *M. plutonius* retain their outer membrane and do not degenerate into a viscous gel that forms a rope when probed. Rather than presenting as pearly-white, c-shaped larvae, larvae infected by *M. plutonius* when found dead, are stretched length-wise. Diseased larvae take on a characteristic pale-yellow or brown color. Tracheal tubes become more apparent in these discolored larvae. Confirmation of an EFB diagnosis usually requires cell culturing, immune-based assay or polymerase chain reaction (PCR)-based techniques. The PCR-based techniques are very sensitive and often can detect sub-clinical levels of EFB in seemingly healthy colonies, but require specialized equipment.

The lack of a spore-forming stage in *M. plutonius* allows for EFB to be managed effectively with antibiotics (specifically oxytetracycline). Any antibiotics applied within the colony should be applied far in advance of any honey harvest to avoid having these chemicals cross into the human food chain. In lieu of or in addition to chemical treatment, the "shook swarm method" is also an effective management practice. In this procedure, all comb is replaced, mimicking the natural behavior of the bees to swarm when faced with overwhelming parasite or disease pressure. EFB is often a disease that reveals underlying stress within a colony. Weakness resulting from depopulation, poor nutrition, and/or environmental stress factors can promote progression of EFB. Further, climactic conditions appear to play a role. EFB is particularly common in certain areas of Asia during the rainy season suggesting a potential connection between the two.

1.5 Pests

Honey bees produce several valuable products the most famous of which being honey; however, wax, propolis, bee bread, and brood are highly coveted by other organisms. There is no environment where their claim to these biochemically expensive products goes uncontested. Their infamous ability to sting is just one response to this constant pressure to defend their produce.

1.5.1 Vertebrate Pests

In addition to predation from human honey-hunters, honey bees in Asia face a variety of avian predators. Bee-eating birds are a common threat to honey bee colonies in Asia and Africa. These colorful, aerial hunters are known to consume all species of *Apis*. Bee-eaters are migratory birds and populations of them often aggregate in one area heavily impacting populations of bees locally. At a coastal watch site in western Malaysia 93.1% of the migrating population was composed of oriental honey buzzards (*Pernis ptilorhynchus*) (Decandido et al. 2004). These birds are known to mount coordinated attacks of *A. dorsata* colonies (Thapa and Wongsiri 2003; Thapa et al. 2000). Honey buzzards have been observed flapping their wings wildly near colonies to incite the pursuit of guard bees. They are able to fly away more quickly than their would-be attackers and only after leading the defensive bees away, do they return to consume honey and brood (Oldroyd et al. 2006; Thapa and Wongsiri 2003). Honey buzzards may also hunt as pairs with one bird functioning as decoy while the other smashes into the comb to dislodge it from the tree or cliff face to which it is attached (Oldroyd et al. 2006).

Certain smaller birds such as honey guides (Indicatoridae) are unique to the bee-eating birds because of their notable ability to make a meal of beeswax. Their digestive system harbors bacteria (*Micrococcus cerolyticus*) that allow them to digest it (Friedmann and Kern 1956). It is not uncommon to observe honey guides sitting in the middle of and consuming the wax remains of an *A. dorsata* colony abandoned by migrating bees. Honey guides are also renowned for their curious habit of actively guiding humans to honey bee colonies where they feed on wax and brood left behind after the humans have effectively destroyed the hive and consumed their fill. This behavior is only seen in honey guides of Africa but is the titular quality of the family nonetheless.

Many bee-eating birds are known to pick off honey bee workers in mid-flight. Several others have evolved as predators of honey bees with varying levels of specialization (the genera *Merops* and *Nyctyornis*, the bee-eaters; *Picus,* Wood peckers; *Dicrurus,* The Drongos; *Martes, Crypsiurus*, Swifts; and *Hirundapus,* Needle-tailed Swifts) (Cervancia 1993; Wongsiri et al. 2005; Akrantanakul 1987). At times, beekeepers in Asia are known to employ net-trapping of offending species or to relocate colonies to limit predation. Other vertebrates ranging from monkeys, toads, martens, bears, and even a tiger have been observed seeking out colonies to consume bees, brood, and/or honey.

1.5.2 Invertebrate Pests

1.5.2.1 Wax Moths

Wax moth larvae attack the combs of honey bees worldwide. The two species *Galleria* mellonella and *Achroia gisella,* the greater and lesser wax moth respectively, are small drab colored moths in the family Pyralidae. Both species have similar life histories as they complete their development from egg to adult in honey bee colonies consuming wax and bits of other hive products (Akrantanakul 1987). In *A. mellifera,* their eggs and larvae are easily removed by workers from strong colonies thus these insects only become a problem when colonies are weakened or otherwise depopulated. They are also pests of stored or unused combs (Pernal and Clay 2013). These parasites can be adequately controlled via nonchemical means. It is recommended that beekeepers maintain adequate colony strength, remove hive debris, and burn or otherwise get rid of unused or infested comb.

In Southeast Asia, wax moths are a major pest of *A. cerana*. Their presence frequently results in the colony absconding (Akratanakul 1987). Recently, larvae of *G. mellonella* infesting *A. cerana japonica* colonies tested positive for IAPV and BQCV (Triyasut et al. 2016). These bee viruses were likely transmitted to the larvae in food or other hive products that they consumed as no viral replication was detected within the larvae. Further work is needed to determine potential connections between wax moth larvae and honey bee viruses.

1.5.2.2 Small Hive Beetles

Aethina tumida (Nitidulidae, Coleoptera) is an emerging pest in Asia as of its first recorded discovery in the Philippines in 2014 (Brion 2015). Hundreds of colonies in Mindanao (southern Philippines) were lost between June 2014 and April 2015 from the parasitic activity of these beetles. They feed on pollen, brood, and honey. As the larvae feed, they produce a viscous slime which, along with their excrement leads to the fermentation of honey and the general fouling of hive products. Like, wax moths, small hive beetle are pests of stored combs as well.

The mild climate of Southern Asia provides excellent conditions for small hive beetles to develop. High temperatures have been shown to shorten larval and pupal development allowing for these insects to have more generations in a year (de Guzman and Frake 2007). In addition, the availability of brood, pollen, and honey year round from multiple bee species promotes heightened fecundity (de Guzman et al. 2015). Considering these qualities, small hive beetles likely pose a threat to stingless bees as well as the native honey bees of Asia. Further study is necessary to determine the impact that these invasive parasites are likely to have and what unique challenges to management will be posed in this new environment.

1.5.2.3 Wasps

Wasps (of the genus *Vespa*) are significant predators of honey bees in Asia (Matsuura 1988). The entrances of honey bee colonies are frequently targeted by large wasps capable of killing several bees in one attack (Cervancia 1993). Such disturbances often result in *A. cerana* absconding. Notably, *A. cerana* and *A. mellifera* have been observed creating a tight ball around hornet species, vibrating their flight muscles to generate enough heat to kill the attacker (Ono et al. 1987; Ken et al. 2005). Some species of wasps are known to specialize in attacking drones in drone congregations and have even been observed mimicking the flight patterns of queen bees to attract their prey (Oldroyd et al. 2006). Asian beekeepers, in an arguably less spectacular fashion, are known to attack wasps with tennis rackets, slippers, and wooden planks (Cervancia 1993).

1.5.2.4 Ants

Ants attack honey bee colonies to obtain food and periodically shelter. Weaver ants (*Oecophylla smaragdina*), black ants (*Monomorium* spp.), fire ants (*Solenopsis* spp.), and field ants (*Formica* spp.) are frequently recorded invading honey bee colonies and can be especially problematic for commercial beekeepers (Akrantanakul 1987). Dwarf honey bee species have to mount an aggressive to *O. smaragdina* as these ants will grab returning foragers and drones directly from the air. Overwhelming or persistent attack on a nest causes absconding in both dwarf bee species. Smaller ants (e.g., *Monomorium* spp.) are more of a challenge for *A. cerana*. Their size makes them more difficult to pick up and remove and they can present overwhelming numbers very quickly. Beekeepers can effectively manage any issues with the application of a grease barrier around colonies (Oldroyd and Wongsiri 2006).

1.5.2.5 Bee Lice

Braula coeca (Braulidae, Diptera) are wingless flies highly specialized to life in honey bee colonies to the extent that they are barely recognizable as flies. These insects are typically mistaken for *Varroa* because of their similarities in color and size but attention to the presence of three body segments (rather than one apparent body segment) and three pairs of legs (rather than four) can easily resolve the identification. Their small larvae consume wax, pollen, and honey as they tunnel through comb. The adults steal nectar and pollen directly from the mouths of worker bees as they feed each other via trophallaxis. It is not uncommon to see *B. coeca* aggregating on queens en masse likely because of the greater frequency of feeding via trophallaxis by her attendants. Though the sight may be alarming, these parasites are of minimal impact to a colony and require no chemical remediation.

1.5.2.6 Black Fungus Beetle

Alphitobius laevigatus (Tenebrionidae, Coleoptera) is a small beetle found recently in association with *A. cerana* colonies in China and Vietnam. These beetles are likely scavenging on various waste products and debris from these colonies and pose no threat to the health of the bees as has been determined thus far.

1.6 Conclusion

Considered keystone species for just their pollination services alone, honey bees are fascinatingly complex creatures wholly worthy of the name superorganism. Asia is unique in its diversity of *Apis* species, and the bees are known to play a vital role in maintaining the local fauna, sustaining the agricultural crops and securing food sources in a region where the population growth is among the highest in the world. Traditional knowledge and beekeeping practices have long been embedded into local cultures and ways of life. This book covers the diversity of honey bees, up to date beekeeping details, and differing beekeeping practices throughout history in the broad diversity of geographic regions comprising the continent of Asia.

Acknowledgement PC would like to acknowledge Thailand research fund ((RSA6080028) and Chiang Mai University fund. SR would like to acknowledge funding provided by the vanEngelsdorp lab at the University of Maryland and the encouragement/support of Dr. Dennis vanEngelsdorp.

References

Abrol DP, Bhat AA (1990) Studies on 'Thai sac brood virus' affecting indigenous honeybee *Apis cerana indica* Fab. colonies - prospects and future strategies – I. J Anim Morphol Physiol 37:101–108

Abrol DP, Putatunda BN (1996) New record of *Tropilaelaps clareae* (Acarina: Mesostigmata) associated with *Xylocopa irridipennis* from India. Sci Cult 62:59

Ai H, Yan X, Han R (2012) Occurrence and prevalence of seven bee viruses in *Apis mellifera* and *Apis cerana* apiaries in China. J Invertebr Pathol 109(1):160–164

Akrantanakul P (1987) Honey bee diseases and enemies in Asia. In: A parasitoidal guide. FAO, Rome

Allen M, Ball B (1996) The incidence and world distribution of honey bee viruses. Bee World 77:141–162. https://doi.org/10.1080/0005772X.1996.11099306

Amr SZ, Shehada SE, Abo-Shehada M et al (1998) Honey bee parasitic arthropods in Jordan. Apiacta 3:78–82

Anderson DL (1984) A comparison of serological techniques for detecting and identifying honeybee viruses. J Invertebr Pathol 44:233–243. https://doi.org/10.1016/0022-2011(84)90019-3

Anderson DL (2004) *Varroa* mites and their host relationships in the Philippines. In: Proceedings of the 7th Asian apicultural association conference and 10th beenet symposium and techno-flora, Laguna, Luzon, Philippines, pp 177–178

Anderson DL, Fuchs S (1998) Two genetically distinct populations of *Varroa jacobsoni* with contrasting reproductive abilities on *Apis mellifera*. J Apic Res 37:69–78. https://doi.org/10.1080/00218839.1998.11100957

Anderson DL, Morgan MJ (2007) Genetic and morphological variation of bee-parasitic *Tropilaelaps* mites (Acari: Laelapidae): new and re-defined species. Exp Appl Acarol 43:1–24. https://doi.org/10.1007/s10493-007-9103-0

Anderson DL, Roberts JMK (2013) Standard methods for *Tropilaelaps* mites research. J Apic Res 52:1–16. https://doi.org/10.3896/IBRA.1.52.4.21

Anderson D, Giacon H, Gibson N (1997) Detection and thermal destruction of the chalkbrood fungus (*Ascosphaera apis*) in honey. J Apic Res 36:163–168. https://doi.org/10.1080/00218839.1997.11100944

Anderson DL, Sukarsih (1996) Changed *Varroa jacobsoni* reproduction in *Apis mellifera* colonies in Java. Apidologie 27:461–466

Anderson DL, Trueman JW (2000) *Varroa jacobsoni* (Acari: Varroidae) is more than one species. Exp Appl Acarol 24:165–189

Ansari MJ, Al-Ghamdi A, Nuru A, Ahmed AM, Ayaad TH, Al-Qarni A, Alattal Y, Al-Waili N (2017) Survey and molecular detection of *Melissococcus plutonius*, the causative agent of European foulbrood in honeybees in Saudi Arabia. Saudi J Biol Sci 24(6):1327–1335. https://doi.org/10.1016/j.sjbs.2016.10.012

Arai R, Tominaga K, Wu M et al (2012) Diversity of *Melissococcus plutonius* from honey bee larvae in Japan and experimental reproduction of European foulbrood with cultured atypical isolates. PLoS One 7:e33708. https://doi.org/10.1371/journal.pone.0033708

Arias MC, Sheppard WS (2005) Phylogenetic relationships of honey bees (Hymenoptera: Apinae: Apini) inferred from nuclear and mitochondrial DNA sequence data. Mol Phylogenet Evol 37:25–35

Aronstein KA, Murray KD (2010) Chalkbrood disease in honey bees. J Invertebr Pathol 103(Suppl 1):S20–S29. https://doi.org/10.1016/j.jip.2009.06.018

Bailey L, Carpenter JM, Woods RD (1982) The diminished incidence of *Acarapis woodi* (Rennie) (Acari: Tarsonelidae) in honeybee, *Apis mellifera* L. (Hymenoptera: Apidae), in Britain. B Entomol Res 72:655–662

Beetsma J, Boot WJ, Calis J (1999) Invasion behaviour of *Varroa jacobsoni* Oud.: from bees into brood cells. Apidologie 30:125–140

Bharadwaj RK (1968) A new record of mite, *Tropilaelaps clareae* from *Apis dorsata*. Bee World 49:115

Boecking O, Spivak M (1999) Behavioral defenses of honey bees against *Varroa jacobsoni* Oud. Apidologie 30:141–158. https://doi.org/10.1051/apido:19990205

Booppha B, Eittsayeam S, Pengpat K et al (2010) Development of bioactive ceramics to control mite and microbial diseases in bee farms. Adv Mater Res 93:553–557

Boot WJ, Tan NQ, Dien PC et al (1997) Reproductive success of *Varroa jacobsoni* in brood of its original host, *Apis cerana*, in comparison to that of its new host, *A. mellifera* (Hymenoptera: Apidae). Bull Entomol Res 87:119–126

Botías C, Anderson DL, Meana A et al (2012) Further evidence of an oriental origin for *Nosema ceranae* (Microsporidia: Nosematidae). J Invertebr Pathol 110:108–113. https://doi.org/10.1016/j.jip.2012.02.014

Brion ACB (2015) Small hive beetle poses threat to bee industry. Available via the Philippine star. https://www.philstar.com/business/agriculture/2015/02/22/1426217/small-hive-beetle-poses-threat-bee-industry. Accessed 22 Nov 2017

Buawangpong N, de Guzman LI, Khongphinitbunjong K et al (2015) Prevalence and reproduction of *Tropilaelaps mercedesae* and *Varroa destructor* in concurrently infested *Apis mellifera* colonies. Apidologie 46:779–786. https://doi.org/10.1007/s13592-015-0368-8

Büchler R, Drescher W, Tornier I (1992) Grooming behaviour of *Apis cerana*, *Apis mellifera* and *Apis dorsata* and its effects on the parasitic mites *Varroa jacobsoni* and *Tropilaelaps clareae*. Exp Appl Acarol 16:313–319

Burgett DM, Kitprasert C (1990) Evaluation of Apistan as a control for *Tropilaelaps clareae* (Acari: Laelapidae), an Asian honey bee brood mite parasite. Am Bee J 130:51–53

Burgett M, Akratanakul P, Morse RA (1983) *Tropilaelaps clareae*: a parasite of honeybees in South-East Asia. Bee World 64:25–28. https://doi.org/10.1080/0005772X.1983.11097904

Camphor ESW, Hashmi AA, Ritter W et al (2005) Seasonal changes in mite (*Tropilaelaps clareae*) and honeybee (*Apis mellifera*) populations in Apistan treated and untreated colonies. Apiacta 40:34–44

Cervancia C (1993) Philippines beekeeping: status of research and development. In: Proc. Beenet Asia, pp 49–63

Chaimanee V, Warrit N, Chantawannakul P (2010) Infections of *Nosema ceranae* in four different honeybee species. J Invertebr Pathol 105:207–210

Chandler D, Davidson G, Pell JK, Ball BV et al (2000) Fungal biocontrol of acari. Biocontrol Sci Technol 10:357–384. https://doi.org/10.1080/09583150050114972

Chantawannakul P, Dancer BN (2001) American foulbrood in honey bees. Bee World 82:168–180

Chantawannakul P, Puchanichanthranon T, Wongsiri S (2005) Inhibitory effects of some medicinal plant extracts on the growth of *Ascosphaera apis*. Acta Hortic 678:183–189

Chantawannakul P, Ward L, Boonham N et al (2006) A scientific note on the detection of honey bee viruses using real-time PCR (TaqMan) in *Varroa* mites collected from a Thai honeybee (*Apis mellifera*) apiary. J Invertebr Pathol 91:69–73. https://doi.org/10.1016/j.jip.2005.11.001

Chantawannakul P, de Guzman LI, Li J et al (2016) Parasites, pathogens, and pests of honeybees in Asia. Apidologie 47:301–324. https://doi.org/10.1007/s13592-015-0407-5

Chen YP, Huang ZY (2010) *Nosema ceranae*, a newly identified pathogen of *Apis mellifera* in the USA and Asia. Apidologie 41:364–374. https://doi.org/10.1051/apido/2010021

Chen YP, Siede R (2007) Honey bee viruses. In: Karl Maramorosch AJS, Frederick AM (eds) Advances in virus research, vol 70. Academic, London, pp 33–80

Chen YW, Wang C, James A et al (2000) Susceptibility of the Asian honey bee, *Apis cerana*, to American foulbrood, *Paenibacillus larvae* larvae. J Apic Res 39:169–175

Chen YP, Zhao Y, Hammond J et al (2004) Multiple virus infections in the honeybee and genome divergence of honeybee viruses. J Invertebr Pathol 87:84–93

Chen YP, Evans J, Feldlaufer M (2006a) Horizontal and vertical transmission of viruses in the honey bee, *Apis mellifera*. J Invertebr Pathol 92:152–159. https://doi.org/10.1016/j.jip.2006.03.010

Chen YP, Pettis JS, Collins A et al (2006b) Prevalence and transmission of honeybee viruses. Appl Environ Microbiol 72:606–611. https://doi.org/10.1128/AEM.72.1.606-611.2006

Choe SE, Nguyen LTK, Noh JH et al (2012) Prevalence and distribution of six bee viruses in Korean *Apis cerana* populations. J Invertebr Pathol 109:330–333. https://doi.org/10.1016/j.jip.2012.01.003

Crane E (1990) Bees and beekeeping: science, practice and world resources. In: Bees and beekeeping: science, practice and world resources. pp xvii-+614

Crane E (1999) The world history of beekeeping and honey hunting. Gerald Duckworth & Co, Ltd, London

Dainat B, Ken T, Berthoud H et al (2009) The ectoparasitic mite *Tropilaelaps mercedesae* (Acari, Laelapidae) as a vector of honeybee viruses. Insect Soc 56:40–43. https://doi.org/10.1007/s00040-008-1030-5

Davis C, Ward W (2003) Control of chalkbrood disease with natural products: a report for the Rural Industries Research and Development Corporation. RIRDC publication no. 03/107, Kingston

de Guzman LI, Delfinado-Baker M (1996) A new species of *Varroa* (Acari: Varroidae) associated with *Apis koschevnikovi* (Apidae: Hymenoptera) in Borneo. Int J Acarol 22:23–27. https://doi.org/10.1080/01647959608684077
de Guzman LI, Frake AM (2007) Temperature affects *Aethina tumida* (Coleoptera: Nitidulidae) development. J Apic Res 46:88–93. https://doi.org/10.1080/00218839.2007.11101373
de Guzman LI, Rinderer TE (1998) Distribution of the Japanese and Russian genotypes of *Varroa jacobsoni*. Honey Bee Sci 19:115–119
de Guzman LI, Rinderer TE (1999) Identification and comparison of *Varroa* species infesting honey bees. Apidologie 30:85–95
de Guzman LI, Burgett M, Rinderer TE (2001) Biology and life history of *Acarapis dorsalis* and *Acarapis externus* Mites of the honey bee. Dadant, Hamilton, pp 17–27
de Guzman LI, Rinderer TE, Frake AM (2015) The effects of diet, mating duration, female to male ratios, and temperature on ovary activation, mating success, and fecundity of *Aethina tumida*. Apidologie 46:326–336. https://doi.org/10.1007/s13592-014-0325-y
de Guzman LI, Williams GR, Khongphinitbunjong K et al (2017) Ecology, life history and management of *Tropilaelaps* Mites. J Econ Entomol 110(2):319–332
De Jong D (1988) *Varroa jacobsoni* does reproduce in worker cells of *Apis cerana* in South Korea. Apidologie 19:241–244
de Miranda JR, Bailey L, Ball BV et al (2013) Standard methods for virus research in *Apis mellifera*. J Apic Res 52:1–56. https://doi.org/10.3896/IBRA.1.52.4.22
Decandido R, Allen D, Yosef KR et al (2004) A comparison of spring migration phenology of bee-eaters and oriental honey-buzzards *Pernis ptilorhyncus* at Tanjung Tuan, Malaysia, 2000-01. Ardea 92:169–174
Delfinado MD, Baker EW (1961) *Tropilaelaps*, a new genus of mite from the Philippines (Laelaptidae (s. Lat.): Acarina). Chicago Nat. Hist. Museum
Delfinado MD, Baker EW (1974) Varroidae, a new family of mites on honey bees (Mesostigmata: Acarina). J Wash Acad Sci 64:4–10
Delfinado-Baker M, Aggarwal K (1987) A new *Varroa* (Acari: Varroidae) from the nest of *Apis cerana* (Apidae). Int J Acarol 13:233–237
Delfinado-Baker M, Baker EW (1982) A new species of *Tropilaelaps* parasitic on honey bees. Am Bee J 122:416–417
Delfinado-Baker M, Baker EW, Phoon ACG (1989) Mites (Acari) associated with bees (Apidae) in Asia, with description of a new species. Am Bee J 129:609–613
Eckert JE (1961) Acarapis mites of the honey bee *Apis mellifera* L. J Insect Pathol 3:409–425
Eickwort GC (1990) Biogeography and taxonomy of honey bees. Am Entomol 36:58–59
Flores JM, Spivak M, Gutiérrez I (2005) Spores of *Ascosphaera apis* contained in wax foundation can infect honeybee brood. Vet Microbiol 108:141–144. https://doi.org/10.1016/j.vetmic.2005.03.005
Food and Agriculture Organization of the United Nations (FAO) (2014) FAOSTAT, Rome. http://faostat.fao.org
Forsgren E, de Miranda JR, Isaksson M et al (2009) Deformed wing virus associated with *Tropilaelaps mercedesae* infesting European honey bees (*Apis mellifera*). Exp Appl Acarol 47(2):87–97. https://doi.org/10.1007/s10493-008-9204-4
Forsgren E, Wei S, Guiling D et al (2015) Preliminary observations on possible pathogen spill-over from *Apis mellifera to Apis cerana*. Apidologie 46:265–275. https://doi.org/10.1007/s13592-014-0320-3
Friedmann H, Kern J (1956) The problem of cerophagy or wax-eating in the honey-guides. Q Rev Biol 31:19–30
Fries I (1988) Infectivity and multiplication of *Nosema apis* Z. in the ventriculus of the honey bee. Apidologie 19(3):319–328. https://doi.org/10.1051/apido:19880310
Fries I (1993) *Nosema apis*—a parasite in the honey bee colony. Bee World 74:5–19. https://doi.org/10.1080/0005772X.1993.11099149

Fries I, Feng F, da Silva A et al (1996a) *Nosema ceranae* n. sp. (Microspora, Nosematidae), morphological and molecular characterization of a microsporidian parasite of the Asian honey bee *Apis cerana* (Hymenoptera, Apidae). Eur J Protistol 32:356–365. https://doi.org/10.1016/S0932-4739(96)80059-9

Fries I, Wei H, Wei S et al (1996b) Grooming behavior and damaged mites (*Varroa jacobsoni*) in *Apis cerana cerana* and *Apis mellifera ligustica*. Apidologie 27:3–12

Fries I, Hansen H, Imdorf A, Rosenkranz P (2003) Swarming in honey bees (*Apis mellifera*) and *Varroa* mite (*Varroa destructor*) population development in Sweden. Apidologie 34:1–9

Fries I, Chauzat M-P, Chen Y-P et al (2013) Standard methods for *Nosema* research. J Apic Res 52:1–28. https://doi.org/10.3896/IBRA.1.52.1.14

Fuchs S, Long L, Anderson D (2000) A scientific note on the genetic distinctness of *Varroa* mites on *Apis mellifera* L. and on *Apis cerana* Fabr. in North Vietnam. Apidologie 31:456–460

Garcia-Gonzalez E, Genersch E (2013) Honey bee larval peritrophic matrix degradation during infection with *Paenibacillus larvae*, the aetiological agent of American foulbrood of honey bees, is a key step in pathogenesis. Environ Microbiol 15:2894–2901

Garg R, Sharma OP, Dogra GS (1984) Formic acid: an effective acaricide against Tropilaelaps clareae Delfinado and Baker (Laelaptidae: Acarina) and its effect on the brood and longevity of honey bees. Am Bee J 124:736–738

Garrido C, Rosenkranz P, Paxton RJ et al (2003) Temporal changes in *Varroa destructor* fertility and haplotype in Brazil. Apidologie 34:535–541. https://doi.org/10.1051/apido:2003041

Gatehouse HS, Malone LA (1999) Genetic variability among *Nosema apis* isolates. J Apic Res 38:79–85

Genersch E (2010) American Foulbrood in honeybees and its causative agent, *Paenibacillus larvae*. J Invertebr Pathol 103:S10–S19. https://doi.org/10.1016/j.jip.2009.06.015

Gerson U, Dag A, Efrat C, Slabezki Y et al (1994) Tracheal mite, *Acarapis woodi*, comes to Israel. Israel Am Bee J 134:486

Haddad NJ (2014) First detection of *Nosema ceranae* in Jordan. Eur Sci 10(33):1857–7881

Hasemann L (1961) How long can spores of American foulbrood live? Am Bee J 101:298–299

Heath LAF (1982) Chalk brood pathogens: a review. Bee World 63:130–135

Heath LAF (1985) Occurrence and distribution of chalkbrood disease of honeybees. Bee World 66:9–15. https://doi.org/10.1080/0005772X.1985.11098816

Hedtke K, Jensen PM, Jensen AB et al (2011) Evidence for emerging parasites and pathogens influencing outbreaks of stress-related diseases like chalkbrood. J Invertebr Pathol 108:167–173. https://doi.org/10.1016/j.jip.2011.08.006

Hitchcock JD, Christensen M (1972) Occurrence of chalkbrood (*Ascosphaera apis*) in honey bees in the United States. Mycologia 64:1193–1198

Hoppe H, Ritter W, Stephen EWC (1989) The control of parasitic bee mites *Varroa jacobsoni*, *Acarapis woodi* and *Tropilaelaps clareae* with formic acid. Am Bee J 129:739–742

Huang W, Jiang J, Chen Y et al (2007) A *Nosema ceranae* isolate from the honey bee *Apis mellifera*. Apidologie 38:30–37. https://doi.org/10.1051/apido:2006054

Huang W-F, Solter LF, Yau PM et al (2013) *Nosema ceranae* escapes fumagillin control in honey bees. PLoS Pathog 9:e1003185. https://doi.org/10.1371/journal.ppat.1003185

Hunter W, Ellis J, vanEngelsdorp D et al (2010) Large-scale field application of RNAi technology reducing Israeli Acute Paralysis Virus disease in honey bees (*Apis mellifera*, hymenoptera: Apidae). PLoS Pathog 6:e1001160. https://doi.org/10.1371/journal.ppat.1001160

Ibay LA (1989) Biology of the two external Acarapis species of honey bees: *Acarapis Dorsalis* Morgenthaler and *Acarapis Externus* Morgenthaler (Acari: Tarsonemidae). Am Bee J 129:816

Ken T, Hepburn HR, Radloff SE et al (2005) Heat-balling wasps by honeybees. Naturwissenschaften 92:492–495. https://doi.org/10.1007/s00114-005-0026-5

Khongphinitbunjong K, de Guzman LI, Burgett MD et al (2012) Behaviour response underpinning resistance and susceptibility of honey bees to *Tropilaelaps mercedesae*. Apidologie 43(5):590–599

Khongphinitbunjong K, de Guzman LI, Rinderer TE et al (2016) Responses of *Varroa*-resistant honey bees (*Apis mellifera* L.) to deformed wing virus. J Asia Pac Entomol 19:921–927. https://doi.org/10.1016/j.aspen.2016.08.008
Klee J, Besana AM, Genersch E et al (2007) Widespread dispersal of the microsporidian *Nosema ceranae*, an emergent pathogen of the western honey bee, *Apis mellifera*. J Invertebr Pathol 96:1–10. https://doi.org/10.1016/j.jip.2007.02.014
Koeniger N, Koeniger G, De Guzman LI et al (1993) Survival of *Euvarroa sinhai Delfinado* and *Baker* (Acari, Varroidae) on workers of *Apis cerana* Fabr, *Apis florea* Fabr and *Apis mellifera* L in cages. Apidologie 24:403–410
Koeniger G, Koeniger N, Anderson DL et al (2002) Mites from debris and sealed brood cells of *Apis dorsata* colonies in Sabah (Borneo) Malaysia, including a new haplotype of *Varroa jacobsoni*. Apidologie 33:15–24
Kojima Y, Toki T, Morimoto T et al (2011) Infestation of Japanese native honey bees by tracheal mite and virus from non-native European honey bees in Japan. Microb Ecol 62:895–906. https://doi.org/10.1007/s00248-011-9947-z
Kongpitak P, Polgár G, Heine J et al (2008) The efficacy of Bayvarol® and Check Mite® in the control of *Tropilaelaps mercedesae* in the European honey bee (*Apis mellifera*) in Thailand. Apıacta 43:12–16
Kumar NR, Kumar R, Mbaya J et al (1993) Tropilaelaps-clareae found on *Apis mellifera* in Africa. Bee World 74:101–102
Laigo FM, Morse RA (1968) The mite *Tropilaelaps clareae* in *Apis dorsata* colonies in the Philippines. Bee World 49:116–118
Lanzi G, de Miranda JR, Boniotti MB et al (2006) Molecular and biological characterization of deformed wing virus of honeybees (*Apis mellifera* L.) J Virol 80:4998–5009
Li J, Qin H, Wu J, Sadd BM et al (2012a) The prevalence of parasites and pathogens in Asian honeybees *Apis cerana* in China. PLoS One 7:e47955. https://doi.org/10.1371/journal.pone.0047955
Li M, Chen H, Tang J et al (2012b) Characterization of nociceptive responses to bee venom-induced inflammation in neonatal rats. Brain Res 1472:54–62. https://doi.org/10.1016/j.brainres.2012.07.005
Lin RH, Sullivan J, Huang ZY (2001) Earlier foraging in *Nosema*-infected honey bee workers: are host *Corpora allata* needed. In: Abstract # D0616. Entomol Soc Am meeting abstracts. http://esa.confex.com/esa/2001/techprogram/programs.htm. Accessed 29 Sept 2017
Liu TP (1991) Virus-like particles in the tracheal mite *Acarapis woodi* (Rennie). Apidologie 22:213–219
Liu ZL, Ho SH (1999) Bioactivity of the essential oil extracted from *Evodia rutaecarpa* Hook f. et Thomas against the grain storage insects, *Sitophilus zeamais* Motsch. and *Tribolium castaneum* (Herbst). J Stored Prod Res 35:317–328. https://doi.org/10.1016/S0022-474X(99)00015-6
Liu H, Pan G, Song S, Xu J et al (2008) Multiple rDNA units distributed on all chromosomes of *Nosema bombycis*. J Invertebr Pathol 99:235–238. https://doi.org/10.1016/j.jip.2008.06.012
Lo N, Gloag RS, Anderson DL, Oldroyd BP (2010) A molecular phylogeny of the genus *Apis* suggests that the giant Honeybee of the Philippines, *A. breviligula* Maa, and the plains Honeybee of southern India, *A. indica* Fabricius, are valid species. Syst Entomol 35(2):226–233
Locke B (2016) Natural Varroa mite-surviving *Apis mellifera* honeybee populations. Apidologie 47:467–482. https://doi.org/10.1007/s13592-015-0412-8
Mahmood R, Wagchoure ES, Raja S (2011) Effect of thymol and formic acid against ectoparasitic brood mite Tropilaelaps clareae in Apis mellifera colonies. Pak J Zool 43:91–95
Mahmood R, Wagchoure ES, Ul Mohsin A et al (2012) Control of ectoparasitic mites in honeybee (Apis mellifera L.) colonies by using thymol and oxalic acid. Pak J Zool 44:985–989
Martin S (1998) A population model for the ectoparasitic mite *Varroa jacobsoni* in honey bee (*Apis mellifera*) colonies. Ecol Model 109:267–281. https://doi.org/10.1016/S0304-3800(98)00059-3
Martin S, Holland K, Murray M (1997) Non-reproduction in the honeybee mite *Varroa jacobsoni*. Exp Appl Acarol 21:539–549
Martin SJ, Kemp D (1997) Average number of reproductive cycles performed by *Varroa jacobsoni* in honey bee (*Apis mellifera*) colonies. J Apic Res 36:113–123

Martin SJ, Highfield AC, Brettell L et al (2012) Global honey bee viral landscape altered by a parasitic mite. Science 336:1304–1306

Martín-Hernández R, Botías C, Barrios L et al (2011) Comparison of the energetic stress associated with experimental *Nosema ceranae* and *Nosema apis* infection of honeybees (*Apis mellifera*). Parasitol Res 109:605–612. https://doi.org/10.1007/s00436-011-2292-9

Matheson A (1993) World bee health report. Bee World 74:176–212. https://doi.org/10.1080/0005772X.1993.11099183

Mathpal Y (1984) Prehistoric rock paintings of bhimbetka. Abhinav Publications, New Delhi

Matsuura M (1988) Ecological study on vespine wasps (Hymenoptera:Vespidae) attacking honeybee colonies: I. seasonal changes in the frequency of visits to apiaries by vespine wasps and damage inflicted, especially in the absence of artificial protection. Appl Entomol Zool 23(4):428–440. https://doi.org/10.1303/aez.23.428

Meikle WG, Sammataro D, Neumann P et al (2012) Challenges for developing pathogen-based biopesticides against *Varroa destructor* (Mesostigmata: Varroidae). Apidologie 43:501–514. https://doi.org/10.1007/s13592-012-0118-0

Michael DS (1957) Acarine disease found in India. Am Bee J 97:100

Milne PS (1957) Acarine disease in *Apis indica*. Bee World 38:156. https://doi.org/10.1080/0005772X.1957.11094995

Mookhploy W, Kimura K, Disayathanoowat T et al (2015) Capsid gene divergence of Black Queen Cell Virus isolates in Thailand and Japan honey bee species. J Econ Entomol 108:1460–1464. https://doi.org/10.1093/jee/tov102

Mossadegh MS (1990) In vitro observations on ontogenesis of the mite, *Euvarroa Sinhai* Delfinado & Baker (Acari: Varroidae), in drone brood cells of the honeybee, *Apis mellifera* L. J Apic Res 29:230–232. https://doi.org/10.1080/00218839.1990.11101224

Mossadegh MS, Bahreini R (1994) Acarapis mites of honey-bee, *Apis mellifera* in Iran. Exp Appl Acarol 18:503–506. https://doi.org/10.1007/BF00051472

Munoz I, Garrido-Bailon E, Martin-Hernandez R et al (2008) Genetic profile of *Varroa destructor* infesting *Apis mellifera* iberiensis colonies. J Apic Res 47(4):310–313

Murray KD, Aronstein KA, de León JH (2007) Analysis of pMA67, a predicted rolling-circle replicating, mobilizable, tetracycline-resistance plasmid from the honey bee pathogen, *Paenibacillus larvae*. Plasmid 58:89–100. https://doi.org/10.1016/j.plasmid.2007.02.001

National Bureau of Agricultural Commodity and Food Standards (2008) Diagnosis of American foulbrood in bee. National Thai Agricultural Standard TAS 10351-2007, Royal Gazette Vol. 125 Special Section 3 D, 4 January B.E.2551. p 16

Navajas M, Anderson DL, de Guzman LI et al (2010) New Asian types of *Varroa destructor*: a potential new threat for world apiculture. Apidologie 41:181–193

Oldroyd B, Wongsiri S (2006) Asian honey bees: biology, conservation, and human interactions. Harvard University Press, London

Oldroyd BP, Wongsiri S, Seeley TD, Siriwat W (2006) Asian honey bees. (alk. paper) edn. Harvard University Press, Cambridge

Ono M, Okada I, Sasaki M (1987) Heat production by balling in the Japanese honeybee, *Apis cerana japonica* as a defensive behavior against the hornet, *Vespa simillima xanthoptera* (Hymenoptera: Vespidae). Experientia 43:1031–1034. https://doi.org/10.1007/BF01952231

Otis GW, Kralj J (2001) Parasitic mites not present in North America. In: Webster TC, Delaplane KS (eds) Mites of the honey bee. Dadant, Hamilton, pp 251–272

Oudemans AC (1904) On a new genus and species of parasitic Acari. Notes Leyden Mus 24(4):216–222

Peng YS, Fang Y, Xu S et al (1987) The resistance mechanism of the Asian honey bee, *Apis cerana* Fabr., to an ectoparasitic mite, *Varroa jacobsoni* Oudemans. J Invertebr Pathol 49:54–60. https://doi.org/10.1016/0022-2011(87)90125-X

Pernal SF, Clay H (2013) Honey bee diseases and pests. Canadian Association of Professional Apiculturists, Beaverlodge, AB, p 68

Pettis JS et al (2013) A rapid survey technique for Tropilaelaps mite (Mesostigmata: Laelapidae) detection. J Econ Entomol 106:1535–1544

Qin X, Evans JD, Aronstein KA et al (2006) Genome sequences of the honey bee pathogens *Paenibacillus larvae* and *Ascosphaera apis*. Insect Mol Biol 15:715–718. https://doi.org/10.1111/j.1365-2583.2006.00694.x

Raffique MK, Mahmood R, Aslam M, Sarwar G (2012) Control of *Tropilaelaps clareae* mite by using formic acid and thymol in honey bee *Apis mellifera* L. colonies. Pak J Zool 44:1129–1135

Ramsey SD (2018) Elucidation of novel nutritional, developmental, and behavioural adaptations for host exploitation in the mesostigmatid honey bee parasite *Varroa destructor*, Doctoral Dissertation, University of Maryland College Park, US

Rashad SE, Eweis MA, Nour ME (1985) Studies on the infestation of honeybees (*Apis mellifera*) by *Acarapis woodi* in Egypt. pp 152–156

Reynaldi FJ, López AC, Albo GN et al (2003) Differentiation of *Ascosphaera apis* isolates by rep-PCR fingerprinting and determination of chalkbrood incidence in Argentinean honey samples. J Apic Res 42:68–76. https://doi.org/10.1080/00218839.2003.11101096

Rice RN (2001) Nosema disease in honeybees: genetic variation and control. RIRDC

Rosenkranz P, Kirsch R, Renz R (2006) Population dynamics of honey bee colonies and varroa tolerance: a comparison between Uruguay and Germany 7th Encontro sobre Abelhas, USP, Ribeirão Preto, Brazil

Rosenkranz P, Aumeier P, Ziegelmann B (2010) Biology and control of *Varroa destructor*. J Invertebr Pathol 103(Suppl 1):S96–S119. https://doi.org/10.1016/j.jip.2009.07.016

Royce LA, Krantz GW, Ibay LA et al (1988) Some observations on the biology and behavior of *Acarapis woodi* and *Acarapis dorsalis* in Oregon Africanized honey bees and bee mites/editors, Glen R Needham [et al]

Sammataro D, Yoder J (2011) Global status of honey bee mites. In: Honey bee colony health: challenges and sustainable solutions. Contemporary topics in entomology. CRC Press, Boca Raton, pp 37–54

Sanpa S, Chantawannakul P (2009) Survey of six bee viruses using RT-PCR in Northern Thailand. J Invertebr Pathol 100:116–119

Shaw KE, Davidson G, Clark SJ et al (2002) Laboratory bioassays to assess the pathogenicity of mitosporic fungi to *Varroa destructor* (Acari: Mesostigmata), an ectoparasitic mite of the honeybee, *Apis mellifera*. Biol Control 24:266–276. https://doi.org/10.1016/S1049-9644(02)00029-4

Singh S (1961) Appearance of American foulbrood disease in Indian honeybee (*Apis cerana indica* Fabr.) Indian Bee J 27:46–50

Solignac M et al (2005) The invasive Korea and Japan types of *Varroa destructor*, ectoparasitic mites of the Western honeybee (*Apis mellifera*), are two partly isolated clones. Proc Biol Sci 272:411–419. https://doi.org/10.1098/rspb.2004.2853

Steche W (1985) Revision of Zander & Bottcher. Nosematose. In: Krankheiten der Biene, Handbuch der Bienenkunde

Strauss U, Pirk CWW, Crewe RM et al (2015) Impact of *Varroa destructor* on honeybee (*Apis mellifera* Scutellata) colony development in South Africa. Exp Appl Acarol 65:89–106. https://doi.org/10.1007/s10493-014-9842-7

Takamatsu D, Morinishi K, Arai R, Sakamoto A, Okura M, Osaki M (2014) Typing of *Melissococcus plutonius* isolated from European and Japanese honeybees suggests spread of sequence types across borders and between different *Apis* species. Vet Microbiol 171:221–226

Thapa R, Wongsiri S (2003) Flying predators of the giant honey bees; *Apis dorsata* and *Apis laboriosa* in Nepal. Am Bee J 143:540–542

Thapa R, Wongsiri S, Manandhar DN (2000) Current status of predators and diseases of honey bees in Nepal. In: Wongsiri S (ed) 7th IBRA and 5th AAA conf. Chiang Mai, Thailand, 2000. International Bee Research Association, Cardiff

Theantana T, Chantawannakul P (2008) Protease and β-N-acetylglucosaminidase of honey bee chalkbrood pathogen *Ascosphaera apis*. J Apic Res 47:68–76. https://doi.org/10.1080/00218839.2008.11101426

Traiyasut P, Mookhploy W, Kimura K et al (2016) First detection of honey bee viruses in wax moth. Chiang Mai J Sci 43:695–698

Tsevegmid K, Neumann P, Yañez O (2016) The honey bee pathosphere of Mongolia: European viruses in Central Asia. PLoS One 11:e0151164. https://doi.org/10.1371/journal.pone.0151164

Tutkun E, Maden S, Inci A et al (1993) General situation of chalkbrood disease in honeybees in Turkey. Turk Entomol Derg 17(2):65–68

van der Zee R, Pisa L, Andonov S et al (2012) Managed honey bee colony losses in Canada, China, Europe, Israel and Turkey, for the winters of 2008–9 and 2009–10. J Apic Res 51:100–114. https://doi.org/10.3896/IBRA.1.51.1.12

Walter DE, Proctor HC (1999) Mites: ecology, evolution and behaviour (No. 639.089 W34). UNSW Press, Sydney

Warrit N, Deborah Roan S, Chariya L (2006) Genetic subpopulations of *Varroa* mites and their *Apis cerana* hosts in Thailand. Apidologie 37:19–30

Whitaker J, Szalanski AL, Kence M (2011) Molecular detection of *Nosema ceranae* and *N. apis* from Turkish honey bees. Apidologie 42:174–180. https://doi.org/10.1051/apido/2010045

White GF (1913) Sacbrood, a disease of bees. US Department of Agriculture, Bureau of Entomology, Washington, DC

Williams GR, Sampson MA, Shutler D et al (2008) Does fumagillin control the recently detected invasive parasite *Nosema ceranae* in Western honey bees (*Apis mellifera*)? J Invertebr Pathol 99:342–344. https://doi.org/10.1016/j.jip.2008.04.005

Williams GR, Shutler D, Little CM et al (2011) The microsporidian *Nosema ceranae*, the antibiotic Fumagilin-B®, and Western honey bee, *Apis mellifera* colony strength. Apidologie 42(1):15–22

Wongsiri S, Thapa R, Chantawannakul P et al (2005) Bee eating birds and honey bee predation in Thailand. Am Bee J 145:419–422

Woyke J (1984) Survival and prophylactic control of *Tropilaelaps clareae* infesting *Apis mellifera* colonies in Afghanistan. Apidologie 15:421–434

Woyke J (1985) Further investigations into control of the parasite bee mite *Tropilaelaps clareae* without medication. J Apic Res 24:250–254. https://doi.org/10.1080/00218839.1985.11100681

Woyke J (1987) Length of stay of the parasitic mite *Tropilaelaps clareae* outside sealed honeybee brood cells as a basis for its effective control. J Apic Res 26:104–109

Wu J, Li J, Li JK (2006) Major honey plants and their utilization in China. Am Bee J 2:153–157

Yang X, Cox-Foster DL (2005) Impact of an ectoparasite on the immunity and pathology of an invertebrate: evidence for host immunosuppression and viral amplification. Proc Natl Acad Sci U S A 102:7470–7475. https://doi.org/10.1073/pnas.0501860102

Yang B, Peng G, Li T, Kadowaki T (2013) Molecular and phylogenetic characterization of honey bee viruses, *Nosema* microsporidia, protozoan parasites, and parasitic mites in China. Ecol Evol 3:298–311. https://doi.org/10.1002/ece3.464

Yen DF, Chyn LC (1971) Studies on a bacterial disease of honey bee in Taiwan. Plant Prot Bull 13:12–17

Yoo KH, Lee Y, Gretzel U et al (2009) Trust in travel-related consumer generated media. In: Hopken W, Gretzel U, Law R (eds) Information and communication technologies in tourism 2009. Springer, New York, p 49e6

Yoshiyama M, Kimura K (2011) Distribution of Nosema ceranae in the European honeybee, Apis mellifera in Japan. J Invertebr Pathol 106:263–267. https://doi.org/10.1016/j.jip.2010.10.010

Zander E (1909) Tierische Parasiten als Krankenheitserreger bei der Biene. Münch Bienenzeitung 31:196–204

Zhang X, He SY, Evans JD, Pettis JS, Yin GF, Chen YP (2012) New evidence that deformed wing virus and black queen cell virus are multi-host pathogens. J Invertebr Pathol 109:156–159. https://doi.org/10.1016/j.jip.2011.09.010

Zhou T, Anderson D, Huang ZSH et al (2004) Identification of *Varroa* mites (Acari: Varroidae) infesting *Apis cerana* and *Apis mellifera* in China. Apidologie 35:645–654

Zioni N, Soroker V, Chejanovsky N (2011) Replication of *Varroa destructor* virus 1 (VDV-1) and a *Varroa destructor* virus 1–deformed wing virus recombinant (VDV-1–DWV) in the head of the honey bee. Virology 417:106–112

Chapter 2
Beekeeping in Turkey: Bridging Asia and Europe

Aslı Özkırım

Abstract Turkey contains the second largest number of honeybee colonies in the world. Beekeeping is managed by beekeepers affiliated with 79 provincial beekeeper associations and the national Turkish Central Beekeepers Association (TAB) (www.tab.org.tr). Turkey is located at the geographic crossroads of Europe, Asia, and the Middle East, and encompasses a wide range of climates and habitats within its borders. Not surprisingly, honeybees of Turkey are also very diverse. Genetic variations among honeybee populations may be attributed by two factors: (1) adaptation to their local environments and (2) movement by humans. Migratory beekeeping has become widespread in Turkey within the last 20–30 years although beekeeping in Turkey is known since 8000–7000 BC. Like other countries, Turkish beekeepers face several challenges to producing their unique, highly respected honeys.

Keywords Turkey · Beekeeping · Apis mellifera · Pollination · Honeybee · Bee products · Pesticides · Bumblebee · History · Nectar plant

2.1 Introduction

Beekeeping is considered a traditional lifestyle in Turkey, particularly in rural areas. Currently, there are approximately 7,709,000 colonies and 83,500 beekeepers in Turkey (Turkish Statistical Institute 2016; Apicultural Registration System 2016), which means only China maintains more colonies at a country level. There are 79 beekeeper associations in the country—one for each province—and one nationwide group, the Turkish Central Beekeepers Association (TAB), that works with all provincial associations (www.tab.org.tr). Beekeepers that maintain more than 30 colonies must become a member of the Beekeepers Association in order to get extra

A. Özkırım
Faculty of Science, Department of Biology, Hacettepe University, Ankara, Turkey

Hacettepe University Bee and Bee Products Applied and Research Center, Ankara, Turkey
e-mail: ozkirim@hacettepe.edu.tr

P. Chantawannakul et al. (eds.), *Asian Beekeeping in the 21st Century*,
https://doi.org/10.1007/978-981-10-8222-1_2

promotions from the government and to be recorded in the Apiculture Registration System (ARS)—a joint program between the Turkish Central Beekeepers Association and The Turkish Ministry of Food, Agriculture and Livestock (MinFAL). Member beekeepers have personal identity cards and ID numbers and are provided with colony-specific barcodes that must be adhered to the hive (Fig. 2.1). Each packaged honey product must contain the same ID number, thus enabling maximum traceability. These efforts improve reporting about colony losses and health, as well as bee product quality. Beekeepers in Turkey must follow European Union regulations for packing and labeling. In total, 56,000 beekeepers are registered with the Turkish Central Beekeepers Association; average operation size is 113 colonies. These colonies produce ~17 kg/per colony.

2.2 Bee Diversity

The natural range of *Apis mellifera* L. includes most of Africa, Europe, and parts of the Middle East. Every beekeeper and honeybee biologist is aware that bees from different locations may differ in traits, such as size, color, wing venation, disease resistance, tendency to swarm, and colony defensiveness (Smith 2002). The bees of Turkey are particularly exciting for studies of honeybee biogeography. Turkey is at the geographic crossroads of Europe, Asia, and the Middle East, and encompasses a wide range of climates and habitats within its borders. Not surprisingly, the honeybees of Turkey are also very diverse, as shown by morphology (e.g., Güler 1996; Güler et al. 1999), behavior, ecology, and physiology (e.g., Adam 1954, 1964, 1977, 1983; Ruttner 1988, 1992). Variation among honeybee populations may be attributed to adaptation of honeybee populations to their diverse local environments.

Ruttner (1988) claimed that southwest Asia is a zone of high morphological diversification and evolution for honeybees. Many clearly distinct races have evolved within this region, which includes a diversity of habitats. Asia Minor, including Anatolia, appears to be the genetic center for these honeybee subspecies according to analysis of morphometric data. Honeybee races in this region include the subspecies *Apis mellifera anatoliaca*, *A. m. caucasica*, *A. m. meda*, and *A. m. syriaca*, which were considered by Ruttner to form a basal branch (O) of the species. Another subspecies that is found in the European part of Turkey, such as Thrace, may be *A. m. carnica*, which belongs to branch C of Ruttner's classification. Migratory beekeeping has become widespread in Turkey within the last 20–30 years. Thousands of colonies are overwintered in the Mediterranean and Aegean regions, and then moved to central and eastern Anatolia during the summer and fall. These practices might promote gene flow among different races and result in homogenization of the gene pool of Anatolian honeybees (Kandemir et al. 2000). Despite the apparent importance of Anatolia for the evolution of honeybees, very little work has been done on the morphological and genetic diversity of Anatolian honeybees (Kandemir and Kence 1995; Smith et al. 1997).

Mitochondrial DNA: Study of the mtDNA of Turkish honeybees shows that *A. m. anatoliaca* and *A. m. caucasica* belong to the Eastern or "C" mitochondrial lineage

Fig. 2.1 Plaques (**a**) and barcodes (**b**) are placed on hives to provide maximum traceability of honey (Photo: Bahri Yılmaz)

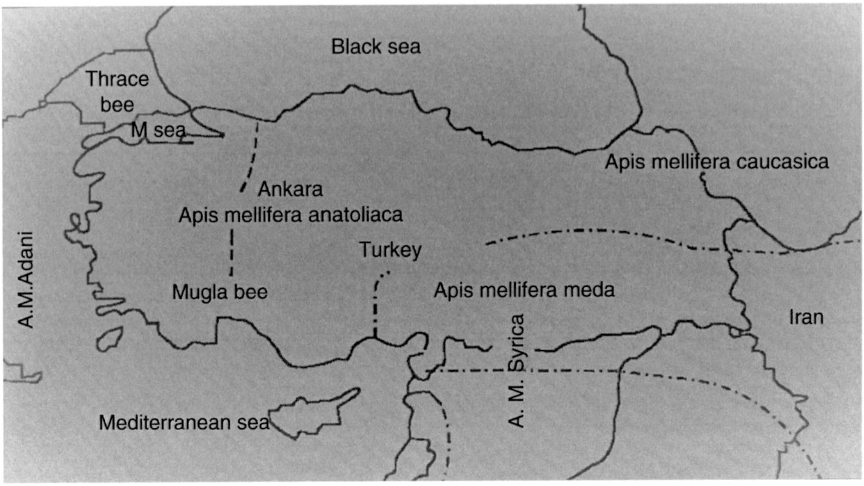

Fig. 2.2 Distribution areas of Turkish Honeybees races and ecotypes (Akyol et al. 2006)

(Smith et al. 1997; Palmer et al. 2000), as do *A. m. carnica* and *A. m. ligustica.* This differs from Ruttner's placement of *A. m. caucasica* in a separate Oriental group of subspecies along with *A. m. armeniaca, A. m.anatoliaca,* and *A. m. adami* (Ruttner 1992; Akyol et al. 2006) (Fig. 2.2). Within the Eastern group, *A. m. anatoliaca* from Thrace differed slightly from the *A. m. anatoliaca* from Anatolia. Some of the bees of Thrace share mtDNA types with *A. m. carnica* from Austria, Slovenia and Croatia (Smith and Brown 1990; Meixner et al. 1993). This suggests some interbreeding among the honeybees of Thrace, the Balkans and southern Austria. It is possible that the honeybees of Thrace form a bridge between the Balkan and Anatolian bees. Since Thrace mtDNA variation was not found in any bees from Anatolia, there could be a barrier to breeding between these two regions; more sampling in northwest Anatolia is needed to understand this. The "homeland" of *A. m. caucasica* is in the Caucasus Mountains, southern valleys of the Caucasus, and the higher reaches of the Little Caucasus mountains (Ruttner 1988), which are primarily located in Georgia and neighboring republics. There is some debate over the range of this bee in Turkey. Ruttner (1988) states that bees resembling *A. m. caucasica* were found along the Black Sea coast of Anatolia as far as Samsun. A detailed look at the mtDNA of *A. m. anatoliaca* and *A. m. caucasica* showed DNA sequence differences between the two. As expected, colonies sampled near the border with Georgia showed a high frequency (77%) of *A. m. caucasica* mtDNA. However, *A. m. caucasica* mtDNA also occurred far to the south of the proposed range of *A. m.caucasica.* Twenty-nine percent of colonies sampled near Erzurum and 25% of colonies sampled around Lake Van had *A. m. caucasica* mtDNA. On the other hand, no evidence of the Caucasian mtDNA along the Turkish Black Sea coast was found although sample size was small. If the mtDNA sequence was (Estoup et al. 1995) found in *A. m. caucasica* is truly characteristic of all *A. m. caucasica*, then results indicate a wide zone of overlap between *A. m. caucasica* and *A. m. anatoliaca*, at least from

Lake Van to the Georgian border. This could be due to natural gene flow and dispersal or to transportation of *A. m. caucasica* by humans. More extensive collections from the north-eastern area of Turkey and the heart of the *A. m. caucasica* range in Georgia would enable us to determine if DNA sequence called "Caucasian" is characteristic of all or most *A. m. caucasica* and will help to determine the true range of *A. m. caucasica* in Turkey. The mtDNA data do not suggest that the bees from the Lake Van region, supposedly in the range of *A. m. meda*, were in anyway distinct from other Anatolian populations (Özkırım et al., unpublished data). In southeast Anatolia, 57% of colonies sampled from Hatay were found to have a type of mtDNA that, at that time, had not been found in any other honeybees. They are called as the Middle Eastern or "M" lineage (Palmer et al. 2000). The honeybees from Hatay appear to be *A. m. syriaca*, whose range is reported to include the eastern coast of the Mediterranean north of the Negev desert, including parts of Israel, Jordan, Syria, and Lebanon. The M type of mtDNA has subsequently been found in Israel and Jordan (Smith and Shafir, unpublished data). Little genetic work has been carried out on the native honeybees of the Middle East, principally because imported *A. m. ligustica*, *A. m. carnica,* and other races have largely replaced the native honeybee where modern apiculture is practiced (Y. Lensky, pers. com.).

Allozymes: A high frequency of the Mdh1100 (or fast) allele is sometimes considered typical of African honeybees (e.g., Nunamaker et al. 1984). However, this allele is also present in high frequency in many Eastern subspecies, including the bees of Turkey. The Mdh1100 allele frequencies reported in the Eastern group of honeybee subspecies range from a low of approximately 23% in north Italian *A. m. ligustica* (Badino et al. 1985; Sheppard and Berlocher 1985) to nearly 100% in *A. m. caucasica* and *A. m. anatoliaca* in Turkey (Asal et al. 1995; Kandemir and Kence 1995; Palmer et al. 2000). We have no information yet on allozyme frequencies in Middle Eastern bees such as *A. m. syriaca*.

In the future, techniques employing microsatellites and DNA fingerprinting (e.g., Estoup et al. 1995) may show a more detailed picture of the diversity that exists among Turkish honeybees. Some of this variation may prove to be of economic importance. Local populations of honeybees are generally adapted to the ecological conditions found in their native habitat. For example, Adam (1983) noted that the Anatolian honeybee, *A. m. anatoliaca*, shows superb resistance to the cold winters and hot dry summers of central Anatolia. The biggest and extensive project of preserving Muğla races has been carried out by the Muğla beekeepers association. In addition, different races or subspecies of bees may differ in resistance to diseases, predators, and parasites. With modern transportation, new parasites and diseases can spread quickly around the world. One has only to look at the spread of the *Varroa destructor* mite. It is difficult to predict which bee races will be most resistant to new the next bee diseases. Genetic variation provides a potential arsenal of countermeasures against invading pests. New genetic techniques for mapping the chromosomes of animals (Hunt and Page Jr 1995) may soon enable apicultural scientists to locate and identify genes involved with particular traits, such as resistance to a particular pathogen. However, much of the genetic diversity of Turkish bee populations is in danger of being lost before it can be identified and studied. Some common aspects of modern beekeeping, such as re-queening colonies from one or few genetic stocks, may have immediate, short-term economic benefits for the beekeeper. However, it may also lead to long-term problems, including the introduction of bees that are not suited to local climates, introduction of new diseases, and loss of native genetic variation. Genetic studies of Turkish honeybee populations will be important for detecting, maintaining, and employing the genetic diversity of Turkey's bees.

2.3 History of Beekeeping in Anatolia and Recent Developments of Turkish Beekeeping

2.3.1 Ancient Period in Anatolia

Since the Middle-East region is the gene center of honeybees, beekeeping has a long history in Anatolia, starting back to 8000–7000 BC. Paintings at Çatalhöyük in Anatolia dating from around 7000 BC have been interpreted to depict honey combs and bees foraging on flowers (Crane 1999). Rock art on the wall of caves, as well as honeybee fossils from thousands of years ago, show that bee biology has not undergone major changes in their recent history. Primitive men had taken honey from bees by killing them. Apart from the history of honeybees, the first clear evidence of the history of beekeeping comes from Hittites (Altman 2010). This evidence takes the form of tablets in cuneiform writing (Fig. 2.3). The Neolithic site of Çatalhöyük was discovered in the 1950s by James Mellaart and was excavated between 1961 and 1965. The site rapidly became famous internationally due to the large size and dense occupation of the settlement, as well as the spectacular wall paintings and other art that was uncovered inside the houses. The site has been widely recognized as being of unique international significance. The spectacular art reflects the life of 8000 years ago and enables us to understand the first examples of agricultural activities and civilization. Çatalhöyük was also the first city in Anatolia possessing domesticated animals: the inhabitants kept dogs, goats, sheep, cattle and also hunted aurochs (wild oxen), pigs, and three kinds of deer. Honey and beeswax were also prevalent. Written laws about beekeeping and honey trade have also been found on tablets (Hoffner Jr 1974, 1997). For example, Hittite sources reveal a common designation used for bread in the texts on the tablets as "NINDA LAL" (Honeybread/cake) (Fig. 2.4) (Akkaya and Alkan 2007). Hittites was a beekeeping country and had been since the earliest times of recorded history. The bee "NIM LAL" also figured in the oldest known myths from the Hattian culture (Fig. 2.5). However, some other written terms such as hive "E NIM LAL," beekeeper "LU NIM LAL" have been found on the tablets of Maştiga Ritual (Anatolian Civilisations Museum).

Fig. 2.3 History of beekeeping in Turkey is coming from Hittite tablets in cuneiform writing (in Anadolu Civilisation Museum; Photo: Bahri Yılmaz)

Fig. 2.4 From the text "NINDA LAL" (Honeybread/cake) a common designation for bread (in Anadolu Civilisation Museum; Photo: Bahri Yılmaz)

Fig. 2.5 Myths on which honeybees' figures were used from the Hattian culture in the Anadolu Civilisation Museum (Photo: Bahri Yılmaz)

2.3.2 *Traditional Apiculture Period*

Beekeeping has frequently been popular in Turkey—for example, during ancient Anatolian civilizations, as well as during Seljuk's State, Anatolian Turks Principalities, and the Ottoman State. Both Seljuk's and Ottoman soldiers ate honey when they went to war and over 3000 kg of honey per year was consumed in Topkapı Palace. Anzer honey was produced only for Ottoman Sultans (Tutkun 2000; Sariöz 2006). The history of Turkish beekeeping can be divided into two parts. The "traditional apiculture period" occurred from ancient times, up to the Ottoman State (1299–1923) in Anatolia. The "modern apiculture period" began after World War I and continues to the present (Kandemir 2015). Turkey has a short history of beekeeping after the introduction of modern apiculture.

2.3.3 *Modern Apiculture Period*

Just after the founding of the Turkish Republic between 1923 and 1950, beekeeping was a successful sector for agricultural development because few resources were necessary for its activities (e.g., agricultural land, funds, expertise, etc.). The beekeeping sector developed very quickly after the World War II in Turkey, primarily because beekeeping was possible within all seven geographical regions of the country (Akbay 1986). The first study about Turkish beekeeping was made by the German named Bodenheimer, who studied the demographics of Turkish beekeeping (i.e., numbers of traditional and modern type of hives, bee products, and bee species) (Bodenheimer 1942). The results of this study were published as a book entitled "Studies on the Honey Bee and Beekeeping in Turkey." Later, Brother Adam visited Turkey three times, in 1954, 1962, and 1972, to observe bee behavior, bee species, and colony morphologies. He took Anatolian bees adapted to difficult climatic conditions that collected nectar quickly back to England for breeding new hybrid bees, such as the Buckfast bee (Adam 1983; Kandemir 2015).

In 1950s, a beekeeping institute was founded by the Ministry of Food, Agriculture and Livestock in Ankara. In 1969, the Turkish Development Foundation (TKV) started beekeeping projects in several regions of Turkey. The largest of this work, called the Integrated Beekeeping Project, focused on distributing equipment, hives, beeswax, and honeybees to young beekeepers in Kazan/Ankara.

During the 1980s and 1990s, modern beekeeping techniques spread among Turkish beekeepers. In 2000s, The Turkish Beekeepers Central Association (TAB) and other Province's Associations were founded. This resulted in most beekeepers with more than 30 colonies, being registered. This resulted in the TAB developing a strong network with the Ministry of Food, Agriculture and Livestock, as well as numerous universities, which continue to this day.

2.3.4 *Current Status of Turkish Beekeeping*

Many people in Turkey make a living from bees. For example, beekeeping is the only income source for more than 150,000 families in Turkey (www.tab.org). However, most Turkish beekeepers come from diverse professional backgrounds, such as teachers, farmers, places of worship, that maintain only a few hives because they enjoy working with these fascinating and useful insects (Kayral and Kayral 1983).

2.3.4.1 Famous Apicultural Areas

Turkey has a number of beekeeping facilities spread across the country to account for climatic conditions and bee flora. The provinces where most professional beekeepers are located in Turkey are (listed from highest to lowest): Muğla, Ordu,

Adana, İzmir, Antalya, Aydın, Erzurum, Sivas, Konya, Kars, İçel, and Ankara (Kumova 2000). The highest production of honey (i.e., 90 kg per hive) comes from the high plateaus of Kars, Erzurum, Bitlis, Kayseri, Tunceli, Ağrı, Yozgat, Şırnak, Batman, Hakkari, Bingöl, Van, Muş, Gümüşhane, Artvin, Giresun, Rize, Erzincan, Bayburt, Adıyaman, Malatya, Sivas provinces, Çukurova, Harran, Ergene and Menderes plains, Karacadağ, Anzer, Ovit, Sultanmurat, Alucra, Çamoluk, Zigana, Sahara, Sarıbulut, Santa, Toros mountainous areas, and the northern area of Thrace (Kayral and Kayral 1983; Şenocak 1988).

2.3.4.2 Amount and Type of Bee Hives

Today, beekeepers in Turkey maintain over seven million beehives. Turkey, with about seven bee colonies per km^2, has one of the highest densities of beehives in the world (Sıralı 2002). The numbers of honeybee colonies in northern and western areas are larger than other regions of Turkey. Only about 14% of total beehives in Turkey are located in Muğla province (Aegean region) and 8% in Ordu province (Black Sea region) (Gökçe 2001). Nearly half of all colonies are used for migratory beekeeping (Genç 1993). Langstroth type hives are commonly used in Turkey, with only about 4.2% of total colonies maintained in traditional hives. Many beekeepers of the Central Anatolian region generally use cylindrical hives made from wood branches covered with mud. In the northern region of Anatolia, beekeepers still use traditional log hives (Sıralı 2002). Wood box hives are used in many districts of South-Eastern Anatolia.

In many districts of Marmara and Central Anatolian, some people still enjoy traditional beekeeping, wherein bees are kept in ancient-style beehives made from *Salix* ssp. branches covered by mud with straw and manure. The production of honey is typically between 3 and 5 kg/hive (Sıralı 2002). All traditional beehives are called Turkish "karakovan" or "sepetkovan" (primitive hive) in Turkish. As a result, the honey yields of traditional beehives in Turkey are low generally. However, products of all these traditional hives are more expensive than modern bee hives (Kayral and Kayral 1989; Genç 1993).

2.3.4.3 Migratory Beekeeping

Migratory beekeeping has become common in Turkey. This means that hives are moved from one location to other in search for good quality honey in reasonable quantities (Gökçe 2001). They will usually move from north to south, and from east to west, following the blooming of honey plants. Sometimes, colonies are moved three times a year. However, some hives are maintained stationary because of the aging population of beekeepers, decreasing honey plants, colony transportation issues, stress on bees, and economic problems (Genç 1993). Honeybee colonies are generally transferred in spring to citrus groves and thyme areas, in June to the fir forests, in summer to cotton, clover, and sunflowers areas, and in August, September, and October to the vast pine forests (Santas 1990). Professional beekeepers that have 100–400 bee colonies, and sometimes up to 1000, transport their hives extensively during the year. In some cases, they may cover up to 2000–4000 km (Sıralı 2002).

Migratory beekeepers often move to wintering areas in the south of the country, where the climate is warmer. As a result, honeybees work almost the entire year. These factors may affect the life span of honeybees, as well as the vitality of colonies. Honeybees can be poisoned by different honey plants, especially when colonies are transferred to a region for the first time. Because of the history of migratory beekeeping in Turkey, most beekeepers follow special routes that avoid poisonous plants. Additionally, migratory beekeeping is preferred because in case one season is not productive, the follow-up season may be if the beekeeper attempts to follow the flora. This allows honey harvesting 2 to 3 times a year. Ultimately, beekeepers must weigh the pros and cons to migratory beekeeping in their attempt to maintain productive colonies.

2.3.4.4 Queen Bee Breeding

Over 100 beekeepers certificated by the Turkish Central Beekeepers Association (TAB) produce 500,000 fertilized queen bees per year (Fig. 2.6). Two locations as gene pools of homozygous honeybees were identified and that maintain strict import regulations are Posof/Ardahan and Macahel/Artvin. They are used for producing Caucasian queens and for performing artificial insemination. Considering that seven million colonies are managed in Turkey, 500,000 queens produced by the Turkish Central Beekeeping Association (TAB) are insufficient. Güler et al. reported that the main reason for the low honey yield is due to insufficient queen bee production in the country (Güler et al. 1999).

2.3.4.5 Colony Management

Colony management is scheduled around natural nectar flows in Turkey. Beekeepers want their colonies to reach maximum strength before the nectar flows begin. Nectar flows are very different between the north and south, or the east and west of Turkey (Delaplane 1999).

By mid-February, the hives are ready for detailed inspection. Queens resume laying eggs in February, after which brood production accelerates rapidly to provide the spring work force. Colonies stores may fall dangerously low in late winter when brood production has started but plants are not yet producing nectar or pollen. As a result, some colonies need supplemental feeding or sugar syrup (Genç 1993; Sıralı 2002).

In Turkey, the Government controls sugar prices. Because they are set lower than honey, beekeepers can afford to feed enough sugar syrup to their bees. This promotes colony productivity and prevents the robbing of weaker hives by strong ones.

Fig. 2.6 Queen bee breeding activities. Over 100 beekeepers who are certificated by the Turkish Central Beekeepers Association (TAB) produce approximately 500,000 fertilized queen bee per year (Photo: Bahri Yılmaz)

This may also prevent colonies from spreading disease and harmful mites. Bees in central, northern, and eastern areas spend five to six winter months without hibernation, and so need large quantities of honey. The amount of honey stored by Turkish bees differs according to the size of their colonies, but should be at least 5–15 kg at the time of winter preparation (Sasaki 1999).

2.3.4.6 Turkish Beekeeping Economy

The contribution of beekeeping to the Turkish economy is approximately $542 million per year. The value of pollination by bees is 10–15 fold this amount, or about $5.4 billion (TUİK 2015). The Ministry of Food, Agriculture and Livestock is aware of the importance of beekeeping to the country's economy. As a result, the government provides some financial support. For example, the T.C Ziraat Bank provides an apiculture credit, with the government paying out 50% of the interest.

Universities, beekeepers associations, and the Ministry of Agriculture work together to prepare projects that encourage beekeepers to produce more bee colonies, honey, and other bee products. For this, new bee equipment, colonies, pollen traps, and education are distributed to new beekeepers. The Ministry of Food, Agriculture and Livestock organizes beekeeper seminars regularly in towns and villages throughout the country in collaboration with various university scholars and the Turkish Association of Beekeepers.

2.3.4.7 Beekeeping Science and Technology

Performance of honeybee colonies, pathogens and pests, bees and bee products are analyzed in several well-equipped university and government laboratories. Furthermore, universities are supported by the government for their bee research endeavors. Many researchers and Ph.D. students from several scientific field in Turkey, such as biology, veterinary medicine, agriculture engineering, food engineering, and chemistry, collaborate with each other, as well as with researchers internationally. The government also maintains an expert list of scientists, beekeepers, representatives of bee companies, and foundations for consultation, and also organizes special meetings to progress the industry.

2.3.4.8 Beekeeping Communication in Turkey

Among the various communication channels, TV programs are the most effective to contact not only Turkish beekeepers, but also Turkish honey consumers. The primary three TV channels—TRT, Bereket TV, Köy TV—highlight practical applications in beekeeping. Other kinds of broadcasts may have invited experts to provide current information about beekeeping. Several scientific books (e.g., Doğaroğlu 2012; Sorkun 2008; Korkmaz and Bacaksız 2016; Korkmaz 2015) have been published about different subjects of Turkish beekeeping. Furthermore, *The Mellifera* journal is a scientific journal that is indexed by EBSCO, Zoological Records, Biological Abstracts, CAB, and SCI Master List; it has been published since 1996. The Uludağ Bee Journal that is published by the Uludağ Beekeeping Association aims to be a common source of information for beekeepers and scientists (http://www.uludag.edu.tr/dosyalar/agam). The two biggest regional beekeepers associations in Turkey, from Ordu and Muğla, also publish beekeeping journals (Oray-Bir'in Sesi ve Arıcının Sesi).

The largest international beekeeping congress is organized by the Muğla Province Beekeepers Association every 2 years. This year, the "5th International Muğla Beekeeping and Pine Honey Congress" will be held in Fethiye/Dead Sea-Muğla (www.muglacongress.org). The congress is a great opportunity to meet and discuss beekeeping issues with beekeepers and scientists from Turkey and abroad.

2.3.4.9 Turkish Beekeeping and Kids

Both Turkish scientists and beekeepers know that the future of beekeeping rests on the shoulders of the next generation. Therefore, mini-bee seminars have been organized by Hacettepe University's Bee and Bee Products Applied and Research Center (HARÜM) for children ranging from kindergarten to high school to explain the natural history and benefits of honeybees in a pedagogical way. The first Turkish story book of honeybees, called the "Honey Bee Castle," was published by the Turkish Scientific Council in 2012 (Özkırım 2011, 2012). The book was later translated to English and published by the International Bee Research Association (IBRA). Furthermore, the Turkish Scientific Council (TÜBİTAK) previously supported the project entitled "We are honey bee, we have honey too," which was led by Assist. Prof. Dr. Meral Kekeçoğlu (Düzce University) for 4 years in order to teach children about bees.

Lastly, collaboration among universities and beekeeper associations has allowed for the organization of "The World Bee Day" for the past 7 years. This event sees short conferences consisting of popular people from the bee sector, honey competitions, roundtables, drawing competitions among primary schools, theater plays about bees, and other activities throughout the country.

2.4 Bee Forage Resources

Turkey has 10,000 native plant species—3900 of them are endemic—contained within its three different phytogeography zones: European-Siberian, Mediterranean, and Iranian-Turonian. This region serves as a land bridge between Europe and Asia. Nearly 500 plants are known to provide nectar and pollen, with 50 classified as dominant nectar plants (Davis 1965–1988; Sorkun 2008).

2.4.1 Nectar Plants/Flower Honey

Turkish bees produce flower honey mainly from cultivated plants, such as oranges (in Muğla, İzmir, Antalya, Mersin, Adana, Hatay provinces), cotton (in Mardin, Diyarbakır, Şanlıurfa, Gaziantep, Adana, Aydın, Nazilli, İzmir provinces), sunflower (in İstanbul, Tekirdağ, Kırklareli, Edirne, Balıkesir, Çanakkale, Samsun, Aksaray, Yozgat, Adana provinces), heather (in Çanakkale, Tekirdağ, Kırklareli, İstanbul, Mersin, Tarsus provinces), chestnut (in Trabzon, Giresun, Samsun, Rize, Sinop, Kastamonu, Bolu, Bursa, Çanakkale, Balıkesir provinces), linden trees (in Tekirdağ, Kırklareli, İstanbul provinces), as well as a variety of orchards. Nectar yielding wild plants throughout Turkey include clovers, acacia, raspberries, strawberry, and bee balm (Ekim 1987; Sorkun and Doğan 1994; Sorkun 2008). Pollen yielding plants are abundant throughout Turkey too. Of special importance are almond trees (*Amigdalus*), *Castanea sativa*, *Castanea vulgaris*, *Salix alba*, *Robinia pseudoacacia* and *Erica*. The latter plant is strongly reputed by some honey lovers to cause improving health (Infandites 1990; Sorkun 2008).

2.4.2 Pine Trees/Pine Honey

Turkey leads the world in pine honey production. Over 90% of total production belongs to Turkey, whereas the remainder is primarily produced in Greece. Pine honey is produced from honeydew of the insect *Marchalina hellenica* (*Monophlebus hellenicus*) (Ülgentürk et al. 2012). Because of the enormous economic interest, bee colonies are transported from all parts of the Turkish country to the lavish pine (*Pinus brutia*) forests found mainly on the Marmara's islands (Büyükada and Heybeliada), on the peninsula of Aegean (Muğla, Fethiye, Denizli, İzmir, Edremit), and in the Mediterranean region (Antalya). September and October are the most plentiful months (Genç 1993). *Marchalina hellenica*, a scale insect that produces honeydew is protected in Turkey. Pine honey primarily consists of honeydew, a highly sugary secretion produced by *M. hellenica* during pine tree sap feeding on the surface of the tree trunk. The pine species on which *M. hellenica* can be found are mainly the "Turkish Pine" (*Pinus brutia*). Insects hide in the cracks and under the scales of the bark of these trees and produce a white cotton-like wax (Yeşil et al. 2005; Özkök et al. 2010). This type of honey brings a unique taste from pine forests that meet the Mediterranean coast at Turkey's Muğla region, where most of the world's pine honey is produced. Harvest occurs between August and December.

Compared to polyflora honey, pine honey has a unique aroma, is darker in color, and rarely crystallizes without any density loss. Pine honey therefore has many uses in the food sector. Nutritionally, honeydew is distinguished from other types of honey because it contains more enzymes, amino acids, and minerals. For these reasons, it is often offered to children. Furthermore, pine honey has substantial amounts of hyphae and spores that are only found in the pine forests of Turkey. Unlike polyflora honey, honeydew has lower levels of glucose and fructose (Özkök et al. 2010; Silici 2011). In western parts of Anatolia, pine honey has been the best possible antiseptic to treat wounds as part of the healing heritage from ancient generations. As a result, pine honey is major export honey of Turkey because of its uniqueness.

2.4.3 Forest Trees/Forest Honey

Colonies foraging on forest plants, especially on *Pinus* trees (the Germans call as "Waldhonig"), yield about 20% of Turkey's produced honey (Kumova 2000). Significant amounts of honeydew honey are also derived from *Papaver, Carduus, Rosa, Tilia, Salix, Quercus, Castenea, Populus, Betula, Tamarix, Ulmus, Picea, Prunus, Pyrus, Cedrus, Abies, and Malus* plants (Genç 1993). Unfortunately, there has been no extensive investigations about properties of these honeydews, insects that produce honey dew on these trees, or the physical, chemical, and microbiological characteristics of these kinds of honeydew honeys so far. Small amounts of honeydew are derived from *Abies* trees in the Marmara region and in the Uludağ mountains situated about 500 km north east of Ankara and 300 km south of Istanbul. This honey is being produced from honeydew of the insect *Lachnus* ssp. (Başak 1991; Çakmak 1999).

2.4.4 Special Flora/Well-Known Turkish Honey

Turkey's world famous thyme honey, renowned its excellent flavor and richness in enzymes, constitutes a rather limited percentage of the annual honey production in Turkey (Infandites 1990). It comes from the Central Anatolia, Aegean, Black Sea, and Marmara regions (Ekim 1987; Sorkun and Doğan 1994).

The most popular and most precious variety of honey in the Black Sea region is made from the nectar of high plateau plants of Anzer. This product, called Anzer honey, is more expensive than honey's produced elsewhere. It has a specific color, and a strong and elegant fragrance of Anzer high plateau plants. These plants only flowers at the end of May and beginning of August (Şekerden et al. 1992; Gökçe 2001). Anzer honey has also high antibacterial activities against some bacteria (Özkırım and Keskin 2001).

In addition, toxic honey named "Deli bal" or "Mad honey" is produced from *Rhododendron* spp. in Turkey during September and October in the northern mountains areas (Kayral and Kayral 1989; Genç 1993). This area stretches from east to west, parallel to the Black Sea, and is limited to the east by the Central Anatolia. Mad honey is widely used in indigenous medicine, especially for the treatment of hypertension and sexual dysfunction. However, the consumption of this honey can result in intoxication. The diagnosis of honey poisoning and a full understanding of its treatment is important for both effective and immediate treatment, and for the prevention of unnecessary medical costs. Evaluation of approximately 34 years of case reports between 1981 and 2014 suggest that poisoning was more frequently reported in males (75.17%), and between the ages 41 and 65. The most common complaints related to honey poisoning were dizziness, nausea, and presyncope. ECG tests revealed sinus bradycardia (79.58%), complete atrioventricular block (45.83%), atrioventricular block (30.91%), ST-segment elevation (22.63%), and nodal rhythm (11.27%). No deaths were report by any of the 1199 cases. The patients were most frequently treated with 0.5 mg atropine (37.79%), 1 mg atropine (49.73%), and saline (iv fluid) (65.35%). Patients were generally discharged within 24 h of recovery (Silici and Atayoğlu 2015).

2.5 Local Knowledge on Honeybees

Beekeeping is a traditional agricultural activity that is performed in almost every region of Turkey. Twenty percent of the world's 25 honeybee subspecies can be found in Turkey. Due to this diversity, bee farmers are encouraged to breed honeybee species native to their region instead of using commercial bees originating elsewhere (Fig. 2.6). Some Turkish beekeepers, especially in Black Sea region, still using pot hives hung on high trees, whereas some in the eastern part put frames underground (Fig. 2.7). Some even try to adapt traditional techniques to modern beekeeping. This has resulted in innovative circle frame hives in pots (Fig. 2.8). On the other hand, pot honey is the most expensive in Turkey. This is economic incentive that encourage traditional beekeeping continues today. In addition to food, honey has been used to treat wounds for millennia. More recently, Turkish people have adopted other bee products, like beeswax to make candles in the caverns or old houses, propolis for the treatment of many superficial and deep internal or external wounds infections (Stangaciu 1998; Gençay and Salih 2009; Silici and Kutluca 2005; Gençay Çelemli et al. 2013; Özkırım et al. 2014). Bee pollen is also used as a nutritional supplement and royal jelly for dignitaries like Ottoman Sultans (Silici and Kutluca 2005). However, honey and pollen is by far the most commonly used products in Turkish culture. Only in the last decades, science and modern technologies, has

Fig. 2.7 Pot hive on the high trees (**a**) and the frames put underground (**b**) in Turkey

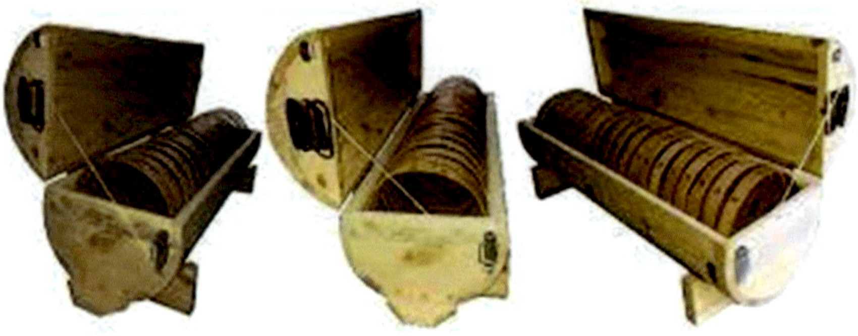

Fig. 2.8 Innovative circle frame hives like pots (http://www.karakovansatis.com)

promoted the production, collection, and use of other important bee products such as bee venom (Stangaciu 2015). In recent years, apitherapy was acknowledged by the Turkish Ministry of Health's Traditional Medical Science Department in 2012. Ultimately, local knowledge, beliefs, and culture of Turkish beekeeping would be integrated with Turkish science (Tanyüksel 2015).

2.6 Pests and Pathogens of Honeybees in Turkey

Honeybees in Turkey are threatened by pests and pathogens like the rest of the world. For example, some regional colony losses were similar to losses in North America and Europe. Over winter colony losses was 28% in 2006, 29.2% in 2007, 13.4% in 2008, 12.9% in 2009, and 18% in 2010 (van Der Zee et al. 2012). The causes of colony losses have been investigated by national and international projects, and in collaboration with the honeybee researcher network, COLOSS (www.coloss.org). Data suggest that losses are the result of several factors, including pests and new pathogens, pesticides, climatic change, poisoning, lack of biodiversity, negative effects on honeybee microbiota, and improper colony management (Muz 2008; Neumann and Özkırım 2011; Özkırım 2015). The Turkish Ministry of Food, Agriculture and Livestock (MinFAL) created a commission in 2013 to discuss reasons and to make a risk assessment for the future.

2.7 Pests of Honeybees in Turkey

2.7.1 Small Hive Beetle Aethina tumida *(SHB)*

There is currently no record of SHB in Turkey. After the first record of SHB in Italy (www.agriculture.gov.ie), *Aethina tumida* was listed as a mandatory pest in the country (www.tarim.gov.tr). Introduction of SHB to Turkey appears possible because of the proximity of Turkey to Italy.

2.7.2 Galleria mellonella *and* Achroia grisella

The greater wax-moth or honeycomb moth *Galleria mellonella* from the family Pyralidae is the only member of the genus *Galleria*. It is found in most of the world, including Europe and adjacent parts of Eurasia like Turkey. In Turkey, the greater wax-moth is an important pest that affects honeybee colony productivity by damaging stored honey combs and wax. While adults and pupae of *G. mellonella* do not damage the combs, larvae do so, particularly when comb is stored in dark, warm, and poorly ventilated areas. Physical, chemical, biological, and cultural methods have been used to control greater wax-moth damage in Turkey. Some of methods of *G. mellonella* control have low efficacy, while others are not easy to use, unsafe, or expensive. Because of these reasons, Turkish beekeepers are unwilling to use many of them (Akyol and Korkmaz 2008). *G. mellonella* has been found in southern regions of Turkey where the weather is hot and moisture levels are high. If the honeybee population decreases in a hive, adult *G. mellonella* are more likely to establish because there are not enough adult bees in the hive; this suggests colony mismanagement (Özkırım, 2000-Oral presentation in beekeepers' seminar). Moths cannot cause damage to strong, healthy colonies, but are dangerous to weak, unhealthy ones. *G. mellonella* is a close relative to the lesser wax-moth *Achroia grisella*, which is a member of tribe Galleriini of the pyralid subfamily Galleriinae. It has been observed in the same areas as *G. mellonella*. Chemical treatment is preferable by Turkish beekeepers to keep *Galleria mellonella* and *Achroia grisella* adults out of wax and honeycombs; however, freezing or cold-shock (−20 °C/6 h or +5 °C/20 h to destroy the eggs) has become popular treatment (Akyol and Korkmaz 2008).

2.8 Pathogens of Honeybees in Turkey

2.8.1 From Varroa jacobsoni *to* Varroa destructor

Varroa destructor is a major problem in Turkey, like many countries in the world. *Varroa* spp. were not known to occur in Turkey prior to 1977, and thereafter were only known in the far western region of the country (İlikler and Yüzbaşı 1981). Soon after, *Varroa* spp. mites spread to all regions of Turkey due to the large migratory beekeeping industry. Upwards of 600,000 colonies were reported to have been lost because of this parasite within the first 4 years of its introduction (Temizer 1983; Güler and Demir 2005). Studies on mtDNA Co-I gene sequences and morphological characters of many populations attributed to *V. jacobsoni* from different parts of the world showed that it was a complex of two species: *V. jacobsoni* and *V. destructor* (Anderson and Trueman 2000; Zhang 2000). This new analysis revealed that *V. jacobsoni* infests *A. cerana* in the Malaysia-Indonesia region only, and that *V. destructor* infests *A. cerana* in Asia and *A. mellifera* worldwide.

Immediately after the reclassification of *V. jacobsoni*, female *Varroa* spp. were collected from worker honeybees from 118 colonies in 24 apiaries that were located in intensive beekeeping areas of 17 Turkish provinces. Later, it was reported that the

V. destructor samples from the eight Black Sea provinces belonged to the economically important Korean strain (Aydın et al. 2007). Morphometric analysis was performed on collected *Varroa* spp. specimens from various regions in Turkey, and they were further identified to be *V. destructor* (Aydın et al. 2007).

Since its introduction, *V. destructor* has infested 98.9% of honeybee colonies in Turkey. Although infestation rates can vary by year, they do not fluctuate significantly. Therefore, some levels of resistance to *V. destructor* may have developed in the endemic populations over time. Regardless, *V. destructor* is widespread in Turkey (Özkırım and Keskin 2003; Çakmak et al. 2003; Warrit et al. 2004). In addition, *V. destructor* is the most damaging parasite of honeybee colonies; without effective chemical control tools, many colonies die off due to varroosis. Beekeeping would be neither profitable nor enjoyable in many areas of Turkey without effective treatment against *V. destructor* (Çakmak et al. 2003). The beekeeping market, with more than seven million colonies in Turkey, is large enough to induce manufacturers to generate a variety of control tools if current registered products fail. Chemical treatments employing Flumethrin, Tau-Fluvalinate, Amitraz, Coumaphos (Girişgin and Aydın 2010), organic acids like lactic, formic and oxalic acids (Akyol and Yeninar 2009), thymol and other essential oils (Akyol and Yeninar 2008), and powdered sugar applications (Çakmak unpublished data) are used against *V. destructor*. Pollen traps and drone bee combs are also used as cultural barriers to *V. destructor*. Currently, the Ministry of Food, Agriculture and Livestock is promoting Turkish beekeepers to use organic products that not to leave residues in hives and to use cultural barriers against *V. destructor*. Furthermore, many provincial beekeeper associations urge the chemical treatment coordinated; this allows all colonies to be treated at the same time by the same active ingredient in order to solve effectiveness and resistance problems of the products.

2.8.2 **Nosema ceranae *and* Nosema apis**

Nosemosis is a common worldwide disease of adult honeybees (*Apis mellifera* L.) that is caused by microsporidia (Fries et al. 1996; Paxton et al. 2007; Fries 2010). Nosemosis has become a major problem in Turkey in recent years (Özkırım and Keskin 2001; Aydın and Çakmak 2005; Özkırım et al. 2006). *Nosema apis* was the only agent known to produce this disease in *A. mellifera* L. until *Nosema ceranae* was identified in this host in 2005 in Europe (Higes et al. 2006, 2009). Investigations that have been performed in Turkey reveal that *N. apis* is widespread in the north and the Black Sea region, where the weather is cool and humidity is high (Özkırım and Keskin 2001; Aydın and Çakmak 2005; Huang et al. 2008). *N. cerenae* seems to be more prevalent than *N. apis* since it has been reported in many regions. Because the geography of Turkey varies to include forest, mountain, steppe, wetland, and coastal regions, it is difficult to discern at present their influence on driving completion between *N. apis* and *N.ceranae* in the different regions. Since 2009, bee samples have been collected from the different provinces of Turkey to understand seasonal pathogenicity and disease levels for both *N. apis* and *N. ceranae*. The results show that *N.apis* pathogenicity directly depends on weather conditions, particularly temperature and humidity. On the contrary, *N. cerenae* has the colonies throughout the country exceedingly quickly, without regard for climatic conditions. Although still poorly understood, Turkish scientists have reported Nosemosis to be associated with Deformed Wing Virus, lack of winter cluster, and queen loss (Muz 2008; Muz and Muz 2009; Muz et al. 2010). As a result, Nosemosis, especially caused by *N. ceranae*, is considered to be one of the main causes of colony loss in Turkey (Muz and Doğaroğlu 2011).

Nosema spores can also be found in Turkish honey, pollen, and beeswax (Özkırım and Keskin 2001). This kind of contamination represents a high risk for healthy bee colonies in Turkey. Turkish beekeepers leave some amount of honey in hives for winter. However, they prefer to leave low-quality honey and to transfer frame honey from weak colonies when combining them. Furthermore, they may also use contaminated honey and pollen to make cake (bee bread) for feeding colonies. Because of the risk of spreading disease, the Ministry of Food, Agriculture and Livestock, universities, and beekeeper associations suggest that beekeepers feed their colonies using their own honey, pollen, and cakes Ministry of Agriculture Risk Assessment Report (2013).

2.8.3 Bee Viruses

Honeybee viruses were the first discovered in Turkish apiaries experiencing elevated colony mortality in 2008 (Muz and Muz 2008). Further studies revealed that bee viruses, including IAPV (Israeli Acute Bee Paralysis Virus), occur in Turkey (Gülmez et al. 2009; Okursoy et al. 2010; Özkırım and Schiesser 2013). Viral infections are mostly correlated with *Varroa destructor*, Nosema spp., and *Ascosphaera apis* (Chalkbrood) infections (Muz and Muz 2008, 2010; Gülmez et al. 2009). Because no viral treatment is available, Turkish beekeepers are advised to reduce *V.destructor* and *Nosema* spp. levels and maintain strong colonies.

2.8.4 Foulbrood Diseases

American foulbrood (AFB) caused by *Paenibacillus larvae* is a mandatory notifiable disease in Turkey. Infected colonies, as well as colonies with radius of 7 km, are placed and are burnt; no compensation is offered by the government. Because of this, Turkish beekeepers often do not notify the Ministry of Food, Agriculture and Livestock of the disease and instead attempt to treat their infected colonies themselves. As a result, colonies often remain as disease reservoirs. The prevalence of AFB in the country is believed to be between 2 and 6% (Özkırım and Keskin 2002; Özkırım et al. 2006, 2012, 2014; Özkırım and Yalçınkaya 2012). European foulbrood (EFB), caused by *Melissococcus plutonius* and associated bacteria, is not widespread in Turkey. In some cases, secondary infection agents *Bacillus laterosporus, Bacillus alvei,* and *Enterococcus* sp. cause some problems in weak honeybee colonies (Özkırım and Keskin 2002).

Foulbrood caused by *Bacillus* spp., *Streptococcus* spp., and *Enterococcus* spp., are also observed in Turkey. Prevalence levels vary tremendously throughout the season, but approximately 21.8% in whole country (Özkırım and Keskin 2002; Özkırım and Yalçınkaya 2012; Özkırım et al. 2014). The main cause of ordinary foulbrood is the large range of day and night temperatures, especially during spring and autumn. During these periods, eggs laid by the queen on the outer areas of frames are often left unattended as the colony contracts during cool temperatures. Ordinary foulbrood should not be confused with AFB or EFB, which are caused by different bacteria that produce different symptoms and require their own control methods. Colonies exhibiting signs of foulbrood disease can result in reduced bee populations and decreased production of bee products.

2.9 Other Problems of Beekeeping in Turkey

Turkish beekeeping faces similar problems as other places in the world. Beekeeping is a traditional agricultural activity that is carried out in almost every region of Turkey. Twenty percent of the world's 25 honeybee *A. mellifera* subspecies can be found in Turkey. Due to this diversity, bee farmers are encouraged to breed bee species native to their region instead of purchasing commercial bees from elsewhere. However, Turkish breeders face many challenges, particularly erratic weather changes. Because of this, they request support from the Ministry of Food, Agriculture and Livestock to accommodate for unexpected conditions. These conditions also affect scale insect *Marchalina hellenica*, which is the most important insect for pine honey and honeydew production. Since 2006, *M. hellenica* has been included in the European and Mediterranean Plant Protection alert list. In addition, the Turkish Ministry of Forestry has sought to protect the forests, mostly in Muğla Province, where this insect lives.

Approximately 70% of plants in Turkey are floristic. Therefore, apicultural production has great potential in Turkey if stress factors can be mitigated (www.tab.org). The major problems of beekeeping in Turkey can be listed as below:

- The environment, introduced parasites, beekeeping techniques, the quality of products, problems in marketing and trading (Nakamura 1999).
- Several issues fervently remain, including the use of old and unproductive queens in colonies, inappropriate apiculture equipment and supplies, and lack of education of beekeepers (Şahinler and Şahinler 1996).
- One critical setback to Turkish beekeeping is that beekeepers must often pay fees to farmers or landowners to use their land, unlike other regions of the world where beekeepers are paid for pollination services (Özkırım, unpublished data).
- Bee products that are collected, but not stored and marketed in appropriate storage conditions cause the misusing of bee products. However, some tricks are used specially on honey, royal jelly, and pollen. Well-educated beekeepers are

important for improving product quality and distribution and will contribute to the development of better beekeeping technology (Tolon 1999).

- Bee wax and comb foundations are the most dangerous reservoir for bacterial and fungal diseases. Unfortunately, beeswax from multiple beekeepers is often consolidated, rendered, and then spread throughout the country (Hornitzky and Wills 1983; Plessis et al. 1985; Machova 1993; Özkırım and Keskin 2005, 2010; Haddad 2011; Ekici and Kırgız 2012). The Ministry of Food, Agriculture and Livestock and Turkish Central Beekeepers Association organized a roundtable to discuss the problem in 2013. They decided to recommend boiling beeswax individually as much as possible. However, bee wax and comb foundation are still major sources of *P. larvae* spread.
- Imidacloprid, a nicotine-derived systemic insecticide belonging to a group of pesticides called neonicotinoids, has been another problem for Turkish beekeeping (Girolami et al. 2009). Colony losses in Thrace and South-East regions, where cotton and sunflowers are produced, were evaluated for imidacloprid in bee colonies by the Ministry of Food, Agriculture and Livestock. As a result, the Ministry of Food, Agriculture and Livestock banned the use of imidacloprid for cotton and sunflower seeds in Turkey.
- Resistance and residue (RR) problems are also an issue in the Turkish beekeeping sector. Turkish beekeepers are keen to overdose medical treatments against any kind of parasite or microorganism. Even if colonies are healthy, they will still use these chemicals prophylactically. As a result, pathogen resistance and chemical residues may occur in the beekeeping sector. In some cases, honeybees can be poisoned by misusing chemical treatments. In the hopes of reducing environmental contamination, the use of powdered pesticides and aerial pesticide applications have been forbidden by the Ministry of Food, Agriculture and Livestock. Despite farmer education, the application of agricultural insecticides during day still occurs (Bahri Yılmaz; personal notes).

2.10 Other Non-*Apis* Species Found in Turkey

Forty-eight species of bumblebee are found in Turkey. For example, *Alpigenobombus wurfleini* is a relatively rare mountain species and appears in north-eastern Anatolia above 1800 mt (Özbek 1997). Ranges of *Bombias handlirschianus* and *B. shaposhnikov* overlap, as both are mountain species (1800–3000 mt) that prefer open areas. In general, they are abundant species, but their populations have been in decline since the 1980s. *Bombus terrestris* and *B. lucorum* are widespread and recorded in the same district but not at the same altitudes. Contrary to *B. lucorum*, *B. terrestris* prefers warmer areas, low altitudes, and open spaces. *B. cryptarum* is restricted to north east Anatolia and prefers open area, high altitudes (1500–3000 mt) and colder climates. With the exception of *B. terrestris*, the other species are relatively cold-adapted insects. Long-term records and observations show that populations of these bumblebee species have declined over the past two decades.

Similar to honeybees, Turkey contains a high diversity of wasps (approximately 1000 species) (Özbek 1997; Aytekin 2009; Güler et al. 2014).

2.11 Future Perspectives

Turkey has a great beekeeping potential because of its very rich flora and suitable ecology; however, its potential has not been reached yet. Total honey production, honey yield, and productions of other bee products are rather low. It is likely that it can be increased two- or threefold. Some of the reasons for low yields are bee diseases and parasites, insufficient queen production, and low educational level, and knowledge of beekeepers. It is essential to harvest process and marketing of bee products such as honey, bee wax, pollen, royal jelly, propolis, and bee venom. In order to develop advantage of Turkey's strong honey industry, bee product standards should be obeyed to ensure that these products are. This is especially important when they are exported.

Acknowledgement First of all, I would like to devote this chapter to the memory of Mr. Bahri YILMAZ, the former president of Turkish Central Beekeepers Association, who passed away 5 months ago. He was the largest and most experienced beekeeper in Turkey. I tried my best to reflect his great works for Turkish Beekeeping through this chapter. I hope that this book will keep alive his efforts forever. I would also like to thank the scientists that have supplied data for this special chapter. Finally, a special thanks goes to all Turkish scientists and Turkish beekeepers that are helping the country's beekeeping sector move forward.

References

Adam M (1954) In search of the best strains of bees: second journey. Bee World 35(12):233–244. https://doi.org/10.1080/0005772X.1954.11094853

Adam M (1964) In search of the best strains of bee: concluding journeys. Bee World 45(3):104–118. https://doi.org/10.1080/0005772X.1964.11097059

Adam M (1977) In search of the best strains of bee: supplementary journey to Asia Minor, 1973. Bee World 58(2):57–66. https://doi.org/10.1080/0005772X.1977.11097644

Adam M (1983) In search of the best strains of bees. Northern Bee Books, Hebden Bridge, West Yorkshire, and Dandant and Sons, Hamilton

Akbay R (1986) Arı ve İpekböcegi Yetistirme. Ankara Üniv. Zir. Fak. Yay. No: 956, Ders Kitabı: 276. Ankara, p 308

Akkaya H, Alkan S (2007) Beekeeping in Anatolia from the Hittites to the present day. J Apic Res 46:120–124

Akyol E, Korkmaz A (2008) The effect of–5°c cold application to protect honeycombs against greater wax-moth *Galleria mellonella* damage. Uludağ Bee J 8(1):26–29

Akyol E, Yeninar H (2008) Controlling *Varroa destructor* (Acari: Varroidae) in honeybee *Apis mellifera* (Hymenoptera: Apidae) colonies by using Thymovar® and BeeVital®. Ital J Anim Sci 7:237–242

Akyol E, Yeninar H (2009) Use of oxalic acid to control *Varroa destructor* in honeybee *(Apis mellifera* L.) colonies. Turk J Vet Anim Sci 33(4):285–288

Akyol E, Şahinler N, Özkök D (2006) Honeybee (*Apis mellifera* L.) races, ecotypes and their general characteristics in Turkey. J Anim Vet Adv 5(9):771–774
Altman N (2010) The honey prescription: the amazing power of honey as medicine. Healing Arts Press, Toronto
Anderson DL, Trueman JWH (2000) *Varroa jacobsoni* (Acari; Varroidae) is more than one species. Exp Appl Acarol 24:165–189
Apicultural Registration System (2016) The ARS register. http://www.tab.org.tr. Accessed 11 Sept 2017
Asal S, Kocabaş S, Elmacı C et al (1995) Enzyme polymorphism in honey bee (*Apis mellifera* L.) from Anatolia. Turk J Zool 19:153–156
Aydın L, Çakmak İ (2005) Honey bee Nosema disease in the Republic of Turkey. J Apic Res 44(4):196–197
Aydın L, Güleğen E, Çakmak İ et al (2007) Occurrence of *Varroa destructor* Anderson and Trueman, 2000 on Honey Bees (*Apis mellifera*) in Turkey. Turk J Vet Anim Sci 31:189–191
Aytekin M (2009) Atlas of bumblebees in Turkey. http://www.atlashymenoptera.net/. Accessed 11 Sept 2017
Badino G, Celebrano G, Manino A et al (1985) Enzyme polymorphism in the Sicilian honeybee. Experientia 41:752–754
Başak M (1991) Haseratın Arıcılıktaki Rolü. Teknik Arıcılık. Sayı 31. Ankara
Bodenheimer FS (1942) Studies on the honey bee and beekeeping in Turkey. Merkez Ziraat Mücadele Enstitüsü, Ankara
Çakmak İ (1999) Honeybees and agriculture. May Agro-Tek Bursa 3(7):7–9
Çakmak İ, Aydın L, Güleğen E et al (2003) Varroa *(Varroa destructor)* and tracheal mite *(Acarapis woodi)* incidence in the Republic of Turkey. J Apic Res 42:57–60
Crane E (1999) The world history of beekeeping and honey hunting. Taylor and Francis Group, New York
Davis PH (1965–1988) Flora of Turkey, vol I–X. Edinburg University Press, Edinburg
Delaplane KS (1999) Honey bees and beekeeping. The University of Georgia, College of Agricultural and Environmental Sciences, Athens
Doğaroğlu M (2012) Modern Arıcılık Teknikleri. Doğa Arıcılık, p 326
Ekici M, Kırgız Y (2012) Arıcılıkta Kullanılan Mum ve Temel Petekte Yaşanan Sorunlar ve Çözüm Önerileri. Paper presented at 3. Uluslararası Muğla Beekeeping and Pine Honey Congress, Muğla, 1-4 Kasım 2012
Ekim T (1987) Arıcılıkta Önem Tasıyan Bitkiler ve Bunların Yurdumuzdaki Durumu. In: Türkiye 1. Arıcılık Kongresi (22-24 Ocak 1980). Tarım ve Köyisleri Bak. Yay. No: 154. Ankara, pp 53–64
Estoup A, Garnery L, Solignac M et al (1995) Microsatellite variation in honey bee (*Apis mellifera* L.) populations: hierarchical genetic structure and test of the infinite allele and stepwise mutation models. Genetics 140:679–695
Fries I (2010) *Nosema ceranae* in European honey bees (*Apis mellifera* L.) J Invertebr Pathol 103:573–579
Fries I, Feng F, Silva AD et al (1996) *Nosema ceranae* n. sp. (Microspora, Nosematidae), morphological and molecular characterization of a microsporidian parasite of the Asian honey bee *Apis cerana* (Hymenoptera, Apidae). Eur J Protistol 32:356–365
Genç F (1993) Arıcılıgın Temel Esasları. Atatürk Üniversitesi Ziraat Fakültesi Yayın No 149. Erzurum, pp 138–185
Gençay Çelemli Ö, Hatjina F, Charistos L et al (2013) More insight into the chemical composition of Greek propolis; Differences and similarities with Turkish Propolis. Z Naturforsch C 68:429–438
Gençay Ö, Salih B (2009) GC-MS analysis of propolis samples from 17 different Regions of Turkey, Four different regions of Brazil and one from Japan. Mellifera 9(17):19–28
Girişgin O, Aydın L (2010) Efficacies of formic, oxalic and lactic acids against *Varroa destructor* in naturally infested honeybee (*Apis mellifera* L.) colonies in Turkey. Kafkas Univ Vet Fak Dergi 16(6):941–945

Girolami V, Mazzon L, Squartini A et al (2009) Translocation of neonicotinoid insecticides from coated seeds to seedling guttation drops: a novel way of intoxication for bees. J Econ Entomol 102(5):1808–1815
Gökçe M (2001) Arıcılığın Genel Durumu Sorunları ve Çözüm Önerileri. Paper presented at the Beekeeping Symposium Ordu, 16 Sept 2001
Güler A (1996) Morphometric characteristics and performances of honeybee (*Apis mellifera* L.) races and ecotypes in Turkey. (Ph. D. Thesis) Çukurova Üniversitesi Fen Bilimleri Enstitüsü, Adana
Güler A, Demir M (2005) Beekeeping potential in Turkey. Bee World 86(4):114–118
Güler A, Kaftanoğlu O, Bek Y, Yeninar H (1999) Türkiye`deki çesitli balarısı (*Apis mellifera*) ırk ve ekotiplerinin morfolojik karakterler açısından ilişkilerinin discriminant analiz yöntemi ile saptanması. TÜBİTAK Doga 23(4):337–344
Güler Y, Dikmen F, Töre D, Aytekin AM (2014) Contributions on the current knowledge of the diversity of the Megachilidae (Apoidea: Hymenoptera) fauna in the Mediterranean Region of Turkey. Turk J Entomol 38(3):255–278
Gülmez Y, Bursalı A, Tekin S (2009) Molecular detection and characterization of deformed wing virus (DWV) in honeybees (*Apis mellifera* L.) and mite (*Varroa destructor*) in Turkey. Afr J Biotechnol 8(16):3698–3702
Haddad N (2011) Honey bee vıruses, dıseases and hıve management ın The Mıddle East and theır relatıon to the colony collapse dısorder and bee losses. U Bee J 11(1):17–24
Higes M, Martin R, Meana A (2006) *Nosema ceranae*, a new microsporidian parasite in honey bees in Europe. J Invertebr Pathol 92:93–95
Higes M, Martin-Hernandez R, Garrido-Ballion E et al (2009) Honeybee colony collapse due to *Nosema ceranae* in professional apiaries. Environ Microbiol Rep 1(2):110–113
Hoffner HA Jr (1974) Legal and social institutions in Hittite Anatolia. In: Sasson JM (ed) Civilizations of the Ancient Near East, vol I
Hoffner HA Jr (1997) Legal and social institutions in Hittite Anatolia. In: Sasson JM (ed) Civilizations of the Ancient Near East, vol I
Hornitzky MAZ, Wills PA (1983) Gamma radiation inactivation of *B. larvae* to control American foulbrood. J Apic Res 22(3):196–199
Huang WF, Bocoquet M, Lee KS et al (2008) The comparison of rDNA spacer regions of *Nosema ceranae* isolates from different hosts and locations. J Invertebr Pathol 97:9–13
Hunt GJ, Page RE Jr (1995) Linkage map of the honey bee, *Apis mellifera,* based on RAPD Markers. Genetics 139:1371–1382
İlikler I, Yüzbaşı A (1981) Ege Bölgesi Balarılarında (*Apis mellifera* L.) Arı Zararlıları. Tarım ve Orman Bakanlı¤ı, Zirai Mücadele ve Zirai Karantina Genel Müdürlü¤ü. Bölge Zirai Mücadele Araştırma Enstitüsü, Bornova, İzmir, p 28
Infandites M (1990) Beekeeping in Greece. Bienenstich. No 17. Ciba-Geigy AG, Basel
Kandemir İ (2015) Bal Arısı ve Biyolojik Özellikleri. Multidisipliner Yaklaşımlı Biyolojik Temelli Doğal Tedaviler-Biyoterapi. Apiterapi Chapter Turkiye Parazitoloji Derneği Yayın No: 25 İzmir, p 462
Kandemir İ, Kence A (1995) Allozyme variation in a central Anatolian honeybee (*Apis mellifera L.*) population. Apidologie 26:503–510
Kandemir İ, Kence M, Kence A (2000) Genetic and morphometric variation in honeybee (*Apis mellifera* L.) populations of Turkey. Apidologie 31:343–356
Kayral N, Kayral G (1983) Son Sistem Arıcılık. Arı İs Yayın No: 1. İstanbul, p 160
Kayral N, Kayral G (1989) Yeni Teknik Arıcılık. İnkılap Kitapevi. İstanbul, p 722
Korkmaz A (2015) Anlaşılabilir Arıcılık. Samsun Gıda Tarım ve Hayvancılık İl Müdürlüğü, p 330
Korkmaz A, Bacaksız D (2016) Ana Arı Yetiştiriciliği. Saybir, p 30
Kumova U (2000) Ülke Arıcılıgını Çagdaslastırma Konusunda Öneriler. Teknik Arıcılık. Sayı 70. Ankara, pp 5–10
Machova M (1993) Resistance of *Bacillus larvae* in beeswax. Apidologie 24:25–31
Meixner MD, Sheppard WS, Poklukar J (1993) Asymmetrical distribution of a mitochondrial DNA polymorphism between 2 introgressing honey bee subspecies. Apidologie 24:147–153

Ministry of Agriculture Risk Assessment Report (2013) published by Ministry of Food, Agriculture and Livestock http://www.tarim.gov.tr.
Muz MN (2008) Bal arılarında ani koloni sönmesi (AKS). Türk Parazitol Derg 32(3):271–275
Muz MN, Doğaroğlu M (2011) *Nosema ceranae* ve arı kolonisine neden olduğu çöküşler. Maybir Dergisi 3:21–24
Muz MN, Muz D (2008) Occurrence and genetic analysis of DWV (Deformed Wing Virus), ABPV (Acute Bee Paralysis Virus) and VDV-1 (*Varroa destructor* -1-Virus) in the Turkey. Paper presented at Apimedica and Apiquality International Conference, University of Rome, Italy
Muz MN, Muz D (2009) *N. ceranae* ve *N. apis*'in PCR-RFLP ile ilk tespiti. In 16. Ulusal Parazitoloji Congress Proceedings, Adana, p 311, 1–7 Nov 2009
Muz MN, Muz D (2010) Co-infections of *Nosema ceranae* with sacbrood and chalkbrood infections in collapsed apiaries of Turkey. In: VI. COLOSS meeting proceedings COST action FA0803, Ankara–Turkey
Muz MN, Girişgin AO, Muz D et al (2010) Molecular detection of *N. ceranae* and *N. apis* in CCD apiaries of Turkey. J Apic Res 49(4):342–344
Nakamura J (1999) Problems of beekeeping in Asia. Farm Jpn 33(3):15–17
Neumann P, Özkırım A (2011) The COLOSS Network "Prevention of colony losses". Mellifera 11(21–22):5–9
Nunamaker RA, Wilson WT, Haley BE (1984) Electrophoretic detection of Africanized honey bees (*Apis mellifera scutellata*) in Guatemala and Mexico based on malate dehydrogenase allozyme patterns. J Kansas Entomol Soc 57:622–631
Okursoy S, Albayrak H, Kurt M, Yazıcı Z (2010) Prevalence of three honey bee viruses in Turkey. Veterınarskı Arhıv 80(6):779–785
Özbek H (1997) Bumblebees Fauna of Turkey with distribution maps. Türk Entomol Dergisi 21(1):37–56
Özkırım A (2011) Honey bee castle. International bee research association press, p 99
Özkırım A (2012) Arılar Şatosu. TÜBİTAK Press, p 99
Özkırım A (2015) Dünyada ve Türkiye'de Arı Kayıpları ve Nedenleri. Multidisipliner Yaklaşımlı Biyolojik Temelli Doğal Tedaviler-Biyoterapi. Apiterapi Chapter Turkiye Parazitoloji Derneği Yayın No: 25 İzmir, p 462
Özkırım A, Keskin N (2001) A survey of *Nosema apis* of honey bees (*Apis mellifera* L.) producing the famous anzer honey in Turkey, Zeitschrift für Naturforsch Verlag der Zeitschrift für Naturforschung Tübingen (ISI), pp 918–919
Özkırım A, Keskin N (2002) Distribution of the major bacterial brood diseases diagnosed in Ankara and its surroundings. Mellifera, pp 40–44
Özkırım A, Keskin N (2003) The occurence of *Varroa jacobsoni* (Acari: Varroidae) and *Acarapis woodi* (Acari: Tarsonemidae) in Capital City of Turkey, Hacettepe. J Biol Chem 32:7–13
Özkırım A, Keskin N (2005) The culture of *Bacillus* spp. from comb foundation. Hacet J Biol Chem 34:37–41
Özkırım A, Keskin N (2010) Türkiye'de Üretilen Hazır Petek ve Eski Peteklerin Arı Hastalıkları ve Arı Ölümlerinin Yayılımındaki Etkisinin Karşılaştırmalı Olarak Biyokimyasal ve Mikrobiyolojik Yöntemlerle Epidemiyolojik Açıdan Araştırılması (Yardımcı Araştırmacı) Project No:0801601004 Supported by: Hacettepe Üniversitesi Bilimsel Araştırmalar Birimi 2008-2010
Özkırım A, Schiesser A (2013) Israeli acute paralyses virus (IAPV) in Turkish Bees. J Apic Res 52(2):56–57. https://doi.org/10.3896/IBRA.1.52.2.09
Özkırım A, Yalçınkaya A (2012) Preventive and antimicrobial activities of alkoxysilane against the American foulbrood pathogen *Paenibacillus larvae* in *Apis mellifera* L. J Apic Res (ISI) 51(1):85–90. https://doi.org/10.3896/IBRA.1.51.1.11
Özkırım A, Keskin N, Sorkun K (2006) The determination of a parasite of honey bee (*Apis mellifera* L.) "*Nosema apis* Z." in honey. Mellifera 6(10–12):2–6
Özkırım A, Keskin N, Kürkçüoğlu M et al (2012) Evaluation of some essential oils as alternative antibiotics against American foulbrood agent *Paenibacillus larvae* on honey bees *Apis mellifera* L. J Essent Oil Res 24(5):465–470. https://doi.org/10.1080/10412905.2012.703504
Özkırım A, Gençay Çelemli Ö, Schiesser A et al (2014) A comparison of the activities of Greek and Turkish propolis against *Paenibacillus larvae*. J Apic Res 53(5):528–536. https://doi.org/10.3896/IBRA.1.53.5.01

Özkök A, D'arcy B, Sorkun K (2010) Total phenolic acid and total flavonoid content of Turkish pine honey dew honey. J ApiProd ApiMed Sci 2(2):65–71
Palmer MR, Smith DR, Kaftanoğlu O (2000) Turkish honeybees: genetic variation and evidence for a fourth lineage of *Apis mellifera* mtDNA. J Hered 91:42–46
Paxton RJ, Klee S, Korpela S et al (2007) *Nosema ceranae* has infected *Apis mellifera* in Europe since at least 1998 and may be more virulent than *Nosema apis*. Apidologie 38:558–565
Plessis TA, Du-Rensburg HJ, Van-Johannsmehier MF (1985) The gamma sterilization of beewax. S Afr Bee J 57(3):54–57
Ruttner F (1988) Biogeography and taxonomy of honeybees. Springer, Berlin, Heildelberg
Ruttner F (1992) Naturgeschichte der Honigbienen. Ehrenwirth, München, p 357
Şahinler N, Şahinler S (1996) A study on the present situation and problems of apiculture and some recommendation in Hatay province. J Agric Fac Mustafa Kemal Univ 1(1):17–28
Santas LA (1990) The dangerous mite plagues grecien bees, varroa spread rapidly. Bienenstich. No 17. Ciba-Geigy AG, Basel
Sariöz P (2006) Arı Biziz Bal Bizdedir, Dünden Bugüne Türkiye'de Arıcılık. İstanbul Altıparmak Pazarlama, p 78
Sasaki M (1999) Ecological characteristics of Asiatic hive bee and its possibility as apicultural species. Farm Jpn 33(3):18–22
Şekerden Ö, Kılıç M, Kaplan ÖM (1992) Türkiye'de Anzer Balı Üretim Bölgesinin Florası, Cografik ve İklimsel Özellikleri ile Bu Şartlarda Arıcılıgın Yapılma Sekli. Doğu Anadolu Bölgesi In 1. Livestock Symposium, Erzurum, 3–4 June 1992
Şenocak K (1988) Modern Arıcılık–Arı Sırrının Çözümü. Remzi Kitabevi, İstanbul, p 222
Sheppard WS, Berlocher SH (1985) New allozyme variability in Italian honeybees. J Hered 76:45–48
Silici S (2011) Determination of volatile compounds of pine honeys. Turk J Biol 35:641–645
Silici S, Atayoğlu AT (2015) Mad honey intoxication: a systematic review on the 1199 cases. Food Chem Toxicol 86:282–290. https://doi.org/10.1016/j.fct.2015.10.018
Silici S, Kutluca S (2005) Chemical composition and antibacterial activity of propolis collected by three different races of honeybees in the same region. J Ethnopharmacol 99:69–73
Sıralı R (2002) General beekeeping structure of Turkey. Uludağ Bee J Kasım 2:31–40
Smith D (2002) Genetic diversity in Turkish honey bees. Uludağ Bee J 2:10–16
Smith DR, Brown WM (1990) Restriction endonuclease cleavage site and length polymorphisms in mitochondrial DNA of *Apis mellifera* and *A. m. carnica* (Hymenoptera: Apidae). Ann Entomol Soc Am 83:81–87
Smith DR, Slaymaker A, Palmer M, Kaftanoğlu O (1997) Turkish honeybees belong to the east Mediterranean lineage. Apidologie 28:269–274
Sorkun K (2008) Türkiye'nin Nektarlı Bitkileri Polenleri ve Balları. Palme Yayıncılık, p 341
Sorkun K, Doğan C (1994) Nektarlı Bitkilerin Çiçeklenme Dönemleri ve Yayılış Alanları Teknik Arıcılık. Sayı 44. Ankara, pp 2–11
Stangaciu S (1998) A guide to the composition and properties of propolis. Apiacta 33:71–77
Stangaciu S (2015) Apiteraphy in the world, past, present and perspectives. Multidisipliner Yaklaşımlı Biyolojik Temelli Doğal Tedaviler-Biyoterapi. Apiterapi Chapter Turkiye Parazitoloji Derneği Yayın No: 25 İzmir, p 462
Tanyüksel M (2015) Dünyada ve Türkiye'de Biyoterapide Güncel Durum, Sorunlar ve Gelecek Hedefler. Multidisipliner Yaklaşımlı Biyolojik Temelli Doğal Tedaviler-Biyoterapi. Apiterapi Chapter Turkiye Parazitoloji Derneği Yayın No: 25 İzmir, p 462
Temizer İ (1983) Folbex-VA İlacının Varroa Parazitine Karflı Etkinliğinin Saptanması Üzerinde Araştırmalar. Tarım ve Orman Bakanlığı, Ziraat İşleri Genel Müdürlüğü. Ege Bölge Zirai Araştırma Enstitüsü Yayın 3, İzmir
Tolon B (1999) Bee product standardization for production and export. Uluslararası Hayvancılık '99 Kongresi (21-24 Eylül 1999). İzmir, pp 519–524
Turkish Statistical Institute (2015) http://www.tuik.gov.tr. Accessed 21 Sept 2015
Turkish Statistical Institute (2016) http://www.tuik.gov.tr. Accessed 14 Sept 2017

Tutkun E (2000) Teknik Arıcılık El Kitabı. Türkiye Kalkınma Vakfı, p 98

Ülgentürk S, Kıran K, Ayhan B et al (2012) Türkiye'de *Marchalina hellenica* Gennadius (Hemiptera: Marchalinidae) ile ilişkili karınca Üzerine Bir Çalısma. (Hymenoptera: Formicidae) türleri. Türk Entomol Bült 2(4):263–270

van Der Zee R, Pisa L, Andonov RS et al (2012) Managed honey bee colony losses in Canada, China, Europe, Israel and Turkey, for the winters of 2008-9 and 2009-10. J Apic Res 51:100–114. https://doi.org/10.3896/IBRA.1.51.1.12

Warrit N, Hagen TAR, Smith DR, Çakmak İ (2004) A survey of *Varroa destructor* strains on *Apis mellifera* in Turkey. J Apic Res 43:190–191

Yeşil A, Gürkan B, Saraçoğlu Ö et al (2005) Effect of the pest *Marchalina hellenica* on the growth parameters on *Pinus brutia* in Muğla Region (Turkey). Pol J Ecol 53(3):451–458

Zhang ZQ (2000) Notes on *Varroa destructor* (Acari; Varroidae) parasitic on honeybees in New Zealand. Syst Appl Acarol Spec Publ 5:9–14

Chapter 3
Beekeeping in Parts of the Levant Region

Nizar Haddad and Lisa Horth

Abstract Beekeeping in the Levant region, a historical geographic area that lies east of the Mediterranean Sea and now comprises several countries, has a very long and colorful apiculture history. In ancient times, bees were revered by humans. They have been included in hieroglyphic writings and were often found associated with royalty. A variety of *Apis mellifera* subspecies currently persist in the Levant region. Historically and presently, these bees have been tended by beekeepers using an assortment of ancient and modern home types. This relationship has allowed humans to reap the benefits of bees including honey, wax, and other valuable products, in addition to pollination services. Some of the subspecies in the region now appear to be locally adapted to harsh environmental conditions and may warrant conservation. Today, genetic introgression among subspecies, as well as the purchase of queens from outside of the region may facilitate the transmission of novel diseases to susceptible colonies and to new geographic regions. Preserving natural honey bee genetic assemblages across the region may prove beneficial not only from a conservation standpoint, but also from an evolutionary and ecological one, when considering the value of locally adapted traits, stable community structure, and species diversity. The reverence for this majestic creature may need to be rekindled if we are to continue to tend honey bees for the economic benefits they provide to our society. Otherwise, they may continue to decline in abundance, perhaps as a result of a disease, movement induced by humans, migration, queen importation, and a myriad of additional disruptive anthropogenic factors that negatively impact their health and welfare.

Keywords Honey bees · Jordan · Palestine · Lebanon · Syria · Disease
Pollination · Agriculture · Genetic · *Apis*

N. Haddad (✉)
Bee Research Department, National Center for Agricultural Research and Extension, Baqa'a, Jordan
e-mail: Director@NCARE.gov.jo

L. Horth
Department of Biological Science, Old Dominion University, Norfolk, VA, USA
e-mail: lhorth@odu.edu

P. Chantawannakul et al. (eds.), *Asian Beekeeping in the 21st Century*,
https://doi.org/10.1007/978-981-10-8222-1_3

3.1 Diversity

3.1.1 The Evolution of Apis mellifera *Subspecies' Classification in the Levant*

Comprehensive morphometric and genetic studies of western honey bee (*Apis mellifera*) subspecies have provided valuable information on the extensive phenotypic and genetic diversity of this group, especially in the Levant (Ruttner 1988; Ruttner et al. 1978; Alburaki et al. 2011, 2013; Franck et al. 2000). We now know that of the 29 honey bee subspecies originally classified by Ruttner using only morphometric data, *A. m. syriaca* is the primary subspecies found in major parts of the geographic region called the Levant.

The Levant native distribution extended throughout much of the region including Syria, Lebanon, Palestine, Jordan, and parts of Iraq. The Levant is a geopolitical region that comprises lands found east of the Mediterranean Sea. It is roughly crescent shaped; the inside edge of the crescent hugs the Mediterranean Sea, and the region is loosely bounded to the northwest by the Taurus Mountain in Turkey, and to the south by the Arabian Desert. The Levant currently comprises parts of the Middle East, and includes Syria, Lebanon, Palestine, the Gaza Strip, Jordan, and Iraq. Some descriptors also include Turkey and Cypress. The area has a long, rich geopolitical history, and served historically as the land bridge between ancient Mesopotamia, subsequent empires, and Egypt.

Historically, three major honey bee (*A. mellifera*) evolutionary lineages were originally identified, based upon morphometric traits, with a fourth lineage added subsequently. The four lineages conventionally addressed include the "O" lineage (or "Oriental" lineage, in the Near and Middle East), "M" (North African/West European or West Mediterranean), "A" (South/Central African), and "C" (North Mediterranean/Eastern Europe) (Ruttner et al. 1978; Franck et al. 2000). Based on morphometry, *A.m. syriaca* was originally classified as part of the "A" lineage (Ruttner et al. 1978), though later molecular work on mitochondrial DNA clustered these subspecies with the "O" lineage (Franck et al. 2000). This shift, however, may have resulted from the import of queens into the range of *A. m. syriaca*, and subsequent genetic mixing with other subspecies.

Recent genetic work on mitochondrial DNA has again placed Syrian bees of the subspecies *A.m. syriaca* in the "A" lineage (Ruttner 1988; Alburaki et al. 2011). An exception occurs for the bees analyzed from northern Syria, near the Turkish border. These populations clustered genetically with a different lineage, including *A. m. anatoliaca* and *A. m. caucasica* (Whitfield et al. 2006), which, along with *A. m. meda* are considered Turkish bees, and have traditionally been classified as members of the "O" mitochondrial lineage, not the "A" lineage (Smith et al. 1997, and Ruttner 1988, who also included *A. m. syriaca* as inhabiting Turkey, and indicated that they were found along Turkey's southernmost border, with Syria). Migratory beekeeping is common in Turkey, with *A. m. caucasica* being the most valued subspecies by these beekeepers (Palmer et al. 2000). Recently, bees in parts of northern

Syria, in the areas around Tartus and Al-Hasakah, have also demonstrated evidence of introgression with *A. m. caucasica* and *A. m. anatolica,* which may have been from Turkey and Cyprus (Alburaki et al. 2011, 2013). Finally, bees in the eastern part of Syria, near the border with Iraq, are morphologically similar to the *A. m. meda* that are found in Iran (Ftayeh et al. 1994). Southern Syria, however, still retains the indigenous *A. m. syriaca.*

Mitochondrial DNA work has recently shown that the Syrian and Lebanese populations of *A. m. syriaca* do not appear to be genetically distinct from one another. A fifth independent lineage, inclusive of these bees, has recently been proposed (Alburaki et al. 2013). Syrian honey bees that appear related to the "O" lineage are confirmed to occur only near the Turkish and Iraqi borders (Alburaki et al. 2013). *A. m. syriaca* now appears to be substantially different from the other three classical morphometric lineages that Ruttner et al. (1978) described (Alburaki et al. 2011, 2013).

A. m. meda is considered native to the majority of the additional parts of Iraq, where *A. m. syriaca* was the ancestral subspecies (Ruttner 1988; Ftayeh et al. 1994; Haddad and Fuchs 2004). Unlike the eastern Syrian populations of *A. m. syriaca* that are morphometrically similar to *A. m. meda* near Iraq (Ftayeh et al. 1994), populations of *A. m. meda* from northern Iraq, for example in Mosul, do not differ from the "O" lineage based upon a principal components analysis of nuclear (microsatellite) DNA data (Alburaki et al. 2013). The continual evolution of the story of which subspecies are located where, and when, in this region implies that we do not yet fully comprehend all aspects of relatedness of the subspecies of bees in this region.

The third major subspecies found in Iraq is *A. m. caucasia* (Saeed 2013). This subspecies is ancestrally native to the valleys of the Caucasus Mountains. These bees appear to be poorly adapted to overwintering in cold climates, and adults tend to demonstrate high susceptibility to *Nosema* infections during this time (Ruttner 1988). Populations become larger in size in the spring, but this growth does not occur rapidly (Ruttner 1988; Winston 1991). *A. m. caucasia* produces prolific amounts of honey, which could relate to its particularly long proboscis that allows it to reach nectar that other subspecies cannot easily exploit (Hermani 2012).

A. m. caucasica and *A. m. anatolica* are present in Lebanon and Syria, too, but in these countries, *A. m. ligustica* rounds out the list of major subspecies identified (Alburaki et al. 2013; Kadamani 2015; El-Obeid 2015). In Palestine, there are three major subspecies found in addition to *A.m. syriaca*, meaning two subspecies differ here from the trio found in Lebanon and Syria. The subspecies in Palestine are *A. m. ligustica*, *A. m. carnica* (Pollmann), and *A. m. lamarckii*, while in Gaza, it is sometimes possible to find *A. m. lamarckii.* This is a result of bee importation from Egypt (Albaba 2010).

Some important genetic differences have been revealed over the last few decades through the use of mitochondrial DNA analysis and microsatellite markers employed to study the biodiversity and genetic relationships of honey bees in the region (Franck et al. 2001; Haddad et al. 2009a, b, c; Alburaki et al. 2013). However, by and large, these genetic studies have actually provided confirmation of the major results found in prior work that was based upon morphological trait assessment. Namely, *A. m. syriaca* is the native subspecies of the region. Even more recently, single nucleotide polymorphism (SNP) analysis of worker bee genomes from around the world has contributed to our improved understanding of global variation in honey bee subspecies' relatedness, as well as provided a deeper understanding of the local adaptation of the Levant region's honey bees to their environment (Wallberg et al. 2014).

Most recently, it appears that the distribution of *A. m. syriaca* is currently being impacted by the importation of commercial *Apis* breeds from many other countries around the world into the Levant region (Haddad et al. 2009a, b, c). This influx of new bees includes a number of *A. mellifera* subspecies, with potentially infectious agents that are novel to *A. m. syriaca* (Haddad and Albaba 2014). This is a cause for concern when honey bee colonies are diseased and dying at high frequency in many regions of the world, including the Levant, while globally we continue to truly understand the causation of these population declines (Haddad 2011a; Lee et al. 2015).

In the Levant region, *A. m. syriaca* is the most south-westerly geographic inhabitant of all of the near-east honey bee races. Neighboring races include *A. m. meda* in the north and north east (Ftayeh et al. 1994), *A. m. lamarckii* and *A. m. yemenitica* in the south west. When considering differences in the races, the subspecies historical geographic ranges suggest that *A. m. syriaca* occupies a particularly interesting place in honey bee biogeography. This is also emphasized in phylogenies of the relationships found among the honey bee subspecies. In genetic relatedness trees, *A. m. syriaca* is found near the origin of the four primary phylogenetic branches that are recognized in principal component analysis when evaluating honey bee intraspecific variability (Ruttner 1988).

In a recent study, high gene flow (as high as 18%) was shown to occur between populations of *A. m. syriaca* on the African and Asian continents. Some of the bees used included *A. m. syriaca* samples collected from Jordan. These were compared with samples collected from other regions, including South Africa (Wallberg et al. 2014). The high rate of gene flow identified across continents is especially interesting, and curious, given that South African bees are not imported by humans into Jordan, specifically, or even into the Levant region in general. This suggests that gene flow occurs through some other mechanism besides a direct anthropogenic effect.

3.1.2 Apis mellifera syriaca *Behavioral Plasticity*

The behavioral manifestations of *A. m. syriaca* include a large number and type of particular traits. These bees have what has been described as a nervous temperament and individuals are known to have the ability to transition from one activity to another, as needed (Ruttner et al. 1978). Another similar, equally valuable trait that *A. m. syriaca* express that is also a response related to the environment, is the ability to preserve the colony through exhibiting reduced brood rearing during the hottest months of summer (Zaitoun et al. 2000; Haddad et al. 2016). During this time, they also demonstrate an increase in swarming tendency. They can produce an excessive number of swarm cells, which is especially relevant because it results in the survival of several virgin queens in the colonies. This occurs until a mated queen is present, and potentially safeguards against major risks associated with queen loss. When a mated queen returns to the colony, alternative virgin queens are killed off (Ruttner 1988).

Similar to African honey bee subspecies (Hepburn and Radloff 1998), *A. m. syriaca* also experience frequent absconding from nest sites; the underlying reasons for this behavior in *A. m. syriaca* warrant further investigation (Bodenheimer and Ben-Nerya 1937). Parasites and diseases are one possible cause for absconding, but this would be somewhat ironic for these particular bees since they are quite good at maintaining hygienic hives (Zakour et al. 2012). In fact, colonies are typically found to be free from pests, such as the pollen beetle, *Cryptophagus hexagonalis* (Haddad et al. 2008a). This pest-free state is particularly relevant given that these hygienic *A. m. syriaca* hives are found in many places, including in the exact same geographic locations and apiaries where *A. m. ligustica* are simultaneously being infected (Haddad et al. 2008a). In fact, *A. m. syriaca* has the strongest hygienic behavior and the lowest *Varroa destructor* mite infestation levels of all subspecies when compared to *A. m. anatoliaca, A. m. carnica,* and *A. m. caucasica* (Kence et al. 2013). These observations lend support to the hypothesis suggested recently that (sub) tropical bees have greater defenses against parasites than other bees (Kence et al. 2013a).

A. m. syriaca have also adapted to withstand attacks from some predators, including oriental hornets (*Vespa orientalis*) and green bee-eater birds (*Merops orientalis*). They do this by cessation of flight activity whenever colonies are besieged by these predators (Blum 1956; Kalman 1973; Ruttner 1988; Haddad and Fuchs 2004; Haddad et al. 2007). In contrast to intensive research efforts towards *A. m. syriaca* genetics and behavior, basic information related to the particular traits and characteristics that may be adaptive for other subspecies like *A. m. meda,* found in Iraq, appears to be limited.

3.1.3 **Apis mellifera syriaca** *Genome Sequencing and Important Candidate Genes*

The genome of *A.m. syriaca* has been sequenced as a prelude to identifying genes that may be associated with beneficial traits of this subspecies. Approximately 600 candidate genes were found based on the pathways in which they were involved (e.g., microbe recognition, antiviral activity, and more). Specific genes found related to heat tolerance and immunity-related traits. Several genes were identified that are of great interest, based on our knowledge about their operation in immunity from other species. These include functional variants in immune-related embryonic development genes (*Cactus*, *Relish*, *Dorsal*, *Ank*2, and *Baz*) (Haddad et al. 2015c). Similarly, several candidate genes of interest with respect to hygienic activity as it relates to *V. destructor* mites were found to exist (*Norp*A2, *Zasp*, *Lan*A, *Gasp*, and I*mpl*3). Additional candidate genes with functional variants appear to be promising with respect to *V. destructor* resistance (*Pug*, *Pcmt*, elk, elf3-s10, *Dscam2*, *Dhc64C*, *gro*, and *futsch*) (Haddad et al. 2015c).

These functional variants found for genetic differences between *A. mellifera* and *A. m. syriaca* might be used to develop effective molecular tools for bee conservation and breeding programs to improve locally adapted subspecies of *A. m. syriaca* and to utilize their advantageous traits for the benefit of apiculture industry (Haddad et al. 2015c, 2016). In addition to next generation sequencing data, the mitochondrial genome of this subspecies was also recently sequenced (Haddad 2015, 2016).

3.1.4 **Apis mellifera syriaca** *Conservation Strategies and Concerns*

The National Agricultural Research and Extension breeding apiary was established in an effort to conserve *A. m. syriaca*. To evaluate the effectiveness of a conservation program that is currently underway, a comparative genome hybridization study was recently employed. The purpose of this work was to evaluate the genetic differences found between extant populations of *A. m. syriaca* and a historic reference collection. If the conservation program were not effective, we might, for example, anticipate a decrease in genetic variation in extant population samples relative to historical ones. The reference collection was a preserved sample that had been collected in 2001 from the southern region of Jordan. This reference sample showed high similarity to an even older reference sample that was collected several decades earlier, in 1952. The comparative genome study found insignificant genomic differences between the current conservation program population bees and the historic reference collection. This suggests that the ongoing conservation program may be successful in the preservation of important genetic variation; bees in the conservation program have remained genetically similar over time, possibly because they are not mating with alternate, immigrant subspecies.

The functional genomic variation that has been identified in these recent genetic studies may be valuable for improving locally adapted subspecies of *A. m. syriaca*, which often live in harsh environments. This subspecies has a history of being tolerant of major stressors, including high environmental temperatures, scarcity of summertime nectar, predatory wasp attacks, and *V. destructor* mite resistance. The study also confirmed that the conservation of this subspecies may be possible through the use of queens that undertake natural mating flights, combined with the careful use of supplemental artificial insemination, to ensure the genetic purity of future subspecies. The study also concluded that *A. m. syriaca* drones might be more successful at insemination of *A. m. syriaca* queens during mating flights than other, imported bees. This may be true because *A. m. syriaca* males are adapted to the challenging local climatic conditions that are so often experienced by the bees which are native to this geographic region (Haddad et al. 2016).

Unfortunately, the importation of commercial breeder *Apis* lines and the establishment of these lines for honey production is bound to result in the original autochthonous subspecies disappearing over time. As early as in 1952, Brother Adam found the frequent importation of *A. m. ligustica* into Israel noteworthy. He remarked at the time that it might be difficult to find pure *A. m. syriaca* in Israel after the importation of *A. m. ligustica*. The tendency to import commercial lines, as well as those used for breeding purposes, has increased substantially in recent years. These lines include subspecies of Carniolan, Buckfast, and/or Caucasian origin (Slabezki et al. 2000; Hussein 2000; Blum 1956). In combination with increased transport of colonies by humans as a component of modern beekeeping, cross-mating between commercial lines and native subspecies may result in wild, local populations that harbor varying degrees of native bee genes blended with commercial breeder lines of diverse characteristics. These bees may not be as well suited to their environment as those that have evolved in the region.

3.1.5 Invasion of **Apis florea** *into the Levant*

A growing issue that has arisen relatively recently is the movement of invasive bees into the Levant. *Apis florea* was discovered in the Levant region for the first time in Iraq in 1992. The species was found north of Baghdad, near the village Makatu, which is a few kilometers from the central portion of the Iraqi-Iranian border (Glaiim 1992). Kurdish beekeepers and farmers in Iraq try to provide their colonies with shelter to protect them from rain since these bees produce only a very small amount of honey (H. Montaser and I. Majed, pers. comm.). *A. florea* has also expanded its distribution to the south, and can now be found in Iraq, parallel with some known populations located on the Iranian side of the border (Özkan et al. 2009). The subspecies has also recently moved northerly and was discovered near the Sirwan River in Shaikh Lankar village in 2012 living inside small shrubs and trees (Hasnawi Montaser and Ibrahim Majed, pers. comm.). West of Iraq, the subspecies has invaded the Aqaba Gulf region in Jordan, where they were discovered in 2008. Since then, they have quickly become established locally (Haddad et al. 2008b) while simultaneously remaining restricted to this localized area in Jordan. However, in Africa they have also migrated as far southwest as Sudan (Lord and Nagi 1987; Mogga and Ruttner 1988), and there is evidence that they can move to new geographic areas fairly rapidly. These bees are currently expanding their range throughout the country of Sudan (Moritz et al. 2010a) as well as moving east from Sudan, into Ethiopia (Bezabih et al. 2014).

An *A. florea* population was recently identified in Aqaba, a city located on the southwestern tip of coastal Jordan (Haddad et al. 2009b, c). The Aqaba population is an isolated population that appears to be expanding in range. This small cluster of colonies is geographically separated by at least 1500 km from the closest geographical populations, which are found in Sudan. These Sudanese populations were themselves introduced in 1985. The *A. florea* in Aqaba are 2000 km away from the nearest naturally occurring populations, which are found in Oman and Iran (Haddad et al. 2009b, c). Recent morphometric analyses suggest that the

Aqaba population is most similar to *A. florea* in Oman, suggestive that Oman bees are likely the source population for the recent colonization of Aqaba (Haddad et al. 2009b, c). By studying the genetic structure of this invasive population through the use of nine microsatellite DNA markers it was found that the entire regional population can be traced back to just a single nest (Moritz et al. 2010). The very high relatedness of these bees increased suspicion that this group originated as an accidental invasion of a single swarm that arrived by chance with fishing boats (Haddad et al. 2009b, c; Moritz et al. 2010). Such results also strengthen the conclusion that these bees are expanding their distribution by normal migration, as well as through accidental anthropogenic invasions that sometimes occur with global transportation. Such movement also increases the risk of *A. florea* migrating to additional new areas in the region, including the Levant, along all of the Middle East, and even further into additional parts of Africa. This raises grave concern since these bees may serve as competition for native species and, as is true whenever there is novel contact, they bring with them the risk of novel disease transmission.

3.2 Historical Records and Recent Developments of Beekeeping

3.2.1 The Ancient History of Honey Bees in the Levant Region

There is a long, historical relationship in the Levant and Middle East between humans and honey bees. One of the earliest known records of this association can be found in the African part of the Middle East region, and dates back to 2400 BC. In the Sun Temple near Cairo, Egyptian hieroglyphs depict bees and show that they symbolize royalty (Ransome 2004). The earliest archeological findings associated with bees that are located in the Asian part of the Middle East date back to the year 2100 BC. These are found in Sumerian and Babylonian cuneiform writings (Abd Al-Wahed 1985). Even the world's oldest record of a love poem dates back to ancient Mesopotamia, and in this poem, the sweetness of honey is used as an allegory referring to the loving relationship between King Shu-Su'en (2038–2030 BC) and his bride (Kramer 1988).

The Levant region is the land of both the Old and New Testaments of the Bible and is referred to, on several occasions, as the Land of Milk and Honey (e.g., Exodus 33:3). Samson was said to have taken wild honey from inside the corpse of the lion in the Soreq Valley (Judges 14: 8–9). In addition to these examples, several metaphors associated with bees and hive products are found in the Bible, including the description of the judgments of the Lord as "… sweeter than honey and the honey comb." John the Baptist, who lived in the Jordan Valley, was himself a honey hunter (Mathew 3:4). Even Jesus Christ was said to have fed his disciples with honey (Luke 24:42).

In the Islamic religion, which is predominant in the Levant region, the Quran is the primary religious doctrine. Chapter 16 of the Quran is entitled "Sūrat an-Naȟl," which translates in English to “The Bee.” It represents the great value placed upon bees, which serve as inspirational creatures. Part of this admiration stems from the inherent value of their hive products, especially their honey (Alhissnawi 2016).

With the exception of Egypt (Ransome 2004) none of these archeological, historical, and religious records provide any information specifically about the craft of beekeeping in the region. Rather, they only provide information about the bees themselves and most likely honey found in nature. In the last decade, clay beehives were discovered in the city of Rehov, Jordan. These date back to the Iron Age (ninth and tenth century BC) and are considered the oldest evidence of beekeeping practice in the region (Mazar et al. 2008).

3.2.2 Recent Developments in Hive Use

Personal communications with long-term beekeepers across the region combined with field site visits and hive inspections allow us to conclude that the most common hive type throughout the region is very similar to the type historically found in the Jordan Valley, dating back to the Iron Age. The primary shape of the hive is a cylinder (Fig. 3.1).

Individual cylindrical hives are laid in groups of 30–100 per grouping. Sometimes several groups are found in the same location; however, there is more commonly one group of hives per beekeeper. In the mountainous areas surrounding the Yarmouk

Fig. 3.1 Traditional clay mud hive found in the Levant region

River in north Jordan bordering Syria, some beekeepers keep their clay hives in caves. In other locations where clay hives are used, they may be clustered under the same, shared shade cover.

Empty spaces between hives are typically filled with clay-mud mixture of water chalky clay, straw, and cow manure to provide a warm microclimate for the bees in the winter and a cooler one in the hot summers. They also serve to prevent the invasion of empty spaces with invasive insects or small animals.

These cylindrically shaped hives are found throughout the Levant region and have an average length of 65–95 cm and an average internal diameter of between 23 and 30 cm. It was also observed that the longer the hive, the thinner its diameter. Most hives in the Levant region are made primarily from sun dried clay mud. However, in areas close to rivers where there is an abundance of cane, beekeepers make a skeleton of the cylinder hive with the cane, then cover this with mud. On a limited number of occasions in Karak, Jordan, beekeepers even use boxes that are typically used for shipping materials. This practice was limited only to those that had access to the boxes, which came from the area around the Gulf of Aqaba. According to local elders, this practice started in 1945.

In Syria, Lebanon and Iraq, beekeepers primarily use clay mud hives, but in most of the Levant region the traditional hives have almost disappeared now, with the exception of a few locations in Daraa, Syria. Langstroth single brood chambered hives are now most common. The widest practice of the use of the traditional hives has until now occurred in the Kurdish region, in the north of Iraq.

3.3 Local Knowledge on Honey Bees: Cultures and Local Beekeeping

3.3.1 Hive Types and Their Use

The historic hives have been described in Sect. 3.2. The Langstroth hive design is most frequently used among beekeepers in the Levant region. In 1852, Reverand Langstroth in the United States developed the hive, probably with knowledge of similar simultaneous, ongoing work by Dr. J. Dzeierzon. The hive used in this region is like others elsewhere, a standard top-opening, frame-type hive that lures bees to build comb only on the frames. Since Lanstroth hives are portable, they are economically valuable in the region since they can be transported.

3.3.2 Practical Beekeeping by Country

In Iraq, Langstroth hives are common. There are currently ~250,000 working hives in existence today in Iraq. About 150,000 of these are in Kurdistan, and about 100,000 are located in the other districts (Alhissnawi 2016). About 1000 tons of honey are produced annually and about 700 of these are produced in Kurdistan. Unlike many other regions of the world, pollination rental is not yet used commercially in Iraq (Alhissnawi 2016). The environment of Iraq does have great potential to be developed, particularly the fertile area between the Tigris and Euphrates Rivers, as well as the oases and agricultural areas that are irrigated using underground water. However, issues related to instability in Iraq since 1979, such as the war with Iran, and then the first and the second Gulf wars, have served as a primary factor that has limited the development of the beekeeping sector over the last several decades.

In Jordan, official data collected for the 2014 Ministry of Agriculture of Jordan reports indicate that the total number of registered bee colonies is 57,990 modern, Langstroth bee hives. However, it also is estimated that ~14,500 hives are not officially registered. These unregistered hives are owned by hobbyist beekeepers who operate ~10 hives each. The total national production of honey in Jordan was reported to be ~725 tons in 2014. Beekeeping-related imports at this time included 13,897 tons of honey, 24 tons of pollen (for human consumption and bee feeding), 1720 queen bees, and 776 bumblebee colonies. The imported bumblebees from Europe were used in the pollination of tomatoes in greenhouses (Amer 2015).

In Lebanon, Langstroth hives are also common (Kadamani 2015). About 175,000 hives (El-Obeid 2015) are currently functioning in the country. Traditional hives here were made of mud. The National Beekeeping Forum of Lebanon indicates that ~1.8 tons of honey are produced annually in Lebanon. One major producer generates organic honey and this honey comprises ~0.5% of the total local production (El-Obeid 2015; Kadamani 2015). Another commercially valuable hive product produced in Lebanon is royal jelly. About ~150 kg of royal jelly are produced annually (El-Obeid 2015). Queens are typically imported to Lebanon from abroad. As few as 10,000 queens are produced locally (El-Obeid 2015; Kadamani 2015). Here, pollination rental is also not yet commercially available. In fact, there are very limited additional economical activities related to the bees and hives, save a few educational trips for students (El-Obeid 2015).

In Palestine, Langstroth hives are also used. According to a 2014 survey regarding livestock, there are 67,864 hives. The traditional hives found in Palestine are also made of mud and clay. About 1125 traditional hives currently remain in operation (PCBS 2014). Palestine produces ~544 tons of honey, with ~484 tons of which come from the West Bank and ~70 tons from the Gaza strip. Queens are often imported to Palestine from Israel (Albaba 2010). Commercial rental of hives for pollination is not yet available (Albaba 2010).

In Syria, modern Langstroth hives are also in use (Duwwara et al. 2015; AOAD 1995). In 2010, it was estimated that there were ~503,000 of these in use annually; traditional hives here are made of clay or wood. It was estimated that about 127,000 traditional hives are in use each year. Honey production in Syria totals 3000 tons annually, about 650 kg of royal jelly are also produced each year. Queens are imported from abroad and commercial pollination rental does not take place (CBSSYR 2011).

3.3.3 Bee Races and Hive Management

A recent survey was conducted on 138 beekeepers across 14 Arabian countries (Al-Ghamdi et al. 2016). This survey revealed that they preferred indigenous bee races over imported bees. Interestingly, these beekeepers also felt that mismanagement of colonies was indeed the primary cause of colony death, along with multiple other factors (Williams et al. 2010). Al-Ghamdi et al. (2016) believe that the results of their survey also continue to support the concept that extension programs are quite valuable for beekeeper training.

3.4 Pests and Diseases of Honey Bees

3.4.1 Devastating Colony Losses: Disease, Colony Collapse Disorder, or Both?

Losses of managed honey bees have been identified in the last few decades that appear to impact many parts of the world (Neumann and Carreck 2010). Unusual honey bee colony losses that began to occur suddenly, on a grand scale, and without obvious explanation (such as an overabundance of mites and obvious signs of disease), became prominent in the United States in the mid-2000s (VanEngelsdorp et al. 2007). Europeans had already identified dramatic winter losses (Faucon 2002) and were quantifying these, while trying to link them to causative agents, including several viruses (Berthoud et al. 2010). Studies from the Levant region quickly followed (Haddad et al. 2009b, c).

Specific descriptors were articulated to distinguish this important and unique phenomenon from other more typical losses and in the United States. These events were named Colony Collapse Disorder (CCD, VanEngelsdorp et al. 2007, 2009). CCD includes a set of particular events including: rapid loss of adult workers while food remains present in the hive, large numbers of deceased workers not found near the hive or immediate vicinity, the presence of capped brood, the possible presence of a queen and perhaps a few workers, and a lack of high quantities of *V. destructor* mites and/or *Nosema* sp. at the time of hive collapse.

The first case of CCD outside of the United States was reported in 2012 by a Swiss beekeeper, with traditional Swiss hives, who lost several colonies in a few days and 10 of his 16 total colonies within about a month of the first colony loss. Workers in existing colonies were assessed for pathogens from "weak" (few frames of bees present) and seemingly "healthy" colonies (Dainat et al. 2012). Akin to US results, there was no difference in pathogen loads and co-infections of pathogens for the "healthy" and CCD colonies investigated (Dainat et al. 2012). In general, sick hives have now been identified in many regions of the Levant, but the specific CCD assessment has not been undertaken.

While severe losses continue to occur in the Levant, no regional name has yet been given to these events, which may result from a number of factors. Pathogens, including *Nosema* spp., *V. destructor* virus, *V. destructor* mites, and deformed wing virus have been found in multiple subspecies of honey bees in various parts of the Levant, including *A. m. jementica* in Yemen (Haddad et al. 2017), *A. m. intermissa* in Algeria (Adjlane et al. 2012), and *A. mellifera* found across MENA (Haddad et al. 2015a, b).

In the Middle East, losses were reported to be between 10 and 85% of honey bee colonies recently (Haddad et al. 2009b, c; Soroker et al. 2009). In Jordan, in 2008, collapsing honey bee colonies were tested for several viruses (Haddad et al. 2008a). Acute bee paralysis virus and sac brood virus, but not several other viruses, were identified through the use of molecular techniques (RT PCR) and found to be present in sick bees. In Yemen, sick bees were also evaluated and found to have pathogens, but aside from *V. destructor* mites that could serve as disease transmission vehicles, no single pathogen was associated with diseased colonies (Haddad et al. 2017).

While a number of parasites have been found to be associated with sick bees, a single causative agent has not been identified in their demise and it has become clear that particular diseases or parasites that are problematic in one region, or country, may not be equally problematic in another (Cox-Foster et al. 2007; Higes et al. 2009; Dainat et al. 2012). The transport of bees for commercial purposes may contribute to the spread of novel diseases in new places, and transport does occur throughout the Levant. The impact of the migration of bees is something which humans would do well to consider and monitor into and around the Levant, and elsewhere.

Globally, a variety of factors appear relevant or associated with the devastating bee losses happening today. In 2013, the European Commission restricted the use of multiple neonicotinoid pesticides (thiamethoxam, clothianidin, and imidacloprid) and put restrictions on the use of fipronil in hopes of lessening colony losses after the European Food Safety Authority deemed them an unacceptable risk (OJEU 2013). This type of ban has not occurred for the Levant.

3.4.2 Bee Pests, Parasites, and Pathogens in Various Parts of the Levant

The tables included here (Tables 3.1 and 3.2) list the common and scientific name of the pests, predators, and pathogens found in each country of the Levant region. In Iraq, there are a number of pests, including destructive parasitizing insects like wasps and yellow jackets, plus mites such as *V. destructor*. Furthermore several bacterial, viral, and fungal infectious disease agents occur such as bacteria responsible for American foulbrood, and Black queen cell virus and Israeli acute paralysis virus. A similar assemblage of pests and pathogens is listed for the other countries too, including Jordan, Lebanon, Palestine, and Syria. There are an abundance of pests

Table 3.1 Honey bee pests and predators that are found in the Levant region

Name of pathogen		Country in which pathogen was confirmed				
Common name	Scientific name	Iraq	Jordan	Lebanon	Palestine	Syria
Oriental Wasp/Red wasp	*Vespa orientalis*	Sivaram (2004), Glaiim (2009)	Nazer and Rateb (1991), Amr et al. (1998), Haddad et al. (2007)	Hussein (2012)	Hussein (2012)	Hussein (2012), Duwwara et al. (2015)
Yellow jacket	*Vespula germanica*	Sivaram (2004)	Amr et al. (1998), Haddad et al. (2007)	Hussein (2012)	Hussein (2012)	Hussein (2012), Duwwara et al. (2015)
Beewolf	*Philanthus* spp.	Glaiim (2010, 2014)	Amr et al. (1998), Al-Ghzawi et al. (2009)	Hussein (2012)	Hussein (2012)	Hussein (2012), Duwwara et al. (2015)
Ants	*Dorylus fulvus*	AOAD (1995)	Al-Ghzawi et al. (2009)	Hussein (2012)	Hussein (2012)	Hussein (2012), Duwwara et al. (2015)
Lesser wax moth	*Achroia grisella* (Fabricius)	AOAD (1995)	Al-Ghzawi et al. (2009)	Hatoom (1996), Hussein (2012)	Hussein (2012)	Hussein (2012)
Greater wax moth	*Galleria mellonella*	AOAD (1995)	Al-Ghzawi et al. (2009)	Hussein (2012)	Hussein (2012)	Hussein (2012)
Senotaina fly	*Senotainia Tricuspis*	–	Al-Ghzawi et al. (2009), Haddad et al. (2015a, b)	–	–	Hatoom (1996)
Bee louse	*Braula* spp.	AOAD (1995)	Al-Ghzawi et al. (2009)	Hussein (2012)	Hussein (2012)	Hussein (2012)
	*Varroa destructo*r	Glaiim (2010), Hussein (2012)	Nazer and Rateb (1991), Haddad (2011a)	Hussein (2012)	Hussein (2012)	Hussein (2012), Duwwara et al. (2015)
	Acarapis woodi	Glaiim (2010)	Nazer and Rateb (1991), Al-Ghzawi et al. (2009)	Hussein (2012)	Hussein (2012)	Hussein (2012), Duwwara et al. (2015)
Bee-Eater	*Merops* spp.	Glaiim (2014)	Al-Ghzawi et al. (2009)	Hussein (2012)	Hussein (2012)	Hussein (2012), Duwwara et al. (2015)
Honey badger	*Mellivora capensis wilson*	Cheesman (1920)	Al-Ghzawi et al. (2009)	Hussein (2012)	Hussein (2012)	Hussein (2012)

Table 3.2 Honey bee microsporidian, microbial, and viral diseases found in the Levant region

Name of pathogen		Country in which pathogen was identified				
Common name	Scientific name	Iraq	Jordan	Lebanon	Palestine	Syria
Nosema	*Nosema apis*	AOAD (1996), Sivaram (2004)	Haddad (2015)	–	1–7	Duwwara et al. (2015)
	Nosema ceranae	–	Haddad (2015)	–	–	–
American foulbrood	*Paenibacillus larvae*	AOAD (1995), Sivaram (2004), Abass et al. (2010)	AOAD (1995), Haddad et al. (2015a, b)	OIE (2009)	AOAD (1995)	AOAD (1995), Duwwara et al. (2015)
European foulbrood	*Streptococcus pluton/ Melissococcus plutonius*	Abass et al. (2010)	AOAD (1995)	OIE (2009)	AOAD (1995)	AOAD (1995)
Chalkbrood	*Ascosphaera apis*	Sivaram (2004), OIE (2009)	AOAD (1995)	OIE (2009)	AOAD (1995)	AOAD (1995)
Chronic bee paralysis virus	–	–	Al-Abbadi et al. (2014)	Haddad (2016)	Haddad (2016)	Haddad (2016)
Acute bee paralysis virus	–	–	Haddad et al. (2008a), Al-Abbadi et al. (2014)	Haddad (2016)	Haddad (2016)	Haddad (2016)
Israeli *acute* paralysis virus	–		Al-Abbadi et al. (2014)	Haddad (2016)	Haddad (2016)	Haddad (2016)
Sac brood virus	–	–	Haddad et al. (2008a)	Haddad (2016)	Haddad (2016)	Haddad (2016)
Black queen cell virus	–	–	Al-Abbadi et al. (2014)	Haddad (2016)	Haddad (2016)	Haddad (2016)
Deformed wing virus	–	Haddad et al. (2008a, 2015a, b)	Haddad et al. (2008a, 2015a, b), Al-Abbadi et al. (2014)	Haddad et al. (2015a, b)	Haddad et al. (2015a, b)	Haddad et al. (2015a, b)
Macula-like virus	–	–	de Miranda et al. (2015)	–	–	–

and predators found throughout the Levant. Notably, the Senotainia fly is only found in Jordan and Syria. As well, *Nosema ceranae* and macula-like virus are also only found in Jordan. Of course, this may simply represent sampling error, meaning these species have not yet been evaluated in other regions. Another interesting point is that the only location in which *Nosema apis* was not found was Lebanon.

3.5 Other Problems of Beekeeping

3.5.1 Pesticides Are Problematic for Honey Bees

Pesticides are detrimental to bees and in all regions globally, we must be mindful in our actions and employ caution when using pesticides in the vicinity of honey bees, other commercialized bee, as well as native bees. The scientific community has been paying more attention to the effects of sublethal pesticide doses recently, which have been shown to impact many bee behaviors and traits, including foraging, learning, memory, homing failure, reproduction, the functioning of a healthy immune system, and survival in honey bees (Bendahou et al. 1997; Desneuz et al. 2007; Henry et al. 2012). The poisoning of honey bees with pesticides became a crucial problem in the Levant region, especially in the areas where intensive agriculture is practiced (AOAD 1995).

3.5.2 Human Conflict

War can be costly with respect to beekeeping. It creates a major challenge for even the most attentive beekeepers, who may have difficulty accessing their bees and thus maintaining healthy hives in wartime. War can also result in hives that must be abandoned rapidly and without warning.

These grave issues were clearly articulated by over 300 Iraqi extension service agents and beekeepers during training courses that occurred in Jordan between the years 2010 and 2016 (Haddad pers. Comm.). These training courses were funded by several agencies including the United State Agencies For International Development (USAID), the Food and Agriculture Organization (FAO), and the Japanese International Cooperation Agency (JICA).

As a result, beekeepers may experience severe financial losses as a result of war. These losses occur not only for the beekeeper, but for others too when hive commodities like honey, wax, and candles cannot be sold; furthermore, buyers may not be able to afford any products produced. This has been the case in Iraq for over 15 years, and the case in Syria for over 5 years. War also means that for unmanaged hives, disease may set in. This may result in the transmission of contagious diseases, many of which can have grave impacts on honey bee health. Unfortunately, this is a challenging issue to research under the current setting.

3.5.3 Past and Predicted Impacts of Climate Change in the Levant

Climate change is expected to impact the Levant quite seriously because the regions that will get hotter and drier are the most fertile land (largely, the fertile crescent), which is already surrounded by desert. From 1980 to 2008, temperature and precipitation data demonstrated that the land near the Mediterranean Sea in several countries, especially Syria, was already becoming hotter and drier, and will continue to do so, with some of the most extreme thermal and precipitation-related changes observed globally (Lobell 2011). Rainfall changes impact honey bee colonies and some cannot increase their population sizes now, before nectar flow begins (Haddad 2011b). Much of the Levant is desert, which means species like *A. m. syriaca* that appear well adapted to heat may be capable of withstanding harsher environments that some subspecies may not tolerate. Evapo-transpiration, evaporation, and drought have increased in many regional locations (Haddad 2011b). Moreover, a clear shift has occurred in the timing, duration, and geographic locations of rainfall over the last 50 years which is converting semidry areas into dry ones, meaning the available areas for beekeeping are decreasing. Urbanization in areas previously used for agriculture also challenge the beekeeping sector.

3.6 Other Non-*Apis* Species Found in the Region

Due to the huge challenges that are facing honey bees, it is becoming more important to find alternative species to be used for crop pollination, particularly since the cultivation of crops dependent upon pollinating insects has increased in the last several decades (Aizen et al. 2009). Wild bees have been proven to be at least as efficient, or sometimes more efficient, pollinators than *Apis mellifera* for important agricultural crops such as nightshade species (Solanaceae) (O'Toole 1993), blueberries, and forage legumes (Richards 1996). Additionally, just the presence of non-*Apis* species along with *A. mellifera* in almond orchards results in greater pollination effectiveness of individual honey bees in the orchard (Brittain et al. 2013). Interest has grown in finding non-*Apis* pollinators that are potentially useful in orchards. Mason bees *(Osmia* sp.) are being used now successfully in apple orchards and on strawberry farms. In the highlands of Jordan in Ajloun, transects were run to assess bee diversity found in stone-fruit tree blossoms. Fifty-three species were found in a collection of over 1400 individuals. Five families were represented: Apidae, Megachilidae, Halictidae, Andrenidae, and Colletidae. The greatest diversity was found in Apidae and Megachilidae, while the highest abundance was found for the Halictidae (Al-Ghzawi et al. 2006).

There is also a positive relationship between bees and floral abundance (Heithaus 1974; Banaszak and Krzysztofiak 1996). The richness of bee species in Jordan can perhaps be attributed to the high plant biodiversity within the country (Al-Eisawi

1998; Abu-Irmaileh 2000). Studies should be conducted to monitor bee visitors of wild and cultivated plants throughout the year in additional parts of the Levant. Additional research into the pollination value of individual species on target crops is urgently needed. The biology and rearing of the most effective pollinators should be investigated in order to increase their numbers in the region so as to improve fruit tree productivity and other pollination needs where pollination services are lacking. In general, information about non-*Apis* bees in the Levant region is very limited. However, according to the records of the National Center for Agriculture Research and Extension, the use of bumblebees imported from Europe for the pollination of some vegetables in green houses has been practiced in the region since 1994.

3.7 Future Perspectives

There remains much to learn about bees globally, and especially in the Levant region. Progress is currently being made in the region to sequence and identify genes related to economically valuable bee traits, such as disease resistance and heat tolerance in those that are native to the region. The unique ability of *A. m. syriaca* to tolerate exceedingly hot and dry conditions could prove particularly valuable in the face of climate change. Anytime additional species' genomes are sequenced for a given taxa, more information can be gleaned about the importance of specific mutations and their value, which may be especially important in extreme environments where the ability to evolve impacts lineage persistence. Active research is ongoing with respect to understanding gene expression, genome sequence variation, and their relationships to disease, bee health, and the production of strong *Apis* lines for commercial use.

In addition to genetic analyses, the importance of epigenetics, genome methylation, and perhaps their relationship to disease cannot be underestimated. Queens experience epigenetic changes relative to workers to allow them to have a functioning reproductive system and lay an enormous number of eggs and the role of methylation and disease is currently poorly studied beyond this. Since *A. m. syriaca* are resistant to the effects of some diseases and able to tolerate extreme environments, this subspecies is one we all benefit from learning more about.

While the search for high quality queens is ever present, we must be mindful of the potential impacts of introgression between subspecies, impregnating queens for use (instrumental insemination), and moving queens from one habitat or hive, to another. Our understanding of disease in bees is young, and the potential for these factors to impact hive health, the queens themselves, the hives they are moved into, nearby commercial hives, and native bees is concerning and warrants careful study. Some evidence in South Africa has already shown the potential for increased risk to other hives when colonies begin to be migrated by humans across novel regions (Pirk et al. 2014). We must be mindful of our actions and the potential consequences for the future when interfering with the natural dynamics of bees that can interbreed easily, such as the subspecies found in the Levant region.

References

Abass A, Shaher K, Al-Jorany R (2010) The abandance of American foul broad *Paenibacillus larvae* and European foul broad *Melissococcus pluton* on the honey bees colonies in the middle of Iraq. Al-Anbar J Agric Sci 8(4):425–437. (In Arabic)

Abd Al-Wahed F (1985) Iraq civilization. Al Horia Publ. House

Abu-Irmaileh B (2000) Weeds of cultivated fields. Publication of Jordan University, Deanship of Research

Adjlane N, Doumandji SE, Haddad N (2012) Situation de l'apiculture en Algerie: facteurs menacant la survie des colonies d'abeilles locales Apis mellifera intermissa. Cah Agric 21:235–241

Aizen MA, Garibaldi LA, Cunningham SA et al (2009) How much does agriculture depend on pollinators? Lessons from long-term trends in crop production. Ann Bot 103:1579–1588. https://doi.org/10.1093/aob/mcp076

Al-Abbadi A, Ghabeish I, Hassawi D (2014) Molecular detection of three viral diseases in honeybees (*Apis mellifera* L.) colonies infested with mite (*Varroa destructor*) in Jordan. J Biol Sci 14:446–451

Albaba I (2010) Status of bee keeping in Palestine. A paper presented at the first Arab countries conference on honey bee viral diseases, Amman–Jordan, 10–15 Oct 2010

Alburaki M, Moulin S, Legout H et al (2011) Mitochondrial structure of Eastern honeybee populations from Syria, Lebanon and Iraq. Apidologie 42(5):628–641

Alburaki M, Bertrand B, Legout H et al (2013) A fifth major genetic group among honeybees revealed in Syria. BMC Genet 14:117

Al-Eisawi D (1998) Field guide to wild flowers of Jordan and neighbouring countries. Jordan Press Foundation "Al Rai", Amman (In Arabic)

Al-Ghamdi A, Adgaba N, Tadesse Y, Getachew A, Al-Maktary AA (2016) Comparative study on the dynamics and performances of Apis mellifera jemenitica and improved hybrid honeybee colonies in southwestern Saudi Arabia. Saudi J Biol Sci 24:1086–1093

Al-Ghzawi A, Zaitoun S, Mazary S et al (2006) Diversity of bees (*Hymenoptera, Apiformes*) in extensive orchards in the highlands of Jordan. Arxius de Miscel-lània Zoològica 4:42–48

Al-Ghzawi A, Zaiyoun S, Shannag H (2009) Incidence and geographical distribution of honeybee (*Apis mellifera* L.) pests in Jordan. Ann Soc Entomol Fr (N.S.) 45(3):305–308

Alhissnawi M (2016) Spatial analysis of honeybee keeping and its products in the middle Euphrates governorates. Dissertation, Kufa University of Iraq

Amer K (2015) Livestock annual report. Ministry of Agriculture, Jordan

Amr SZ, Shehada SE, Abo-Shehada M et al (1998) Honeybee parasitic arthropods in Jordan. Apiacta 3:78–82

AOAD (Arab Organization for Agricultural Development) (1995) Questionnaire of Arab Organization for Agricultural Development on the present and constrains of Beekeeping. Khartoum, Sudan (In Arabic)

AOAD (Arab Organization for Agricultural Development) (1996) Technical report about the crawling honey bees in Iraq. Khartoum, Sudan (In Arabic)

Banaszak J, Krzysztofiak A (1996) The natural wild bee resources (*Apoidea, Hymenoptera*) of the Wigry National Park. Pol Pismo Ent 65:33–51

Bendahou N, Bournias M, Fleche C (1997) Acute toxicity of cypermethrin and fenitrothion on honeybees *(Apis mellifera mellifera)*, according to age, formulations, and (chronic paralysis virus)/insecticide interaction. J Environ Biol 18:55–65

Berthoud H, Imdorf A, Haueter H et al (2010) Virus infections and winter losses of honey bee colonies *(Apis mellifera)*. J Apic Res 49(1):60–65

Bezabih G, Adgaba N, Hepburn HR, Pirk CWW (2014) The territorial invasion of Apis florea in Africa. Afr Entomol 22:888–890

Blum R (1956) The bees in the Near East. Proc Int Beekeep Congr, Apimondia Wien 16:28

Bodenheimer S, Ben-Nerya A (1937) One–year study on the biology of the honeybee in Palestine. Ann Appl Biol 24:385–403

Brittain C, Williams N, Kremen C et al (2013) Synergistic effects of non-*Apis* bees and honey bees for pollination services. Proc R Soc B 280(1754):21212767. https://doi.org/10.1098/rspb.2012.2767

CBSSYR (2011) Poultry, beehives and boxes silk warms (2006–2010). 2011 Yearbook, Ch. 4. http://www.cbssyr.sy/yearbook/2011/Chapter4-EN.htm. Accessed 1 June 2017

Cheesman E (1920) Report on the mammals of mesopotamia. J Bombay Nat Hist Soc 27:323–346

Cox-Foster DL, Conlan S, Holmes EC et al (2007) A metagenomic survey of microbes in honey-bee colony collapse disorder. Science 318:283–287

Dainat B, Evans JD, Chen YP et al (2012) Dead or alive: deformed wing virus and *Varroa destructor* reduce the lifespan of winter honeybees. Appl Environ Microbiol 78:981–987

de Miranda J, Cornman RS, Evans JD et al (2015) Genome characterization, prevalence and distribution of a Macula-like virus from *Apis mellifera* and *Varroa destructor*. Viruses 7(7):3586–3602

Desneuz NA, Decourtye A, Delpuech JM (2007) The sub lethal effects of pesticides on beneficial arthropods. Annu Rev Entomol 52:81–106

Duwwara M, Saa'b R, Taki K (2015) Study of reality of honey beekeeping and it's economic feasibility in Sweida governorate, South Syria. Jordan J Agric Sci 11(3):827–838

El-Obeid D (2015) Beekeeping course. Lebanese University Faculty of Agriculture and Veterinary Science, LU-Press

Faucon JP (2002) Honey bee winter mortality in France in 1999 and 2000. Bee World 83:14–23

Franck P, Garnery L, Solignac M, Cornuet JM (2000) Molecular confirmation of a fourth lineage in honey beesfrom Near East. Apidologie 31:167–180

Franck P, Garnery L, Louiseau A, Oldroyd BP, Hepburn HR, Solignac CM, Cornu JM (2001) Genetic diversity of the honeybee in Africa: microsatellite and mitochondrial data. Heredity 86:420–430

Ftayeh A, Meixner M, Fuchs S (1994) Morphometrical investigation in Syrian honeybees. Apidologie 25(4):396–401

Glaiim MK (1992) First definitive record of Apis florea in Iraq. Beekeep Dev 24:3

Glaiim K (2009) Hunting behavior of the Oriental hornet, *Vespa orientalis* L., and defense behavior of the honey bee, *Apis mellifera* L. in Iraq. Bull Iraq Nat Hist Mus 10:17–30

Glaiim MK (2010) On the common sharp decline in honey bee, *Apis mellifera* L., activity occurring during parts of summer and fall in central and southern Iraq, with a particular reference to the role of bee-eaters, *Merops* spp. J Kerbala Univ (Iraq) 8(1):244–268

Glaiim K (2014) Occurrence and status of bee-eaters, *Merops* spp. (Coraciiformes: Meropidae), and their attacks on honey bee colonies in Kerbala Province, Iraq. J Apic Res 53(4):478–488

Haddad N (2011a) Bee losses and colony collapse disorder in the Middle East. Uludag Bee J 11(1):17–24

Haddad N (2011b) Climatic changes and Jordanian food security. Aldastoor Press and Publishing House, Jordan. (In Arabic)

Haddad N (2015) Mitochondrial genome of the Levant Region honeybee, *Apis mellifera* syriaca (Hymenoptera: Apidae). Mitochondrial DNA 27(6):4067–4068

Haddad N (2016) Detection of honeybee viruses in samples from the Levant region. Annual Report of National Center for Agricultural Research and Extension Annual report of the year 2015

Haddad N, Albaba I (2014) Example of some Middle East and North African countries quarantine requirements for the import of honey bees. In: Wolfgang (ed) The Bee health and veterinarians. Examples of regional and national regulations. Chapter 3. 24. – Standards and regulations. OIE, pp 285–289

Haddad N, Fuchs S (2004) Honeybee agrobiodiversity: a project in conservation of *Apis mellifera syriaca* in Jordan. Uludag Bee J 2:166–120

Haddad N, Dvořák L, Adwan O et al (2007) New data on Vespid wasp fauna of Jordan (H. Vespidae). Linzer Biol Beitr 39(1):137–142

Haddad N, Brake M, Migdadi H et al (2008a) First detection of honeybee viruses in Jordan by RT-PCR. Jordan J Agric Sci 4(3):242–247

Haddad N, De Miranda J, Batainh A (2008b) The discovery of *Apis florea* in Jordan. J Apic Res 47(2):172–173

Haddad N, Meixner MD, Fuchs S et al (2009a) Mitochondrial DNA support for genetic reserves of *Apis mellifera syriaca* in Jordan. J Apic Res Bee World 48(1):19–22

Haddad N, Fuchs S, Hepburn HR et al (2009b) *Apis florea* in Jordan: source of the founder population. Apidologie 40(4):508–512

Haddad N, Bataeneh A, Albaba I et al (2009c) Status of colony losses in the Middle East. Paper presented at in the 41st Apimondia Cong, Montpellier, France, 15–20 Sept 2009

Haddad N, Adjlane N, Loucif-Ayad W et al (2015a) Presence and infestation rate of *Senotainia tricuspis* (Meigen) (Diptera, Sarcophagidae) on honey bees in the Mediterranean Region. J Apic Res 54(2):121–122

Haddad N, Al-tellawi A, Adjlane N et al (2015b) Diagnosis of *Paenibacillus larvae* honeybees disease in Jordan using microbiological and chemicals techniques. Asian J Anim Sci 9(6):318–329

Haddad N, Batainh MA, Migdadi SO et al (2015c) Next generation sequencing of *Apis mellifera syriaca* identifies genes for *Varroa* resistance and beneficial bee keeping traits. Insect Sci 23(4):579–590. https://doi.org/10.1111/1744-7917.12205

Haddad N, Batainh A, Saini D et al (2016) Evaluation of *Apis mellifera syriaca* Levant region honeybee conservation using comparative genome hybridization. Genetica 144:279–287

Haddad NJ, Noureddine A, Al-Shagour B et al (2017) Distribution and variability of deformed wing virus of honeybees (*Apis mellifera*) in the Middle East and North Africa. Insect Sci 24(1):103–113. https://doi.org/10.1111/1744-7917.12277

Hatoom A (1996) *Senotainia tricuspis* a parasite at Syrian honey bees. Paper presented at The First Inter Arab Apic Cong, American University of Beirut, Lebanon, 17–20 Aug 1996

Heithaus E (1974) The role of plant-pollinator interactions in determining community structure. Ann Mo Bot Gard 61:675–691

Henry M et al (2012) A common pesticide decreases foraging success and survival in honey bees. Science 336:348–350

Hepburn HR, Radloff SE (1998) Honeybees of Africa. Springer, Berlin

Hermani H (ed) (2012). Social insects, vol 1. Elsevier

Higes M, Martín-Hernández M, Garrido-Bailón E et al (2009) Honeybee colony collapse due to *Nosema ceranae* in professional apiaries. Environ Microbiol Rep 1:110–113. https://doi.org/10.1111/j.1758-2229.2009.00014.x

Hussein MH (2000) A review of beekeeping in Arab countries, 2nd edn. Al-Rehab Press, Assiut, Egypt, 641 p (Arabic)

Hussein MH (2012) Beekeeping in Arab countries. Al-Rehab Press, Assiut

Kadamani S (2015) A picture of Lebanese beekeeping, development strategies for Lebanese and Mediterranean beekeeping. Beirut, Lebanon. (In Arabic)

Kalman C (1973) Our formally very aggressive bee. Proc Int Beekeep Congr Apimondia Burkarest 24:306–307

Kence M, Oskay D, Giray T et al (2013) Honeybee colonies from different races show variation in defenses against the varroa mite in a ‘common garden’. Entomol Exp Appl 149(1):36–43

Kramer SN (1988) History begins at Sumer. University of Pennsylvania Press, Philadelphia

Lee K, Steinhauer N, Rennichet K et al (2015) A national survey of managed honey bee 2013–2014 annual colony losses in the USA. Apidologie 46(3):292–305

Lobell DB, Schlenker W, Costa-Roberts J (2011) Climate trends and global crop production since 1980. Science 333(6042):616–620

Lord WG, Nagi SK (1987) Apis florea discovered in Africa. Bee World 68:39–40

Mazar A, Namdar D, Panitz-Cohen N et al (2008) The iron age beehives at Tel Rehov in the Jordan Valley. Antiquity 82:629–639

Mogga JB, Ruttner F (1988) *Apis florea* in Africa: source of the founder population. Bee World 69:100–103

Moritz R, Haddad N, Bataieneh A et al (2010) Invasion of the dwarf honeybee *Apis florea* into the near East. Biol Invasions 12:1093–1099

Nazer I, Rateb M (1991) Status of honey bee diseases and pests in Jordan. In: Ritter W (ed) Proceedings of the international symposium on recent research on bee pathology, Ghent, 5–7 Sept 1990. Rijksstation voor Neematologie en Entomologie on behalf of Apimondia; Merelbeke, Belgium, pp 210–212

Neumann P, Carreck NL (2010) Honey bee colony losses. J Apic Res 49(1):1–6. https://doi.org/10.3896/IBRA.1.49.1.01s

O'toole C (1993) Diversity of native bees and agroecosystems. In: LaSalle J, Gauld ID (eds) Hymenoptera and biodiversity. Cab International, Wallingford, pp 169–195

OIE (2009) World Animal Health Information Database—Version: 1.4. World Animal Health Information Database. Paris, France: World Organization for Animal Health. http://www.oie.int. Accessed 1 June 2017

OJEU (2013) Official Journal of the European Union—Commission Implementing Regulation (EU) 485/2013 139:12–23

Özkan A, Gharleko M, Özden B, Kandemir I (2009) Multivariate morphometric study on Apis florea distributed in Iran. Turk J Zool 33:93–102

Palmer MR, Smith DR, Kaftanoglu O (2000) Turkish honey bees: genetic variation and evidence for a fourth lineage of *Apis mellifera* mtDNA J. Heredity 91:42–46

PCBS (2014) Palestinian Central Bureau of Statistics-Announces the main results for Livestock Survey. http://www.pcbs.gov.ps. Accessed 1 June 2017

Pirk CWW, Human H, Crewe RM, vanEnglesdorp D (2014) A survey of managed honeybee colony losses in the Republic of South Africa—2009 to 2011. J Apic Res 53(1):35–42

Ransome HM (2004) The sacred bee in ancient times and folklore. The Hebrew and the Mohammedens. Dover Publications, pp 65–74

Richards KW (1996) Comparative efficacy of bee species for pollination of legume seed crops. In: Matheson A, Buchmann S, O'Toole C, Westrich P, Williams IH (eds) The conservation of bees. Academic, London, pp 81–103

Ruttner F (1988) Biogeography and taxonomy of honeybees. Springer, Berlin

Ruttner F, Tassencourt L, Louveaux J (1978) Biometrical-statistical analysis of the geographic variability of *Apis mellifera*. Apidologie 9:363–381

Saeed MS (2013) Selecting the best periods of spring, summer, and autumn months to use the wet and dry grafting method when queen rearing honeybees *Apis mellifera* L. (*Apidae Hymenoptera*) by Doolittle method for honey bee hives in the Ninevah region. J Tikrit Univ Agric Sci 13(2):329–335

Sivaram V (2004) Apiculture development in Iraq. Paper presented at the 7th Asian apicultural association conference and 10th BEENET symposium and technofora, University of Philippines Los Banos, 20–22 Feb 2004

Slabezki Y, Efrat H, Dag A, Kamer Y, Yakobson BA, Mozes-Koch R, Gerson U (2000) The effect of honey bee tracheal mite infestation on colony development and honey yield of Buckfast and Italian honey bee strains in Israel. Am Bee J 140(3):231–235

Smith DR, Slaymaker A, Palmer M, Kaftanoglu O (1997) Turkish honey bees belong to the east Mediterranean mitochondrial lineage. Apidologie 28:269–274

Soroker V et al (2009) Colony losses in Israel: incidence of viral infection and beehive populations. Paper presented at the 41st Apimondia Congress, Montpellier, France, 15–20 Sept 2009

VanEngelsdorp D, Underwood R, Caron D et al (2007) An estimate of managed colony losses in the winter of 2006-2007: a report commissioned by the apiary inspectors of America. Am Bee J 147:599–603

VanEngelsdorp D, Frazier M, Frazier J et al (2009) Colony collapse disorder: a descriptive study. PLoS One 4:e6481. https://doi.org/10.1371/journal.pone.0006481

Wallberg A, Han F, Wellhagen G et al (2014) Worldwide survey of genome sequence variation provides insight into the evolutionary history of the honeybee *Apis mellifera*. Nat Genet 46:1081–1088

Whitfield CW, Behura SK, Berlocher SH, Clark AG, Johnston JS, Sheppard WS, Smith DR, Suarez AV, Weaver D, Tsutsui ND (2006) Thrice out of Africa: ancient and recent expansions of the honey bee, *Apis mellifera*. Science 314:642–645

Williams NM, Crone EE, Roulston TH, Minckley RL, Packer L, Potts SG (2010) Ecological and life-history traits predict bee responses to environmental disturbances. Biol Cons 143:2280–2291

Winston ML (1991) The biology of the honey bee. Harvard Univ Press, Cambridge, MA

Zaitoun ST, AL-Ghzawi A-M, Shannag HK (2000) Population dynamics of the Syrian honeybee, *Apis mellifera syriaca*, under semi-arid Mediterranean conditions. Zool Middle East 21:129–132

Zakour M, Ehrhdardt K, Bienefeld K (2012) First estimate of genetic parameters for the Syrian honey bee *Apis mellifera syriaca*. Apidologie 43:600–607

Chapter 4
Apiculture in Israel

Victoria Soroker, Slabezki Yossi, and Nor Chejanovsky

Abstract Honey bee (*Apis mellifera*) has a thousand years of history in this part of Asia. The native honey bee race is considered to be *A. mellifera syriaca*. This subspecies was generally replaced by a more docile subspecies *A. mellifera ligustica*, but members of other subspecies are also occasionally introduced. Only about parts of the country area are suitable for honey bees due to very limited vegetation arid conditions in the southern part of the country. The majority of the colonies are concentrated at the center and north of the country reaching density of more than 14 colonies per square kilometers. To supplement bee forage, *Eucalyptus* trees are planted intensively around the country. Local beekeepers practice modern methods of beekeeping using Langstroth hive boxes. All professional beekeepers usually treat their colonies preventively against *Varroa destructor* mites and foulbrood diseases. Many beekeepers also treat preventively against Nosema disease. Despite regular management, *Varroa* and viruses remain a major obstacle for successful beekeeping.

Keywords European honey bee · *Apis mellifera* · Pests · Bee diseases

4.1 Diversity of Honey bee Races

What are the native bees in the land of Israel is hard to tell. Until recently, it was believed that the native honey bee race is *Apis mellifera syriaca*; however, identification of honey bee remains according to morphometric data revealed that those were not to the local *syriaca* race but *A. m. anatolica* (Mazar and Panitz-Chohen 2007). Could it be that people in the iron age not only cultivated local bees but also imported swarms from at least 500 km from Anatolia that had well-established beekeeping in those times? The historians still do not have answer to this question.

V. Soroker (✉) · N. Chejanovsky
Department of Entomology, The Volcani Center, Agricultural Research Organization, Rishon LeZion, Israel
e-mail: sorokerv@volcani.agri.gov.il; ninar@volcani.agri.gov.il

S. Yossi
Extension Service, Ministry of Agriculture, Rishon LeZion, Israel

P. Chantawannakul et al. (eds.), *Asian Beekeeping in the 21st Century*,
https://doi.org/10.1007/978-981-10-8222-1_4

Fig. 4.1 Traditional honey bee colonies in Israel: (**a**) Abu Kabir Israel; (**b**) Moavia, Israel 1986, photo by Eitan Eilon with permission

Until about 100 years ago, the most abundant strain of honey bees was *A. m. syriaca*. These bees were grown by the local growers in clay cylinders piled one on top of the others as shown in Fig. 4.1. Although this strain is well adapted to local climatic condition, these honey bees behavior is quite aggressive. Thus, with the development of modern honey bee industry, *A. m. syriaca* was substituted by the less aggressive Italian race—*A.m. ligustica*. Later, in the mid-1980s of the last century even the wild bee colonies of *syriaca* became extinct, presumably following the *Varroa* invasion in 1984. Today, *A.m. ligustica* is the main honey bee race cultivated in Israel but it is not the only race grown, as queens from other bee strains are imported for breeding purposes, especially *Buckfast and A. m. caucasica.*

4.2 Historical Records and Recent Developments

Honey bee cultivation in the Middle East is over 3000 years old. Specifically in the land of Israel, there is so far a unique archeological evidence of bee cultivation at the ancient city of Tel Rehov dating back to the sixth–eighth century BC. Nowadays, the bees are usually cultivated in standard Langstroth hives (Fig. 4.2). Most of the breeding is conducted via selective rearing progeny from queens of the most honey producing colonies but the daughters are usually matting freely, resulting in genetically mixed populations. Thus, it is not uncommon to find light brown and almost black drones in the same colony.

The local beekeeping industry consists of about 600 beekeepers that operate approximately 110,000 honey bee colonies, producing 3500 metric tons of honey and totally about 100,000 cycles of pollination services. This results in revenue of about 250 million USD-from honey and pollination services. The number of colonies per beekeeper varies: nearly 35% of beekeepers own more than 100 colonies and can be considered commercial beekeepers, but only 11% possess more than 500 colonies and just a few have several thousands. These large operations play a crucial role in pollination services. Over 90,000 colonies provide pollination services to

Fig. 4.2 Modern small organic apiary in West Galilee

about 20 fruit and vegetable crops. In particular, there is a high demand for honey bee pollination in avocado, almonds, apples, watermelons, and melons. Beekeeping requires registration and license for each apiary.

4.3 Local Culture and Honey bees

The significance of honey in the area is provided by the texts in the Bible (old and new testaments), describing the promised land of Israel as the "land of milk and honey." Honey symbolizes prosperity and as such serves as traditional food eaten with apple slices and as part of pastries during holiday meal on Jewish new years' eve for a successful new year. Honey is also mentioned in the Quran and Hadith especially for its medicinal purposes. Today, it is also popular to use honey as an important natural food component and natural medicine as well as a component in variety of cosmetic products.

4.4 Bee Forage Resources

Honey bee forage is composed of agricultural crops such as cotton, vegetables, and fruit trees as well as wild flowers and eucalyptus. Different species of *Eucalyptus* are planted around the country in attempt to provide bee forage year around. As can be seen in Fig. 4.3, the colonies are not distributed evenly along the country. The majority of them are concentrated at the center and north of the country reaching among the highest density in the world of more than 14 colonies per square kilometers. On the other hand, most of the southern part of the country is quite empty of honey bee colonies. The reason for this diversity in colony density is that areas receiving annually less than 200 mm rain cannot support much vegetation and thus beekeeping. All the growers operating large number of colonies are migrating with their hives

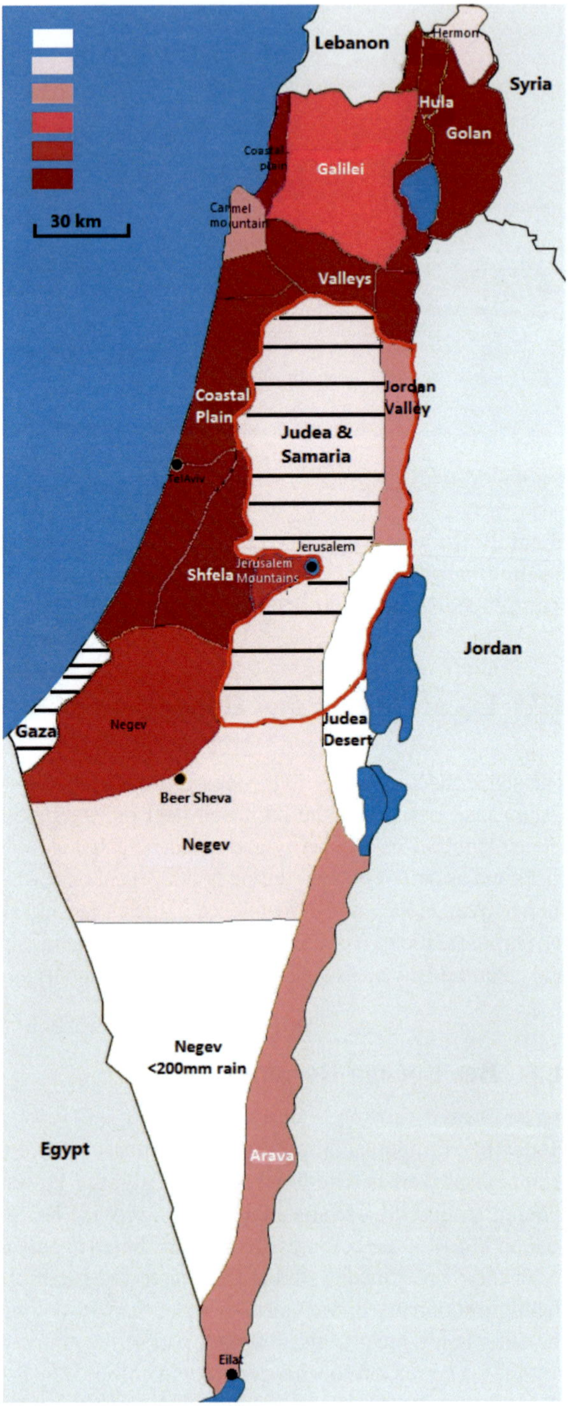

Fig. 4.3 Apiary distribution along the country. Total country area is 22,072 sq. km. Dashed areas indicate regions under Palestinian Authority. The color intensity represents apiary density. At the darkest color density is the highest, with over 14 colonies per sq. km

following the bloom of different flowers and some also going to better over-wintering sites. The large grower may operate more than 30 sites (apiaries) and migrate several times during the year for honey production or pollination services.

4.5 Annual Cycle of Honey bee Colony and Main Timing of Losses

The local climatic conditions are over all comfortable to bees. The winter is mild, temperatures go rarely below zero and not in all places, allowing the bees to have brood continuously along the winter (Fig. 4.4); on the other hand, the breeding is going down, and sometimes even stops at the end of the summer in August–September. This period is usually characterized by nectar and pollen dearth. On the other hand later in autumn, there is an additional flowering season and for beekeepers this is considered a small "spring." This period enables the colonies to get stronger after the summer drought, and beekeepers to make splits and to change queens as a preparation for the winter and the next season. Large commercial beekeepers routinely change their queens annually in autumn. The main period for honey harvest is in early spring. However, summer and autumn harvest are also possible in some locations depending on local nectar sources. Between the sites the colonies are transported by trucks.

4.6 Major Pests and Diseases of Honey Bees

In the last decade, the honey bee pathogens of major concern are common honey bee pathogens such as microsporidia *Nosema* affecting adults, the ectoparasite *Varroa destructor*, bacteria-like *Melissococcus pluton*, and *Paenibacillus larvae* affecting larvae and pupae, and viruses such as deformed wing virus (DWV),

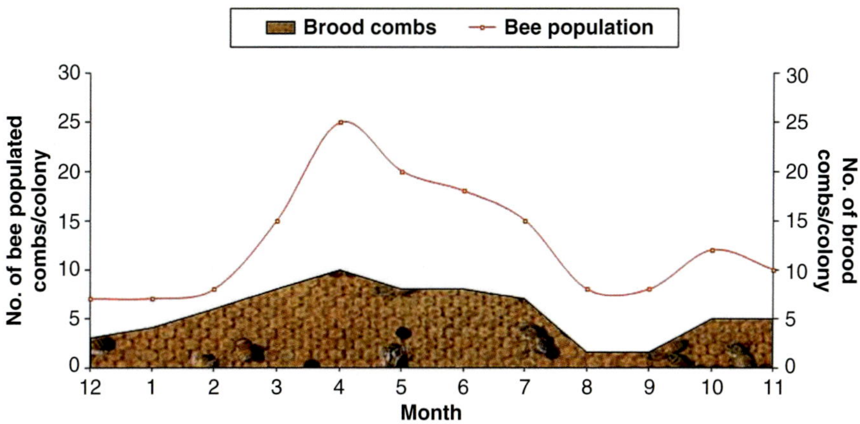

Fig. 4.4 Typical annual honey bee colony cycle

sacbrood virus (SBV), Kashmir bee virus (KBV), black queen cell virus (BQCV), Israeli acute paralysis virus (IAPV), acute paralysis virus (ABPV), and chronic bee paralysis virus (CBPV) (reviewed by Evans and Schwarz 2011). Bacterial diseases are rare, as most beekeepers control these diseases by preventive antibiotic treatment with oxytetracycline at the end of the winter.

4.6.1 Nosemosis

Nosema infections and fecal marks on the hives associated with diarrhea are well known to local beekeepers. *Nosema apis* was believed to be responsible for this phenomena and subsequently weakening of the colony. The recommended treatment against Nosema was and still is: Fumagilin-B (Medivet Pharmaceuticals Ltd.) in three consequent applications of 1 g in 2 L of sugar syrup per hive. Annual monitoring of hives from five to eleven selected apiaries combining microscopic and PCR analysis with primers (synthetic oligonucleotides, specific for both *N. ceranae* and *N. apis*), on over 100 colonies during 2009–2011, have revealed that colonies suspected to be infected by *Nosema apis* by microscopic analysis were in fact infected by *N. ceranae* (Soroker et al. 2011). It is hard to establish accurately when *N. ceranae* appeared in the area replacing *N. apis*. As infection by *N. ceranae* is largely asymptomatic, in the recent years diarrhea is rarely reported by beekeepers and not all of them are treating against Nosemosis (Fig. 4.5). The impact of this change in honey bee management on colony losses is not yet clear.

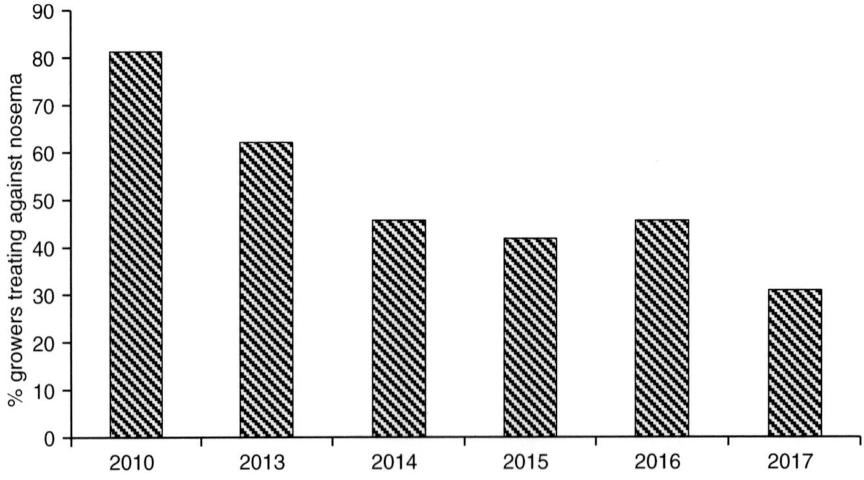

Fig. 4.5 Changes in percentage of growers treating their colonies against Nosema as reported by the beekeepers. It should be noted that although the data represent more than 20% of the total colonies, over the years they do not always represent the same beekeepers. Most of beekeepers that treat Nosemosis apply Fumagilin dissolved in 60% sugar solution

4.6.2 *Varroa*

Varroa destructor Anderson and Trueman (Acari: Varroidae) is an invasive ecto-parasitic mite of honey bees, originally found in the Asian honey bee *Apis cerana*. The mite first appeared in Israel in 1984. High hive density and mobility have contributed to spread of *Varroa*, and the first deaths of colonies due to heavy mite infestations were observed by 1985 (Efrat and Slabezki, unpubl. data). Ever since, *Varroa* has been the main threat to the local apiculture as it is in other areas cultivating European honey bees (Nazzi and Pennacchio 2014), thus requiring regular treatments. At local mild climatic conditions, honey bee colonies usually maintain brood year round, creating conditions for continuous *Varroa* reproduction. Under these conditions, the mite control is extremely challenging. High *Varroa* load over the years is believed to be responsible for the extinction of local feral honey bee colonies *of A. melifera syriaca*. According to our survey conducted among Israeli beekeepers since 2009, the majority of the beekeepers regularly treat against *Varroa* in the late summer and winter seasons. Since *Varroa* invasion, its control has been based mainly on synthetic chemical insecticides. Formamide (Amitraz) was the first chemical used, followed by fluvalinate. After the activity of the latter was lost due to resistance development, it was replaced with Coumaphos (strips of "CheckMite+", Bayer). Coumaphos was used for over 12 years until the mite developed resistance for this chemical in 2011. Today *Varroa* control is conducted mainly by locally produced Amitraz loaded strips (Galbitraz). Only a few beekeepers use other methods. These are mainly oxalic or formic acids. Honey is regularly sampled and inspected for residues according to EU standards.

4.6.3 *Viruses*

Several honey bee viruses were associated with colony losses worldwide and in particular viruses from the DWV-VDV-1-KV clade and those of the ABPV-IAPV-KBV clade (De Miranda et al. 2009) as well as CBPV (Ribière et al. 2010). Viruses from the DWV and ABPV clades are transmitted by *V. destructor*, but not CBPV. In Israel, all the above viruses were present together with black queen cell virus (BQCV) (Leat et al. 2000), sacbrood virus (Chen and Siede 2007), and Lake Sinai virus (Runckel et al. 2011 and NC unpublished results). The seasonal distribution of IAPV, SBV, ABPV, BQCV, VDV-1, and DWV is presented in Fig. 4.6. As can be seen, the most prevalent viruses were IAPV, BQCV, and DWV. Also there were outbreaks of CBPV at the beginning of the summer in localized places (not shown). IAPV prevalence is high in the winter and in the fall while DWV's highest prevalence is in the fall, followed by the summer. BQCV prevalence peaks in the summer but it is high throughout the year.

Most of the viral infections by the viruses in Fig. 4.6 are asymptomatic and can convert to symptomatic due to biological, chemical, and environmental stresses that

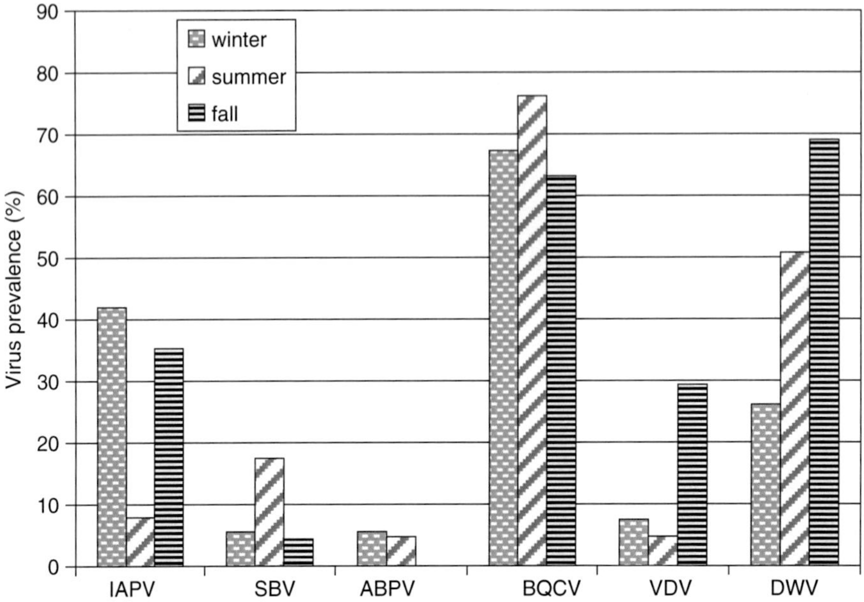

Fig. 4.6 Seasonal prevalence of the most common honey bee viruses in *Apis mellifera* colonies across Israel. The prevalence of the viruses annotated in the *X* axis was determined using reverse-transcription polymerase chain reaction (RT-PCR) and primers specific for each virus (Soroker et al. 2011). CBPV was detected using the primers1-1 5′TCAGACACCGAATCTGATTATTG-3′ and 1-2 5′-ACTACTAGAAACTCGTCGCTTCG-3′ (Colonies $n = 71$)

the colonies become exposed to (Genersch and Aubert 2010; Nazzi and Pennacchio 2014). We have specially monitored colonies for the presence of IAPV throughout several years and most of the time the virus was present in covert infections. However, occasionally we observed outbreaks of IAPV infections following heavy infestations with *V. destructor* (together with other viruses like DWV and VDV-1, YS, NC, unpublished) and in one special case we witnessed collapse of honey bee colonies with the typical symptoms of colony collapse disorder, CCD (Hou et al. 2014). In this case, the virus loads were significantly higher by several orders of magnitude from those of DWV (Hou et al. 2014).

4.6.3.1 A Particularly Interesting Case of Virus-Induced Mortality

During the last two decades, there is in Israel and in other countries an increased shift of crops grown on open fields to cultivation under net-houses and glasshouses. This confers a better protection to crops from insect pest's invasion, from harsh climatic conditions improving water consumption efficiency and the micro-clime, as well as better fecundity. For example, apple, pears, avocado, prune, pomegranate, and other crops are cultivated under nets. In parallel, there is a constant increase in

Fig. 4.7 Honey bee workers collide with the nets

the utilization of *A. mellifera* colonies to pollinate those protected crops (Dag 2008). Beekeepers have observed that many times following the introduction of the colonies under or near the nets they start to show increased mortality at the hive entrance. Usually, this high mortality was attributed to observations that pollen and nectar loaded workers collide with the nets in an attempt to return to the hive (Fig. 4.7). Moreover, the dying bees were hairless, with a dark or almost black appearance; they seemed smaller than healthy bees, having a broader abdomen. They were also shiny, appearing greasy in bright light and with abnormal trembling motion of the wings. To a virologist's eyes, they reminded the typical symptoms of CBPV infections (Ribière et al. 2010). Preliminary screening of paralyzed and dying bees collected at the entrance from such beehives revealed high loads of CBPV.

To investigate the possibility that CBPV infections could be related with the high mortality observed at the entrance of beehives placed for pollination of net-houses, we performed the following experiment in a loquat plantation grown under nets. Twenty uniform colonies were selected. Ten of them were placed at the entrance of loquat grown under the net (open net with entrance and exit sites) at the beginning of the fall (October 2013) just before its flowering and another ten in a locality at 10 km nearby open field crops. The colonies were followed for appearance of bee mortality as well as for estimating the loads of the virus by a molecular method (RT-PCR). Mortality at the entrance of the colonies placed under the nets appeared about 1 month later on November 21st (Table 4.1). Four colonies placed under the net showed medium to severe mortality of adult workers (levels ++ to +++ in Table 4.1) and two colonies placed in the open environment showed low mortality

Table 4.1 Honey bee mortality at the hive entrances

Treatment	Colony	October 23	November 21	December 17
Under the net	1	–	+++	+++
	2	–	+	+++
	3	–	+	+++
	4	–	+	+++
	5	–	+++	+++
	6	–	+++	+++
	7	–	++	+++
	8	–	+	+++
	9	–	+	+++
	10	–	+	+++
Open environment	11	–	–	–
	12	–	+	+/–
	13	–	–	+/–
	14	–	+/–	–
	15	–	–	–
	16	–	–	–
	17	–	–	–
	18	–	–	–
	19	–	–	–
	20	–	–	–

Mortality scale: (–), no dead bees; (±) less than 20 bees dead; (+) about 50 bees dead; (++) between 100 and 400 bees dead; (+++) over 500 bees dead

(levels +/– to +). In the "under the net" group of colonies, we found paralyzed and dead bees displaying symptoms characteristic of CBPV. About 2 months from the beginning of the experiment on December 17th all the colonies under the net showed high mortality and CBPV-like paralysis while only two colonies placed in the open environment showed mild mortality level (+/–, Table 4.1).

To evaluate if the paralysis and mortality we observed in the "under the net" group was due to CBPV infection, we collected samples from the hives that showed dead bees at their entrance and analyzed them by a molecular detection method (reverse-transcription polymerase chain reaction, RT-PCR) with primers (synthetic oligo nucleotides) specific for the viruses IAPV, ABPV, SBV, BQCV, VDV-1, DWV, and CBPV using methods well established in our laboratory (Hou et al. 2014; Soroker et al. 2011). This diagnostic test confirmed that CBPV was present consistently in all the hives that showed paralyzed honey bees and high mortality. None of the other viruses tested showed in a consistent form and only serendipitous positives were obtained for them (not shown). To further confirm our findings, we quantified the number of CBPV present per bees using reverse transcription-quantitative PCR, a method that determines the number of copies of the viral genome (equivalent to the chromosome of eukaryotic organisms). As can be seen in Fig. 4.8, samples from hairless worker bees showing CBPV symptoms or paralyzed workers collected from the entrances of the hives 1, 3, and 5 (under the net hives) bore high

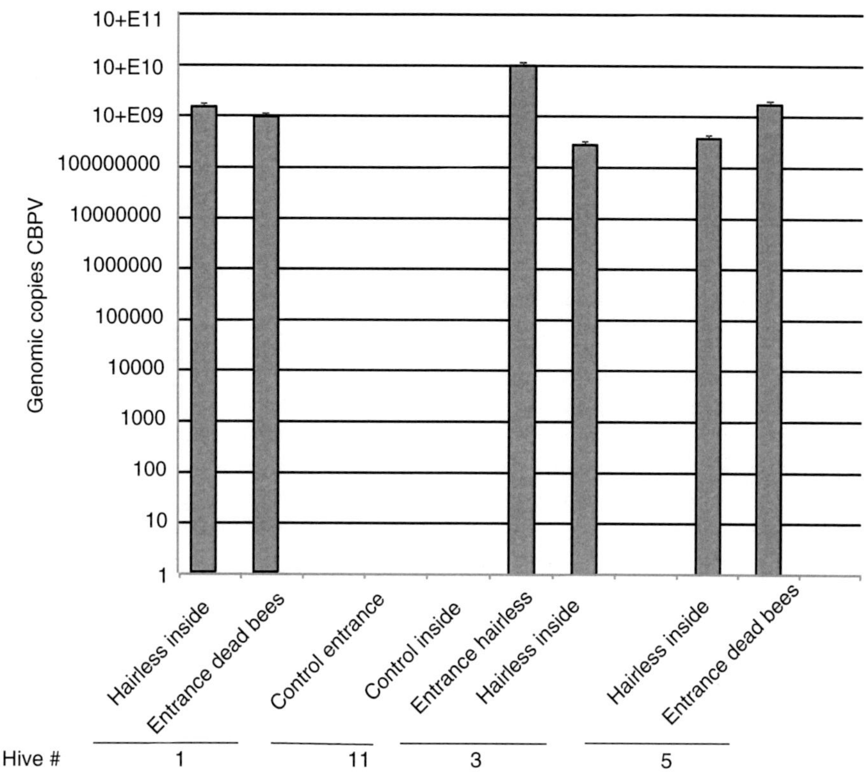

Fig. 4.8 Genomic copies of CBPV of various symptomatic and asymptomatic bees in the hive under the net or in open field. *Y* axis, CBPV genomic copies; *X* axis, symptomatic and asymptomatic bees, the hive number is indicated at the bottom

number of CBPV genomic copies in contrast to workers from hive 11 (open environment) that were negative for CBPV. These results confirmed that individual workers of hives placed under the net contracted CBPV infection. This probably occurred as a consequence of injuries in the cuticle of foragers that collided with the net in their attempt to lift off and return to the colony. It is well known that opportunistic pathogens like viruses can replicate better in their insect body if they reach the hemolymph. For example, lower doses of IAPV, ABPV, and DWV are required to obtain fatal infections in workers if they are directly injected in its body via a natural carrier like *V. destructor* or in laboratory experiments (Boncristiani et al. 2013; Ryabov et al. 2014).

What was the source of CBPV in bee colonies kept under net? That is an intriguing question since it could be present in an undetectable controlled infection in the same colonies and only the mechanical stress and injuries enabled the outbreak of the viral epidemic or alternatively, it could be present in another insect that co-visited the loquat flowers and became the virus transmitter through pollen or nectar for example. Various honey bee viruses were detected in pollen and it is believed that it constitutes a source of transmission of viruses between pollinators (Mazzei et al. 2014; Singh et al. 2010). Also, colonies that are exposed to agricultural monocultures like for in-house pollination of one crop with single source of pollen may experience food nutritional deficiencies due to less balanced diets, and that in turn may increase their susceptibility to pathogens including viruses (Arien et al. 2015).

4.7 Other Problems of Beekeeping

The general problem of beekeeping is relatively high annual losses of colonies that result in drop in honey production that cannot be replaced by colony splitting. Overall annual losses can hardly be determined accurately. Our data based on beekeepers responses to questionnaires, indicate annual loss of about 25%. The losses peak mainly from July up to October. This timing is not surprising as it occurs during the period of forage shortage (pollen dearth) that during the dry season in summer and early autumn. At this period, reproduction is severely reduced, subsequently colonies get smaller, population of bees is thus older, weaker and loaded with pests and pathogens. Experiments with protein supplementation during pollen dearth indicate its significance to colony strength (Soroker et al. 2012).

When beekeepers are asked to grade problems besides pests and pathogens, they point mainly towards lack of forage and insecticide spread. As was numerously shown, good forage not only ensures good honey harvest but may also contribute to colony health. The availability of pollen as a protein source plays a particular role, promoting fast regeneration and increased life expectancy and bee health (Brodschneider and Crailsheim 2010; Genersch and Aubert 2010). Climate changes can be accounted for some problems in foraging. Reduction and or changes in winter rain, dramatically effect blooming and nectar flow of wild flowers and trees. On the other hand, intensive urbanization reduces foraging areas. Changes in agricultural practices and cultivated varieties especially when the crops' requirement do not meet those of the bees, e.g., seedless varieties and management practices (covered crops is an example of such problem as previously described). Insecticide spread is yet another problem. Our recent study revealed that bees are exposed to combinations of various agrochemicals (Soroker et al. 2017). Residues of fungicides, herbicides, and insecticides were detected in bees and pollen stores. The assortment of agrochemicals depended on the site and time of the year. In most of the cases, only part of the colony population suffered acute effects and dyed. In these cases, residues in bodies of dead bees, found next to the hives were often below the LD50 of each of the chemical, thus suggesting that it was the combination of insecticides that had the lethal effect. On the other hand, residues accumulating in the bodies of leaving bees are also expected to have long-term negative effects, e.g., sabotaging bees immune system thus weakening their ability to cope with pest and pathogens, e.g., Nosema (Alaux et al. 2010).

4.8 Other Bee Species

In general, local bee species community is rich (about 800 species) including a eusocial, subsocial, and solitary species. Among eusocial bees, there is dwarf honey bee, *Apis florea* that so far is limited to the south of Arava region. *A. florea* is not native and invaded the region in 2007 (Haddad et al. 2008). It is not considered of agricultural significance, but rather a threat for commercial honey bee population and a public nuisance. On the other hand, another eusocial bee, a non-*Apis* member of Apidae, *Bombus terrestris* has a very high economic significance as a pollinator. This species is native to northern part of the country, but lately expanded due to its intensive cultivation and implementation for pollination especially in green house grown vegetables, such as tomatoes and bell peppers. However, Apidae are not the only bee pollinators in agricultural crops in the Mediterranean habitat and in Israel. In particular, recent studies by Pizanti and coworkers showed contribution of bee species from several genera; primitive eusocial such as *Lasioglossum politum atamarium* and *L. malachurum*, and solitary such as *Osmia* sp., *Colletes,* and *Megachille* to pollination of almond, confection sunflower, and seed watermelon (Pisanty et al. 2014, 2016; Pisanty and Mandelik 2016).

4.9 Conclusions and Future Perspectives

Israel is a country of a very intensive agriculture including beekeeping. The tight link between honey bees and agriculture demands care for well-being of honey bees. Since wild honey bees do not exist, local honey bees are mainly dependent on beekeepers management and wise use of agrochemicals. Efforts are made by the extension and plant protection services to limit usage of bee harming agrochemicals. Despite regular management, *Varroa* and its associated viruses remains a major obstacle for successful beekeeping and a major challenge to bee management. The viruses are another issue of concern even though most of the viral infections are asymptomatic. However, these can convert to symptomatic due to various stressors, e.g., mechanical stress of the nets helps to induce severe CBPV infection of bees foraging under netted conditions. Apparently, CBPV acts as an opportunistic pathogen that gains access inside the bee body through cuticle injuries provoked by the collision of the workers with the nets. To avoid these injuries and thus virus outbreak, is may be possible to utilize nets that do not cause injuries to the workers cuticle. This option requires further study.

The impact of *Nosema* and especially *N. ceranae* on colony survival demands further study. As under intensive agriculture, pollinators including honey bees bees are heavily dependent on crops as food source, efforts are required to promote bee-friendly agricultural management practices.

Acknowledgements The authors thank very much Dr. Eitan Eilon and Mr. Moshe Peer for teaching us about history and culture of beekeeping in the region. To Inna Goldenberg for the figure preparation and assistance in data collection and to all the beekeepers for devoting their time to complete questionnaires providing the data about current practices and colony loss data. NC was supported by a Grant of the Chief Scientist of the Ministry of Agriculture number 131-1723.

References

Alaux C, Brunet J-L, Dussaubat C et al (2010) Interactions between Nosema microspores and a neonicotinoid weaken honeybees (*Apis mellifera*). Environ Microbiol 12:774–782

Arien Y, Dag A, Zarchin S et al (2015) Omega-3 deficiency impairs honey bee learning. Proc Natl Acad Sci 112:15761–15766

Boncristiani HF, Evans JD, Chen Y et al (2013) *In vitro* infection of pupae with Israeli acute paralysis virus suggests disturbance of transcriptional homeostasis in honey bees (*Apis mellifera*). PLoS One 8:e73429

Brodschneider R, Crailsheim K (2010) Nutrition and health in honey bees. Apidologie 41:278–294

Chen YP, Siede R (2007) Honey bee viruses. Adv Virus Res 70:33–80

Dag A (2008) Bee pollination of crop plants under environmental conditions unique to enclosures. J Apic Res 47:162–165

De Miranda JR, Cordoni G, Budge G (2009) The acute bee paralysis virus, Kashmir bee virus, Israeli acute paralysis virus complex. J Invertebr Pathol 103(Suppl):S30–S47

Evans JD, Schwarz RS (2011) Bees brought to their knees: microbes affecting honey bee health. Trends Microbiol 19:614–620

Genersch E, Aubert M (2010) Emerging and re-emerging viruses of the honey bee (*Apis mellifera* L.) Vet Res 41:54

Haddad N, De Miranda JR, Bataena A (2008) Discovery of *Apis florea* in Aqaba, Jordan. J Apic Res 47:173–174

Hou C, Rivkin H, Slabezki Y et al (2014) Dynamics of the presence of Israeli acute paralysis virus in honey bee colonies with colony collapse disorder. Virus 6:2012–2027

Leat N, Ball B, Govan V, Davison S (2000) Analysis of the complete genome sequence of black queen-cell virus, a picorna-like virus of honey bees. J Gen Virol 81:2111–2119

Mazar A, Panitz-Chohen N (2007) It is the land of honey: beekeeping at Tel Rehov. East Archaeol 70:202–219

Mazzei M, Carrozza ML, Luisi E et al (2014) Infectivity of DWV associated to flower pollen: experimental evidence of a horizontal transmission route. Martin SJ, editor. PLoS One 9:e113448

Nazzi F, Pennacchio F (2014) Disentangling multiple interactions in the hive ecosystem. Trends Parasitol 30:556–561

Pisanty G, Mandelik Y (2016) Profiling crop pollinators: life history traits predict habitat use and crop visitation by Mediterranean wild bees. Ecol Appl 25:742–752

Pisanty G, Klein A-M, Mandelik Y (2014) Do wild bees complement honeybee pollination of confection sunflowers in Israel? Apidologie 45:235–247

Pisanty G, Afik O, Wajnberg E et al (2016) Watermelon pollinators exhibit complementarity in both visitation rate and single-visit pollination efficiency. J Appl Ecol 53:360–370

Ribière M, Olivier V, Blanchard P (2010) Chronic bee paralysis: a disease and a virus like no other? J Invertebr Pathol 103(Suppl):S120–S131

Runckel C, Flenniken ML, Engel JC et al (2011) Temporal analysis of the honey bee microbiome reveals four novel viruses and seasonal prevalence of known viruses, *Nosema*, and *Crithidia*. PLoS One 6:e20656

Ryabov EV, Wood GR, Fannon JM et al (2014) A virulent strain of deformed wing virus (DWV) of honeybees (*Apis mellifera*) prevails after *Varroa destructor*-mediated, or in vitro, transmission. PLoS Pathog 10:e1004230

Singh R, Levitt AL, Rajotte EG et al (2010) RNA viruses in hymenopteran pollinators: evidence of inter-taxa virus transmission via pollen and potential impact on non-*Apis* hymenopteran species. PLoS One 5:e14357

Soroker V, Hetzroni A, Yakobson B et al (2011) Evaluation of colony losses in Israel in relation to the incidence of pathogens and pests. Apidologie 42:192–199

Soroker V, Avni D, Slabezki Y et al (2012) Colony losses and their potential causes in Israel. Proceedings of the 19° Congreso Internacional De Actualizaciőn Apícola, Oxaca, Mexico, 2012

Soroker V, Sarig S, Slavetsky Y et al (2017) Leakage of agrochemicals from agriculture to environment: an impact on honeybees. Ecol Environ Hebr 2:16–23

Chapter 5
Beekeeping and Honey Hunting in Nepal: Current Status and Future Perspectives

Ratna Thapa, Sunil Aryal, and Chuleui Jung

Abstract Nepal is rich in honey bee diversity. Four native species of honey bees viz. *Apis laboriosa*, *Apis dorsata*, *Apis florea*, and *Apis cerana* and one exotic species, *Apis mellifera* are found in Nepal. Beekeeping with *A. cerana* was started in 1960, whereas *A. mellifera* in 1990. The Nepal Government and several foreign agencies have been involved in promoting beekeeping for the development of livelihood of deprived communities in rural areas, but beekeeping has been hindered due to lack of modern beekeeping technology, improper management of colonies, and lack of techniques in controlling mites, and brood diseases. The Nepal government and hundreds of international NGOs/INGOs are distributing bee colonies, honey buckets, and some hive tools to villagers for poverty alleviation without any proper beekeeping training. In addition, honey hunting from cliffs is a spectacular event for tourists to enhance the income generation of mountain people. Due to deforestation, and over-harvesting, the cliff-nesting species, *A. laboriosa*, is in an alarming decline. Other *Apis* species and 34 species of bumble bees have been recorded in Nepal, pollinating wild plants, as well as some of the cultivated crops.

Keywords Diversity · Himalayan honey hunting · Honey flora · Mites predators · Ecotourism · Nepal

R. Thapa (✉)
Zoology Department, Tri-Chandra Multiple College, Tribhuvan University, Kirtipur, Kathmandu, Nepal

S. Aryal
Department of Bioresources Sciences, Andong National University, Andong, Republic of Korea

Zoology Department, Amrit Science Campus (ASCOL), TU, Kathmandu, Nepal

C. Jung
Department of Bioresources Sciences, Andong National University, Andong, Republic of Korea

P. Chantawannakul et al. (eds.), *Asian Beekeeping in the 21st Century*,
https://doi.org/10.1007/978-981-10-8222-1_5

5.1 Diversity

Nepal, located in the Himalayan region with an areas 47,181 km^2, is rich in honey bee diversity. In total, four native honey bee species, *Apis laboriosa*, *Apis dorsata*, *Apis florea*, and *Apis cerana*; and one exotic honey bee, *Apis mellifera*, exist from the foothills of the Himalaya to the Terai region (Fig. 5.1a–g). *A. laboriosa*, the Himalayan cliff bee, the largest honey bee in the world builds a single colony in open spaces along rivers (Otis 1996). It is commonly known as the Black Himalayan Giant honey bee (Fig. 5.1a). It is found at high altitudes ranging from 1200 to 3700 m, and forages up to 4100 m (Fig. 5.1e; Roubik et al. 1985; Thapa 2001; Joshi et al. 2004; Woyke et al. 2012). Its average honey yield is 25–60 kg/colony/year. This species plays crucial role in the conservation of Himalayan ecosystem by pollinating wild flowers in the entire Himalayan region, and the star species for ecotourism development in poverty alleviation of

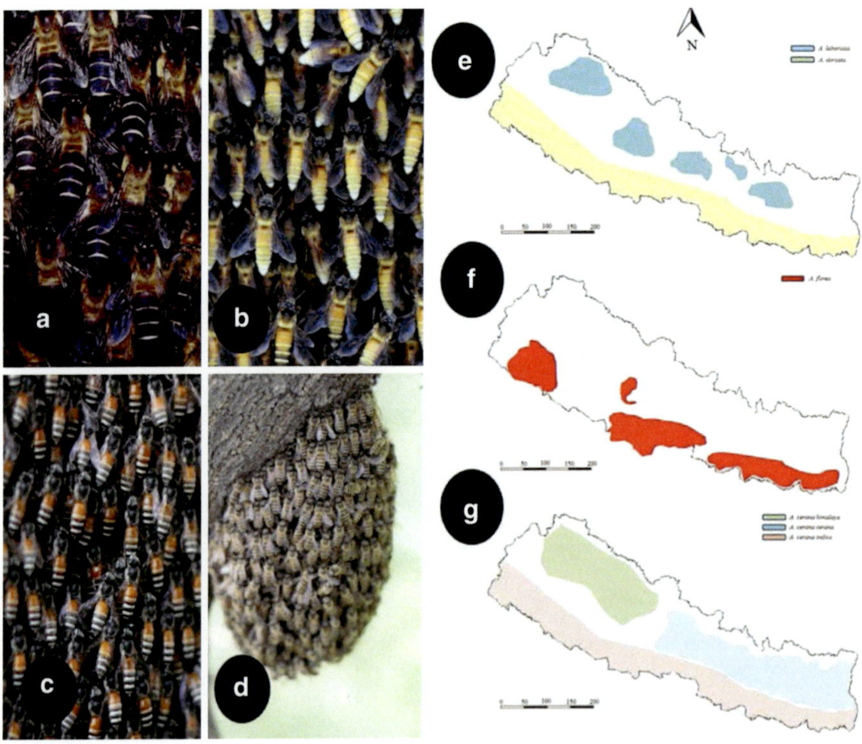

Fig. 5.1 Distribution of native honey bees in Nepal. (**a**) *A. laboriosa*, (**b**) *A. dorsata*, (**c**). *A. florea*, (**d**). *A. cerana*, (a swarm hanging on a tree branch), (**e**) Distribution of *A. laboriosa* (blue), and *A. dorsata* (yellow), (**f**) Distribution of *A. florea*, (**g**) Distribution of *A. cerana himalaya* (upleft part, light green), *A. cerana cerana* (upright part, light blue), and *A. cerana indica* (lower part, light pink color, Photos by R. Thapa)

villagers in Nepal (Thapa 2001). *A. dorsata*, a giant yellow honey bee (Fig. 5.1b), builds a single large comb on the tallest trees, windows, balconies of buildings, and water tanks. *A. dorsata* colonies seasonally migrate between 190 and 1200 m and return to the same old nesting sites (Fig. 5.1e; Thapa 2001; Thapa and Wongsiri 2011). The honey yield is between 15 and 50 kg honey/colony/year. *A. florea*, known as a dwarf honey bee (Fig. 5.1c), builds a single comb in the bushes (Wongsiri et al. 1996; Thapa et al. 2000a). It is found from 350 to 1200 m (Fig. 5.1f). It is an efficient pollinator of fruit and natural flora (Wongsiri et al. 1996). It yields less than 1 kg/colony/year). *A. florea* honey is traditionally used to cure snake bite wounds. *A. cerana*, the Asian hive bee (Fig. 5.1d), is kept in traditional hives such as wall and log hives and modern Newton type-B hives (Newton type-A has a flat roof, whereas Newton type-B has house roof-like structure). There are three sub-species of *A. cerana* viz. *A. c. indica* in plain areas, *A. c. himalaya* in valleys, and *A. c. cerana* in high areas up to 1500 m (Fig. 5.1g; Thapa et al. 2000a). Its honey yield is 8–12 kg/colony/year. *A. mellifera* was introduced to Nepal during the 1990s and has been successfully established up to around 1500 m (Thapa et al. 2000a). The commercial beekeepers seasonally migrate during the honey harvest times. Swarming and absconding tendencies are quite low in *A. mellifera*, but it is highly susceptible to sacbrood disease and the parasitic mite, *Tropilaelaps mercedesae*. The honey production is between 30 and 50 kg/hive/year.

5.2 Historical Records and Recent Development in Beekeeping

The government has had beekeeping training for farmers starting in 1968. In 1975, the Vocational Entomology Section was established to promote and develop beekeeping and sericulture in Nepal. In 1980, the Beekeeping Development Section was specially established for the development and extension of beekeeping in rural areas. The Entomology Division under the Nepal Agricultural Research Council has made it mandatory to conduct research on various aspects of applied entomology including industrial entomology. The beekeeping program is an integral part of governmental policies and pursued by INGO/NGOs in the development of rural and marginal people of Nepal to enhance their income generation.

The Beekeeping Development Section, Directorate of Industrial Entomology Development (DoIED), Department of Agriculture, Nepal government has focused on honey production of domesticated honey bees (*A. cerana* and *A. mellifera*) and wild colonies (*A. laboriosa* and *A. dorsata*) since 2001. It is estimated that the honey production of *A. mellifera*, *A. cerena*, and wild honey bees are 51%, 36%, and 13%, respectively (FNCCI 2006). The number of hives recorded during 2012/13 was 169,000 with the honey yield of 1675 MT (APSD 2013).

5.3 Bee Foraging Sources

A key factor of survival of Himalayan honey bees in fragile Himalayan ecosystems is the diverse bee flora available all year round (Fig. 5.2). The major honey species are *Eupatorium odoratum* (local name (LN): Ban mara; 350–1600 m), *Brassica nigra*, *Brassica butyracea* (LN: Tori; 300–3500 m), *Nyctanthes arbortristis* (LN: Rudhilo; 300–400 m), and *Diploknema butyracea* (LN: Churi or Indian butter trees; 300–1600 m). Various plants, including hundreds of species of agricultural, horticultural, and forage crops, ornamental plants, wild plants, and forest trees are visited by honey bees in different agro-ecozones of Himalaya regions have been reported (Table 5.1).

Bee pollination is important to food production (FAO 2008). Honey bees help increase fruit set 13.0–21.9% in peach and plum (Verma 1992). *A. cerana* is a major pollinator of *Citrus sinensis* and strawberry. Flowers of cauliflower are the most attractive to *A. cerana* followed by broad-leaf mustard and radish (Verma

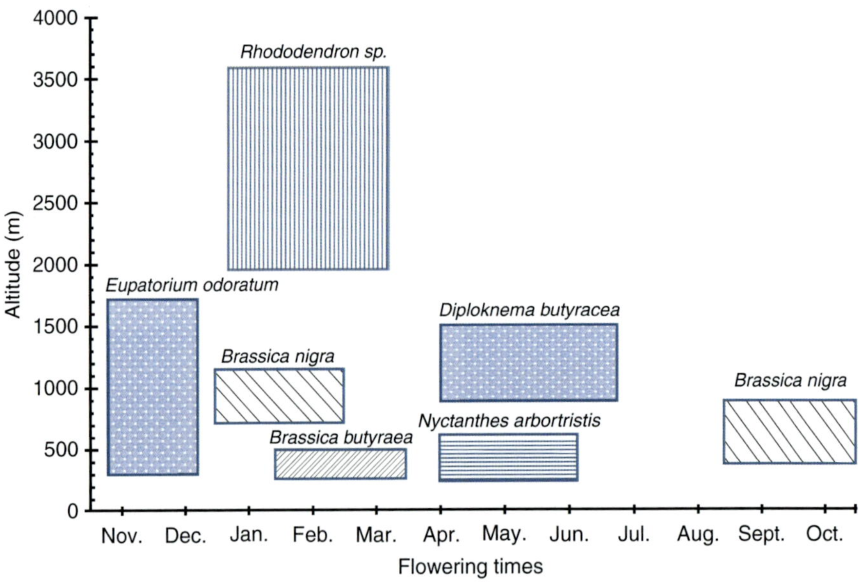

Fig. 5.2 Major honey floral resources and their flowering times based on altitude

Table 5.1 Number of honey bee flora species reported from various parts of Nepal

Location		Total number of plant species	References
Kathmandu	27.7172° N, 85.3240° E	156	Kafle (1984)
		113	Partap and Verma (1996)
Jumla	29.2788° N, 82.1278° E	103	Partap (1997)
Dolkha	27.7784° N, 86.1752° E	119	Bista (2001)
Chitwan	27.5291° N, 84.3542° E	85	Devkota (2003)

and Partap 1994). *A. cerena* starts foraging 56 min after the sunrise and stops at 10 min before sunset with peak foraging time between 11:00 and 14:00 h. Previous studies revealed that the foraging duration of *A. cerana* is significantly longer than *A. mellifera*.

5.4 Local Knowledge on Honey Bees: Local Culture and Beliefs/Practical Guide to Local Beekeeping/Honey Bee and Bee Product Usages Among the Locals

Several types of traditional hives and modified hives are used in Nepal. Rural beekeepers use traditional hives such as log hives and wall hives (Fig. 5.3a–f). Other modified hives for *A. cerana* were introduced by several NGOs/INGOs, Beekeeping Development Section (BDS), Beekeeping Training and Extension Support Project, and International Centre for Integrated Mountain Development (ICIMOD) in their project areas (Fig. 5.3g, h).

Log hives, about 66.75 cm long and 30.2 cm breath, are made by scratching the central rotten part of a semi-dry hollow log from the ends (Fig. 5.3a, b), which are closed with circular wooden plates opened during the honey harvesting time and plastered with a mixture (3:1) of red soil and dried rice coats. The log hives festooned with thorns from wild raspberries are hung in a horizontal position in front balcony-platform to afford protection against yellow-throated martens. House wall hives, about 45 × 45 cm and 30 cm deep, are usually constructed on the second floor by removing six to eight bricks from the wall (Crane 1992; Thapa et al. 2000a). A flat piece of wood, about 5 mm thick, is fixed on the ceiling to provide support for combs construction. The inner surface is varnished with red soil (Fig. 5.3c, d). The front face has a small hole size 1 in., whereas the rear end, size 45 × 45 cm, is covered with *nanglo* (bamboo plate) having a small hole in the middle for viewing the colony size, bee activities, and honey. The rear end is used to harvest honey. In mid and high hills, where the temperature drops below zero degrees centigrade in the winter, house wall hives are frequently used (Kafle 1992; Thapa et al. 2000a; BDS 2002). After Nepal earthquake on April 15, 2015, 90% of wall hives located in old houses were destroyed. Therefore, the wall hives are now vanishing. Cubicle hive, which are similar to wall hives, and use in high hill areas, are constructed using wood remains. The boxes are built without frames, and heavily plastered with a mixture of red soil and cow dung (Fig. 5.3e, f). The front face is covered with a rock slab and plastered with clay. It does not have rear end. During honey harvesting times, the front plate is removed, and villagers can harvest all combs (Opptiz 1990). The Godavari mud hive, a modified form of Newton hive which has brood and honey chambers, was introduced by the Godavari Beekeeping Section, the Nepal government, to replace Newton hives. The mud hive is built on the sloped ground (angle 35–40°) using brick and red soil (Fig. 5.3g). The mud hive is not popular among *A. cerana* beekeepers because this hive is very difficult to

Fig. 5.3 Traditional hives used in Nepal. (**a**) log hive (front view), (**b**) log hive (side view showing where honey is harvested), (**c**) wall hive (front view), (**d**) wall hive (inner view with several combs), (**e**) box hive, (**f**) colony inspection, (**g**) Godavari mud hive, and (**h**) straw hive (Photo by R. Thapa)

clean, and the bottom board is always wet in the rainy season. Straw hives were introduced by ICIMOD to protect *A. cerana* colonies from the winter in Jumla, northwest Nepal (2600 m). The straw hive is constructed using rice, wheat or buckwheat straw, and used wooden frames (Fig. 5.3h). The entrance is made using bamboo pipe to minimize *Vespa* attacks. It is a well-insulated hive for *A. cerana* (Partap et al. 1997). Bee hives make from wheat ad or rice cutting straw use only in Jumla, the project area of ICIMOD, they have not been adopted by traditional *A. cerana* beekeepers. Newton hive type-A and type-B were introduced in 1960. The Newton hives are movable hives with ten foundation-free frames and have two chambers: a honey chamber and brood chamber. The Newton type-A hive is smaller than Newton type-B. The Newton hives are commonly used in Terai and mid hill areas (Kafle 1992; Verma 1992; BDS 2002; Thapa 2002). The Newton type-A hive is smaller than Newton type-B. The Nepal government recommends Newton type-A for higher hilly areas, whereas Newton type-B for low land *A. cerana* beekeeping (BDS 2002). The top bar hive, a design from Kenya, was introduced in 1980 to generate more income for Chepang hill tribe people (Nakamura 1989; Crane 1992). Nevertheless, top bar hives completely fail to keep *A. cerana*.

5.4.1 **A. laboriosa** ***Honey Hunting***

In Nepal, honey from *A. laboriosa* colonies is harvested twice a year. Summer harvests are obtained from the cool temperate regions (2500–3000 m), while winter harvests are from the warm temperate regions (1200 m). Before honey harvesting from the cliffs, a traditional honey harvesting ritual takes place near the cliffs to ask permission from the forest gods to harvest honey from the cliff. Folklore says, if honey hunters do not perform this ceremony before harvesting honey, an accident will definitely occur (for instance, a honey hunter will lose his life). During this ceremony, a cock or goat buck or sheep buck is sacrificed to the forest god, accompanied by cooked rice, fruits, and flowers. After the ritual, a fire is lit at the base of the cliff to smoke the colonies. Honey hunters use rope ladders made from locally available grasses. When collecting honey, one hunter climbs down the rope while two other hunters watch from the base of the cliff to receive honey and brood in a basket. The honey hunter does not wear a bee veil, nor any protective gloves or shoes. The hunter uses his toes to hold the bamboo rungs, and the basket when harvesting honey. After collection, the honey is equally divided among the members of the team (Thapa 2001).

This is a spectacular event to photograph and film. Several documentary films have been made on this topic (Fig. 5.4a, b). Trekkers walking along nearby trails and staying at camp sites can enjoy 2–3 days of honey harvesting activities in summer. Seasonal foreign and local tourist groups or film makers buy a whole cliff for honey or documentary film. The cost of a single cliff depends on the number of colonies present and tourist activities (Table 5.2). The money earned from the honey hunting activities is spent on social welfare of local villages (Thapa 2001).

Fig. 5.4 Colonies of *A. laboriosa*. (**a**) *A. laboriosa* colonies on the rocky cliff, (**b**) Rope fixing to cut honey combs, Barabishey, Tatopani, Kathmandu, Nepal (Photos by R. Thapa)

Table 5.2 Rates for honey hunting exhibitions in Nepal

Activates	Price (US$)
Photograph of honey hunter	5–10
Honey hunting activities	50–100
A single cliff (for honey and brood)	1000
Video and documentary films	1500

5.5 Pests and Diseases of Honey Bees

Nepal, a land of diverse *Apis* and non-*Apis* species, has several brood diseases, pests, and invertebrate and vertebrate predators of honey bees (Table 5.3). Sacbrood disease and *T. mercedesae* are major problems of beekeeping in Nepal.

Sacbrood diseases: In the early 1980s, Thai sacbrood disease killed 95% of *A. cerana* colonies in Nepal. This resulted in great economic loss to *A. cerana* beekeeping all over the country from reduced honey production. Thai sacbrood disease has also been reported from all parts of Nepal where *A. cerana* is present. When resources are scarce, and during the rainy season, *A. cerana* frequently robs *A. mellifera* colonies, as a consequence, sacbrood disease is transmitted to *A. mellifera* (Thapa et al. 2012b). Thai sacbrood disease is an endemic disease in *A. cerana* and *A. mellifera* all over the country. Since sacbrood disease is caused by viruses, antibiotics cannot cure the infested bees. However, there have been attempts to treat the infested colonies with antibiotic drugs (oxytetracyline, already banned) and by requeening. American foulbrood disease has not been reported yet.

Honey bee mites: Beekeepers are facing problems in colony management due to the existence of mites. Thapa et al. (2000b) recorded three honey bee mites, namely, *V. destructor*, *V. underwoodi*, and *T. mercedesae* from *A. cerana* and *A. mellifera* colonies (Fig. 5.5a–c). Often times, honey bee broods of *A. mellifera* were attacked by two mite species, *V. destructor* and *T. mercedesae* (Fig. 5.6a). Around 5% of virgin queens were emerged with wing deformity when infested by *T. mercedesae* (Fig. 5.6b). This suggests that *T. mercedesae* cause considerable damage to *A. mellifera* queen breeding programs more than *V. destructor*. Baldbrood was also observed only in *T. mercedesae* infested *A. mellifera* colonies. *T. clareae* is found in

Table 5.3 Honey bee diseases, pests, bee eaters, and other vertebrate predator of honey bees in Nepal (√ are positives recorded according to author's observation and known clinical symptom)

A. List of diseases	*A. laboriosa*	*A. dorsata*	*A. florea*	*A. cerana*	*A. mellifera*	References
European foulbrood	–	√	–	–	√	Thapa et al. (2000a)
Sacbrood	–	–	–	√	√	
Chalkbrood	–	–	–	–	√	
B. List of pests						
Mites						
Varroa destructor	–	–	–	√	√	Thapa et al. (2015)
Tropilaelaps mercedesae	–	√	–	√	√	
Acarapis woodi	–	–	–	√	√	
Neocypholaelaps spp.	–	–	–	–	√	
Pyemotes naferi	–	–	–	√	–	
Forcellinia faini	–	–	–	√	√	
Other pests						
Pseudoscorpions (*Ellingsenius indicus*)	–	–	–	√	–	Thapa et al. (2013)
Waxmoths	√	√	√	√	√	
Death's head moth (*Acherontia atropos*)	–	–	–	–	√	
Dipetra						
Humpbacked fly (*Megaselia rufipes*)	–	–	–	√	√	
C. List of bee eaters						
Green bee eater (*Merops orientalis*)	–	√	—	–	√	Thapa and Wongsiri (2003)
Drongo (*Dicrurus macrocercus*)	–	√	–	√	√	
Honey buzzard (*Pernis ptilorhynchus*)	–	√	–	–	–	
D. List of other vertebrate predator						
Yellow-throated marten (*Martes flavigula*)		–	–	√	–	

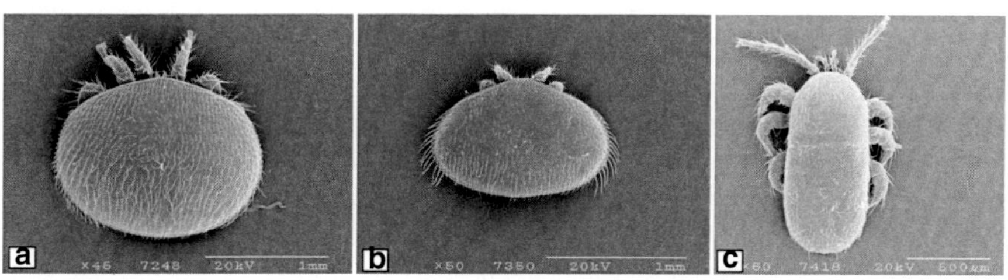

Fig. 5.5 Bee mites found in Nepal (under a scanning electron microscope). (**a**) *V. destructor* from *A. mellifera* colonies, (**b**) *V. underwoodi* from debris of *A. cerana*, and (**c**) *T. mercedesae* from *A. mellifera* brood cells

Fig. 5.6 Mite infestation in *A. mellifera* virgin queens. (**a**) *T. mercedasae* and (**b**) *Tropilaelaps* infected virgin queen of *A. mellifera* (Photo by R. Thapa)

Fig. 5.7 Pseudoscorpions (*Ellingsenius indicus*) from *A. cerana* colonies in Kathmandu, Nepal (Photo by R. Thapa)

A. cerana and *A. dorsata* colonies, whereas *V. destructor* was found in *A. mellifera* (78%) and *A. cerena* (70% of observed colonies) in Chitwan, Nepal (Neupane 2009). Similarly, *Acarapis woodi* has also been observed in *A. cerana* and *A. mellifera* colonies (Thapa et al. 2000a).

Pseudoscorpions: Himalayan *A. cerana*, which live in logs, walls, and modern hives, attracts many organisms ranging from viruses to vertebrates (Morse 1978; Seeley 1985), including pseudoscorpions (*Ellingsenius indicus*, Arachnida, Chelonethi (Fig. 5.7). Pseudoscorpions are small predatory arachnids. They are highly diverse (Harvey 2002) and can be classified as either social, such as *Paratemnoides elongatus* (Atemnidae) that live in groups, share food, and hunt cooperatively (Levi 1953; Brach 1978; Zeh and Zeh 1990), or non-social, such as *Dinocheirus arizonensis*, *Parachelifer hubbardi*, *Cordylochernes scorpioides*, and *Semeiochernes armiger* that show cannibalism (prey on adults and nymphs) and aggressive intraspecific contacts (Weygoldt 1969; Zeh 1987). Thapa et al. (2013)

Fig. 5.8 Death's head moth, *A. atropos* collected from the bottom board of *A. mellifera* hive (Photo by R. Thapa)

demonstrated that pseudoscorpions use abandoned cells for breeding and feed on larvae of *A. cerana*. Therefore, pseudoscorpion are not *Varroa* mite predator. However, psuedoscorpion have not been observed in colonies of *A. mellifera*, *A. dorsata*, *A. laboriosa*, and *A. florea*.

Death's head moth: Geographically, *A. atropos*, only one species from Mediterranean and N. Africa, is found in South and Southeast Asia (Smith 1953). *A. atropos* is 10.5 cm in length and has a wingspan of 11.5 cm. The forewings are mainly black, suffused with brown patches, and small eyespots. The hind wings are bright yellowish with two thick black crosslines. The abdomen has black and orange, yellow bands (Fig. 5.8). Adult *A. atropos* has a short proboscis and cannot obtain nectar from the bottom of tubular flowers, but can feed on sap exuded by trees. It is strong enough to pierce the seal of honey cells and suck the honey. The death's head moth attacks *A. mellifera* colonies at night. This pest has recently been recorded from *A. mellifera* colonies of Kathmandu valley and in *A. dorsata* nests.

Vespa: The major vespa predators of honey bees in Nepal are *Vespa velutina*, *V. tropica*, and *V. magnifica* (Fig. 5.9a–e). *V. velutina* is 18–22 mm long with a brown head and appendages. The first and second abdominal segments are black with orange-brown stripes on first segment, while the third, fourth, and sixth segments are orange-brown (Fig. 5.9a). *V. tropica* is 30–32 mm long with yellow-orange band on first and second metasomal segments. From third abdominal segment to the last abdominal segments are black (Fig. 5.9e). *V. magnifica* is 40–43 mm long and is the largest vespine wasp species in Nepal. The head and antennae are orange (Fig. 5.9c). *V. velutina* hovers in front of the hive entrance, catches only incoming foragers (Fig. 5.9b), *V. tropica* attacks the colonies from cracks by making a hole in a decomposed area of the hive (Fig. 5.9e), and *V. magnifica* attacks guard bees by cutting and breaking their body (Fig. 5.9d; Thapa et al. 2012a).

Bee eaters: There are frequent reports of *A. mellifera* and *A. cerana* colonies being attacked by green bee eaters. The green bee eater and drongo are found all

Fig. 5.9 Vespa species attacking *A. cerana* and *A. mellifera* colonies. (**a**) *V. velutina*, (**b**) *V. velutina* mass attacking *A. cerana* colony, (**c**) *V. magnifica* killing workers of *A. mellifera*, (**d**) Mass of *V. magnifica* killing thousands of *A. mellifera* workers, and (**e**) *V. tropica* attacking *A. mellifera* hive (Photos by R. Thapa)

over the country. They attack flying bees on cloudy and rainy days. The green bee eater and drongo are fed heavily en-mass on foragers of *A. mellifera* in the rainy season, which weakens the *A. mellifera* colonies (Thapa and Wongsiri 2003).

5.6 Other Problems of Beekeeping

Beekeeping practice started a long time ago in Nepal, but *A. cerana* beekeeping has not been commercialized yet. The farmers do not have "know-how" for queen breeding of *A. cerana*. Most *A. cerana* beekeepers collect and/or trap feral colonies in the swarming season. Most feral colonies are headed by old queens or virgin queens with a high tendency towards swarming behavior. Due to such queen behavior, *A. cerana* colonies frequently abscond. Frequent swarming and absconding cause problem with *A. cerana* colony management (Thapa et al. 2000a). Subsequently, commercial *A. mellifera* beekeepers also breed their own queens without knowing the hygienic behavior of the breeder queen colonies. Therefore, most of the commercial beekeepers breed queens that are highly susceptible to the parasitic mites. There have already been attempts to better manage *A. cerana* beekeeping, but *A. cerana* queen breeding programs are still lacking. The current bee research program is not developed well enough to conduct research on most issues of beekeeping in Nepal. Therefore, advanced apicultural research and educational activities to improve beekeeping practice to large commercial scale is needed.

Colony management: Moving *A. mellifera* colonies from lowland to highlands based on flowering seasons gives higher honey yield that of stationary *A. mellifera* beekeeping (Thapa et al. 2000a). Dawadi (2003) suggests the application of Apistan strips one to two times a year to control *T. clareae*. Use of *Metarhizium anisopliae* on bee hives infested by *Galleria mellonella* in laboratory experiment showed that greater wax moth is highly susceptible, and there is no harm to honey bees (Neupane 2005). A study of diet supplements for the honey bee stated that a diet of a low dose sugar syrup combined with a pollen substitute (30 g) was suitable for off-season honey bee colony management and also helped maintain high rates of flight activities. The bee food supplement should be supplied to bees during the cold winter and rainy season (depending upon the weather condition of Khumaltar, Lalitpur, Nepal) to improve the colonys health, produce sufficient population to maintain the colony strength, and ultimately produce more honey yield (Thapa and Pokhrel 2005; Bhusal et al. 2011). However, more research is needed in this area though many efforts are being made to impart existing knowledge to the farmers by training. Sophisticated laboratory to rear queen is lacking; therefore, queen replacement is not practiced by farmers. Pesticide residue and drugs are used to manage diseases; and pests in the hive also deteriorate the quality of honey (FNCCI 2006).

Conservation problems of wild honey bees: Conservation of *A. laboriosa* and *A. dorsata* is another challenge to Nepalese beekeeping due to human activities in burning the colonies and over-harvesting. *A. mellifera* was not a native honey bee of Nepal. It was introduced in 1990 for commercial purposes in Nepal also causes interspecies competition between the native honey bees (*A. cerana* and *A. dorsata*), and the introduced species for nectar, and pollen collection as recorded in the Terai and Inner Terai (Partap and Verma 1996). A rapid decline of *A. laboriosa* colonies has been observed due to deforestation, change in land use pattern, destructive honey hunting practices, and attacks of pests and predators (Thapa 2001). Recently, raising awareness of wild bee conservation has been carried out in *A. laboriosa* hunting areas in the Lamjung and Dolakha districts (Joshi et al. 2004).

Pesticides: Farmers in Nepal use pesticides heavily in crops. There is also lack of legislation to prohibit the use of pesticides, and this has affected honey bees. Pesticide use on agricultural crops also causes colony loss (Thapa and Wongsiri 1999). Most of the farmers spray the pesticides in the morning between 10:00 and 14:00 h, which are the peak hours for foragers to collect pollen and nectar. This suggests that the farmers do not realize the effect of using the pesticide during foraging time of honey bees or the pesticide effects on honey bees and other pollinators. Therefore, educational activities for pesticide application for crops are urgently needed to address the issues.

5.7 Other Non-*Apis* Species Found in the Region

Bumble bees are key pollinators of wild plants and some cultivated crops, but their pollination services have never been studied in the Himalayan regions. Five *Bombus* species were recorded from central parts of Nepal by Thapa (1988), whereas Williams et al. (2010) recorded 29 species from different parts of Nepal (Table 5.4).

Table 5.4 List of *Bombus* species recorded in Nepal

	Species	Location	Altitude (m)	Flowers	Status	References
1	*B. orientalis*	Kathmandu	1200	Wild/ cultivated	Common	Thapa (1988)
2	*B. flavothoracius*	Helumbu	3000	Wild flowers	Rare	
3	*B. festivus*	Helumbu	3000	Wild flowers	Rare	
4	*B. eximius*	Kakani	2148	Wild flowers	Rare	
5	*B. flavescens*	Dunche	1951	Wild flowers	Rare	
6	*B. rotudiceps*	–	1021	–	–	Williams et al. (2010)
7	*B. breviceps*	–	1296	–	–	
8	*B. haemorrhoidalis*	–	1622	–	–	
9	*B. trifasciatus*	–	2016	–	–	
10	*B. luteipes*	–	2162	–	–	
11	*B. lucorum*	–	3291	–	–	
12	*B. lepidus*	–	3327	–	–	
13	*B.cornutus*	–	3436	–	–	
14	*B.asiaticus*	–	3475	–	–	
15	*B.avinoviellus*	–	3513	–	–	
16	*B.pressus*	–	3658	–	–	
17	*B.turneri*	–	2250	–	–	
18	*B.branickii*	–	3700	–	–	
19	*B.nobilis*	–	3751	–	–	
20	*B.abnormis*	–	3788	–	–	
21	*B. funerarius*	–	2696	–	–	
22	*B. lemniscatus*	–	3835	–	–	
23	*B. grahami*	–	2725	–	–	
24	*B. personatus*	–	4867	–	–	
25	*B. tunicatus*	–	2731	–	–	
26	*B. novus*	–	3876	–	–	
27	*B. hypnorum*	–	2825	–	–	
28	*B. kashmirensis*	–	3946	–	–	
29	*B. parthenius*	–	2872	–	–	
30	*B. mirus*	–	4031	–	–	
31	*B. miniatus*	–	3207	–	–	
32	*B. rufofasciatus*	–	4081	–	–	
33	*B. melanurus*	–	3250	–	–	
34	*B. waltoni*	–	4300	–	–	
35	*B. skorikovi*	–	3250	–	–	
36	*B. ladakhensis*	–	4350	–	–	

5.8 Future Perspectives

Grayanotoxin honey: No honey produced in Nepal is toxic except for *A. laboriosa* honey. *A. laboriosa* forages on Ericaceae, *Rhododendron ponticum* and *R. luteum* (Altun et al. 2014). These contain grayanotoxin, a toxic substance, that has been of great concern due to the toxification of Himalayan honey (Jansen et al. 2012). Several case studies suggested that consumption of *A. laboriosa* honey containing the grayanotoxin caused symptoms of blurring of vision, diplopia, nausea, vomiting, and even cardiac depression with low blood pressure (Dubey et al. 2009; Oguzturk et al. 2012; Altun et al. 2014). This type of honey is called "mad honey." Recently, various Asian countries have tried to prevent the import of intoxicating honey from Nepal requiring a grayanotoxin-free certificate. Therefore, there is an urgent need of grayanotoxin content analysis of *A. laboriosa* honey.

Until today, Nepalese beekeepers produce honey from only two species of honey bees, *A. cerana*, and *A. mellifera*, due to lack of proper guidance and equipment for honey production. The other hive products such as propolis, royal jelly, and bee venom are poorly known to Nepalese beekeepers. Therefore, along with low cost honey production technology with proper beekeeping management, disease, and pest control, research on other hive products, and their harvesting, and processing techniques are also needed.

Destruction of forest habitats from urbanization and increase in agricultural activities adversely affect the availability of resources because many staple crops such as rice, wheat, and potatoes are not food sources for honey bees. Therefore, the bee flora deficit area should be identified, and development of a flowering time calendar may help beekeeping management. Therefore, recommended research areas to improve beekeeping operation in Nepal are as follows:

- Genetic improvement and breeding domesticated native *A. cerana* to maintain genetic diversity of *A. cerana*
- Improvement of genetic stocks for colony management practices
- Surveillance and management of honey bee pests
- Honey bee floral identification, bee-pasture mapping, and carrying capacity analysis
- Conservation and utilization of wild *Apis* and non-*Apis* species
- Quality monitoring of honey and other hive products

Acknowledgements This manuscript was largely based on Aryal S, Thapa R, and Jung C 2015. An overview of beekeeping economy and its constraints in Nepal. J. Apic. 30:135–124. The authors are grateful for the permission from the Journal of Apiculture. CJ was supported from PJ00997004, RDA, Korea.

References

Altun İ, Fatih A, Nuri K, Beydilli H, Ethem A (2014) A rare cause of complete atrioventricular block and accelerate nodal rhythm; mad honey poisoning. J Contemp Med 4:41–43

APSD (2013) Statistical information on Nepalese Agriculture 2012/2013(2069/070). Government of Nepal, Ministry of Agricultural Development. Agribusiness Promotion and Statistics Division (APSD). Agriculture Statistics Section (APSD), Nepal Government, p 205

BDS (2002) Annual report of beekeeping, Bee Development Section (BDS). Agricultural Department, Kathmandu, Nepal, p 46

Bhusal SJ, Kafle L, Thapa RB, Shih CJ (2011) Effect of colony strength on the performance honey bees (*Apis mellifera*) in Nepal (Hymenoptera:Apidae). Sociobiology 58(2):1–13

Bista S (2001) Honey bee flora at Kabre, Dolakha District. Nepal Agric Res J 4:18–25

Brach V (1978) Social behavior in the pseudoscorpion, *Paratemnus elongates* (Banks) (Pseudoscorpionida: Atemnidae). Insect Soc 25:3–11

Crane E (1992) Report to beekeeping training and extension support project on a visit to Nepal. British Executive Service Overseas BETREP, Kathmandu, Nepal, p 24

Dawadi DP (2003) Incidence and control of parasitic mite, *Tropilaelaps clareae* Delfinado and Baker in *Apis mellifera* L. Dissertation, Tribhuvan University

Devkota KH (2003) Economic impact of apiculture in Nepal, Chitwan. Dissertation. Tribhuvan University

Dubey L, Maskey A, Regmi S (2009) Bradycardia and severe hypotension caused by wild honey poisoning. Hell J Cardiol 50:426–428

FAO (2008) A contribution to the international initiative for the conservation and sustainable use of pollinators rapid assessment of pollinators' status. In: Monitoring the status and trends of pollinators, FAO, p 52

FNCCI (2006) The study report on trade competitiveness of Nepalese honey. In: Importing the policy environment for agricultural trade, production and marketing in Nepal. Study report prepared by FNCCI/AEC under a cooperative grant agreement project between USAID and FNCCI. Federation of Nepalese Chambers of Commerce and Industry/Agro Enterprise Center (FNCCI), p 43

Harvey MS (2002) The neglected cousins: what do we know about the smaller arachnid orders? J Arachnol 30:357–372

Jansen SA, Kleerkooper I, Hofman ZLM, Kappen FPM, Stary-Weinzinger A, van der Heyden MAG (2012) Grayanotoxin poisoning: "mad honey disease" and beyond. Cardiovasc Toxicol 12:208–225

Joshi SR, Ahmad F, Gurung MB (2004) Status of *Apis laboriosa* populations in Kaski. J Appl Biol 43(4):176–180

Kafle GP (1984) A general survey of bee flora in and around Kathmandu valley. Nepal J Agric 15:89–99

Kafle GP (1992) Silent features of beekeeping in Nepal. In: Verma LR (ed) Honeybees in mountain agriculture. ICIMOD, Kathmandu, pp 155–162

Levi HW (1953) Observations on two species of pseudoscorpions. Can Entomol 85:55–62

Morse RA (1978) Honey bee pests, predators, diseases. Comstock, Cornell University Press, Ithaca, NY

Nakamura J (1989). Intermediate beekeeping in Nepal. In: First Asia Pacific conference of entomology, Chiang Mai, pp 803–809

Neupane BP (2005) Evaluation of the entomopathogenic fungus *Metarhizium anisopliae* (Metsch.) Sorokin on honey bee (*Apis mellifera* L.) and its pest greater wax moth (*Galleria mellonella* L.). Dissertation, Tribuvan University

Neupane KR (2009) Honey bee mites and their population variation in *Apis* spp. during different seasons, Chitwan, Nepal. J Inst Agric Anim Sci 30:159–166

Oguzturk H, Ciftci O, Turtay MG, Yumrutepe S (2012) Complete atrioventricular block caused by mad honey intoxication. Eur Rev Med Pharmacol Sci 16:1748–1750

Opptiz M (1990) Beekeeping and the hunt for wild honey among the Magar of Nepal. Bee World 74(2):165–169

Otis GW (1996) Distributions of recently recognized species of honey bees. (Hymenoptera: Apidae; *Apis*) in Asia. J Kansas Entomol Soc 69(4):311–333

Partap U (1997) Bee flora of the Hindu Kush-Himalayas. Inventory and Management. International Centre for Integrated Mountain Development (ICIMOD), Kathmandu, p 306

Partap U, Verma LR (1996) Impact of food resource availability on growth and development of *Apis cerana* colonies in the Kathmandu valley of Nepal. Ecoprint 3:21–29

Partap U, Joshi SR, Verma LR, Pechhacker H (1997) Straw hive: an eco-friendly moveable frame hive for *Apis cerana*. Ecoviews 4:12–15

Roubik DW, Sakagami SF, Kudo I (1985) A note on distribution and nesting of the Himalayan honey bee *Apis laboriosa* Smith (Hymenoptera:Apidae). J Kansas Entomol Soc 58(4):746–749

Seeley TD (1985) Honeybee ecology. Princeton University Press, Princeton, NJ, p 201
Smith FG (1953) Beekeeping in the tropics. Bee World 34(12):233–245
Thapa R (1988) Bumble bees and large carpenter bees (Order:Hymenoptera) of Kathmandu valley and some parts of Nepal. Dissertation, Tribhuvan University
Thapa R (2001) The Himalayan giant honey bee and its role in ecotourism development in Nepal. Bee World 82(3):139–141. https://doi.org/10.1080/0005772X.2001.11099516
Thapa R (2002) Impact of European honey bee; *Apis mellifera* L. on genetic diversity of Himalayan honey bee; *Apis cerana* F. Int Mt Semin 1:246–250
Thapa R (2003) Integrated pest management strategies for protecting honey bees from pesticides. Integr Pest Manag 1:147–153
Thapa RB, Pokhrel S (2005) Impact of supplement diets on flights of cross breed honey bee (*Apis mellifera* L). J Inst Agric Anim Sci 26:71–76
Thapa R, Wongsiri S (1999) Comparison of toxic effects of neem (*Azadirachta indica*) and some selective pesticides to forager honey bees; *Apis cerana* and *Apis mellifera* in cages field of oilseed rape. Asian Bee J 1:73–79
Thapa R, Wongsiri S (2003) The flying predators of honey bees; *Apis dorsata* and *Apis laboriosa* in Nepal. Am Bee J 1:400–404
Thapa R, Wongsiri S (2011) Wax specks as remnants cues for recognition of old nest sites of immigrant colonies of *Apis dorsata* F. in northern Thailand. Bee World 88:79–81. https://doi.org/10.1080/0005772X.2011.11417432
Thapa R, Shrestha R, Manandhar DN, Kafle B (2000a) Beekeeping in Nepal. In: 7th international bee research & 4th Asian apiculture association conference. Proceeding of 7th inter. conf. tropical bee management and diversity, pp 409–413
Thapa R, Wongsiri S, Manandhar DN (2000b) Current status of predators and disease of honey bees in Nepal. In: Wongsiri S (ed) Seventh international conference on tropical bees management and diversity. International Bee Research Association (IBRA), Cardiff, pp 221–226
Thapa R, Choi SY, Lee ML (2012a) Fight or flee. Defensive fitness of Himalayan *Apis cerena* against Vespine wasps. Bee Word 89:43–47. https://doi.org/10.1080/0005772X.2012.11417472
Thapa R, Kongpitak P, Lee ML, Choi YS (2012b) Recognition of sac brood viral disease of *Apis cerana*. J Apicult 27(3):201–207
Thapa R, Wongsiri S, Lee ML, Choi YS (2013) Predatory behaviour of pseudoscorpion (*Ellingsenius indicus*) associated with Himalayan *Apis cerana*. J Apic Res 52(5):219–226. https://doi.org/10.3896/IBRA.1.52.5.08
Thapa R, Wongsiri S, Lee ML, Choi YS, Kwon HW (2015) SEM observation of Korean Haplotype *Varroa destructor* (Acari: Varroaidae) collected from *Apis mellifera* colonies. J Apicult 30(3):143–153. https://doi.org/10.17519/apiculture.2015.09.30.3.143
Verma LR (ed) (1992) Honeybees in mountain agriculture. Oxford and IBH Publishing Co., New Delhi, p 274
Verma LR, Partap U (1994) Foraging behavior of *Apis cerana* on cauliflower and cabbage and its impact on seed production. J Apic Res 33:231–236
Weygoldt P (1969) The biology of pseudoscorpions. Harvard University Press, Cambridge, p 154
Williams PH, Ito M, Matsumura T, Kudo I (2010) The bumblebees of the Nepal Himalaya (Hymenoptera:Apidae). Insecta Matsumurana Ser Entomol 66:115–151
Wongsiri S, Lakprayoon A, Thapa R, Thirokupt K, Rinderer TE, Sylvester A, Oldroyd BP, Boonchai U (1996) Comparative biology of *Apis andreniformis* and *Apis florea*. Bee World 77(4):23–25
Woyke J, Wilde J, Wilde M (2012) Swarming and migration of *Apis dorsata* and *Apis laboriosa*; honey bees in India, Nepal and Bhutan. J Apic Res 56(1):81–91
Zeh DW (1987) Aggression, density, and sexual dimorphism in chernetid pseudoscorpions (Arachnida: Pseudoscorpionida). Evolution 41(5):1072–1087
Zeh JA, Zeh DW (1990) Cooperative foraging for large prey by *Paratemnus elongatus* (Pseudoscorpionida: Atemnidae). J Arachnol 18:307–311

Chapter 6
Current Status of the Beekeeping Industry in China

Huoqing Zheng, Lianfei Cao, Shaokang Huang, Peter Neumann, and Fuliang Hu

Abstract China is a country with high honeybee diversity, long history of beekeeping and large amount of managed honeybee colonies. Based on the development of a high royal jelly-producing lineage of honeybees and related techniques and tools, China is producing nearly all of royal jelly in the world market. Over the past decade, great development has occurred in Chinese beekeeping industry, e.g., the development of honeybee pollination market, the promotion of beekeeping with *Apis cerana,* and the mechanization of beekeeping. However, there are also challenges facing the industry, e.g., honey adulteration and colony losses caused by biotic and/or non-biotic factors. The exploitation of native bumble bees and stingless bees is still in the very early initial stage. The role of honeybee pollination in agriculture has been recognized by government and farmers. We believe that Chinese beekeeping is entering an era full of opportunities.

Keywords China · Beekeeping industry · Honeybee diversity · History · Royal jelly · Pollination · Pests and diseases · Non-*Apis* species · Perspective

---

H. Zheng (✉) · F. Hu
College of Animal Sciences, Zhejiang University, Hangzhou, P. R. China
e-mail: hqzheng@zju.edu.cn; flhu@zju.edu.cn

L. Cao
Institute of Animal Husbandry and Veterinary Science, Zhejiang Academy of Agricultural Sciences, Hangzhou, P. R. China
e-mail: caolianfei1981@163.com

S. Huang
College of Bee Science, Fujian Agriculture and Forestry University, Fuzhou, Fujian, P. R. China
e-mail: honeysk@qq.com

P. Neumann
Institute of Bee Health, Vetsuisse Faculty, University of Bern, Bern, Switzerland
e-mail: peter.neumann@vetsuisse.unibe.ch

P. Chantawannakul et al. (eds.), *Asian Beekeeping in the 21st Century*,
https://doi.org/10.1007/978-981-10-8222-1_6

6.1 Honeybee Diversity in China

There are five indigenous and one exotic *Apis* species distributed in China (China National Commission of Animal Genetic Resources [CNCAGR] 2011). Among the indigenous species, the two giant honeybees, *Apis dorsata* Fabricius and *Apis laboriosa* Smith, and the two dwarf honeybees, *Apis florea* Fabricius and *Apis andreniformis* Smith, are wild. The eastern honeybee *Apis cerana* Fabricius is both wild and partly managed by beekeepers. At present, the introduced western honeybee *Apis mellifera* Linnaeus dominates the production of bee products in China and is distributed throughout almost all the country.

A. dorsata is distributed in southern China, including Yunnan and Guangxi provinces and Hainan Island. Significant genetic differentiation has been found between the population on Hainan Island and that on the Chinese mainland (Cao et al. 2012a, b).

A. laboriosa is distributed in the mountainous areas of southern Tibet, and western and southern Yunnan Province in China. Related studies focused mainly on the morphometric analyses compared with that of *A. dorsata* (Li et al. 2009; Cao et al. 2012c). In a recent study, *A. laboriosa* from China was compared with *A. dorsata* from China and the giant honeybees from neighboring Asian areas. The comparison was based on mitochondrial DNA sequences, which further confirmed the genetic divergence between *A. laboriosa* and *A. dorsata*, and genetic variation was also found among *A. laboriosa* populations from different areas (Cao et al. 2012b).

A. florea is mainly distributed in Yunnan and Guangxi provinces, whereas *A. andreniformis* is in the Yunnan Province. Few studies have been conducted on dwarf honeybees in China, except for comparative morphometric analyses on the two species (Kuang et al. 1983; Zhang et al. 2009).

A. cerana is widely distributed in the southern and eastern China. *A. cerana* in China has been separated into nine ecological races based on the combination of morphological characteristics, geography, and climate (CNCAGR 2011). In recent years, the Palm Forest and mountain race of Hainan Island, and the Tableland and alpine valley race of Aba were shown to be significantly divergent from others based on molecular studies (Tan et al. 2006; Tian et al. 2014; Zhao et al. 2014).

As the only exotic *Apis* species in China, *A. mellifera* was introduced in the late nineteenth century. *A. mellifera* subspecies in China at present include *A. mellifera ligustica, A. mellifera carnica, A. mellifera carpatica, A. mellifera caucasica*, and *A. mellifera anatolica* (CNCAGR 2011). The descendants from the first three subspecies are prevalent in beekeeping of China and several locally developed breeds have been derived from them. It is noteworthy that a new *A. mellifera* subspecies, *A. mellifera sinisxinyuan* n. ssp., was recently reported in northwestern Xinjiang Province in China, which might extend the natural distribution of *A. mellifera* in the world (Chen et al. 2016).

There is relatively high honeybee diversity in China because of the vast size of the country and its diverse climates. According to the recent version of the Köppen–Geiger climate classification system (Peel et al. 2007), the distribution areas of *A. dorsata*, *A. florae*, and *A. andreniformis* in China belong to the climate type Cwa

(Temperate, dry winter, hot summer), and that of *A. laboriosa* belongs to the climate type Cwb (Temperate, dry winter, warm summer). For the widely distributed species *A. cerana,* climate types include Cfa (temperate, without dry season, hot summer), Cwa (temperate, dry winter, hot summer), and Dwa (cold, dry winter, hot summer).

6.2 Historical Records of Beekeeping

Beekeeping has been practiced since ancient China (CNCAGR 2011). Safeguarding and obtaining honey from *A. cerana* colonies living in tree holes or stone caverns can be dated back to >2000 years ago. In the book "Golden Rules of Business Success" (《Zhi Fu Qi Shu》 or 《致富奇书》 in Chinese) written by Fan Li (or Tao Zhu Gong) during the Spring and Autumn period, there are sections describing the art of beekeeping, stressing the importance of the quality of the wooden box used, and how this can affect the quality of the honey. Beekeeping with a wooden box placed in a courtyard began in the form of vertical type or horizontal type. In the Yuan Dynasty (nearly 1000 years ago), a bee box with adobe block, and brick or bamboo weaving appeared. In the dynasties of Ming and Qing (from about 500 years ago to 100 years ago), professional bee farms were founded to supply honey to the royal court.

Approximately 300,000 colonies of *A. m. ligustica* were imported from Japan between 1928 and 1932, and these bees formed the Chinese founder population (CNCAGR 2011). Since then, *A. m. ligustica* has expanded very fast in China because of its outstanding production performance compared to that of *A. cerana.* After the founding of the People's Republic of China in 1949, other *A. mellifera* subspecies were also introduced for performance improvement of honeybee colonies.

A massive replacement of *A. cerana* occurred after the importation of *A. mellifera*; in some regions, the introduction of *A. mellifera* resulted in the near extinction of the endemic *A. cerana* populations (Yang 2005). Accordingly, several protection zones have been established for *A. cerana* in China (CNCAGR 2011).

6.3 Overview of Beekeeping Industry in China

6.3.1 Beekeeping

There are approximately six million *A. mellifera* colonies in China (CNCAGR 2011). This number is likely close to the actual number because beekeepers with *A. mellifera* are usually registered, and the numbers of colonies they keep are typically stable. The number of *A. cerana* colonies in China is hard to estimate because of the lack of knowledge regarding wild populations and the difficulties in conducting a countrywide survey. The number of *A. cerana* colonies managed by beekeepers was estimated to be approximately two million by CNCAGR (2011). However, according to Luo and Chen (2014), the number was close to four million in 2008 (Luo and Chen 2014). After 2010, the population size of *A. cerana* increased significantly because people realized the important role *A. cerana* played in the local environment, and the price of *A. cerana* honey is increasing. According to Luo and Chen (2014), the number of *A. cerana* colonies managed by beekeepers increased to approximately five million in 2014.

The annual yield of honey is approximately 450,000 metric tons, a fourth to a third of which is exported, mainly to Japan, the UK, Belgium, and Spain (Fang 2015, 2016). Recent nationwide estimates for other bee product include approximately 3000 metric tons of royal jelly, 5000 metric tons of pollen, 6000 metric tons of beeswax, and 400 metric tons of propolis (Fang 2015, 2016; Chen 2017).

The distribution of colonies over the country is not uniform. The Province of Zhejiang, which is located in China's eastern coastal region, is certainly a province of important status in the beekeeping industry. There are 1.1 million beehives registered to 15,000 households in the province, which accounts for one eighth of the total number nationwide (Chen 2014). These hives contribute one fourth of the honey and 60% of the royal jelly production for the entire country (Chen 2014). In some other provinces, however, especially in western China, the beekeeping industry is still in its infancy (lv et al. 2014). Local government and associations are making efforts to promote its development to help the rural economy.

The number of *A. mellifera* colonies per apiary usually ranges between 50 and 150 with up to 140 thousand apiaries in the entire country. Most apiaries are run by couples occasionally with the help of one or two assistants. Figure 6.1 shows *A. mellifera* apiaries in Zhejiang Province, China.

Fig. 6.1 Apiaries with *Apis mellifera* colonies in Zhejiang Province, China. (**a**) In summer, the hives are covered with rice straw for shading. (**b**) In autumn, the apiaries have been recently moved to a location to harvest tea pollen

6.3.2 *Consumption of Bee Products*

Bee products are considered healthy food and widely used in traditional Chinese medicinal recipes. For example, in the "Compendium of Materia Medica" (《Bencao Gangmu》 or 《本草纲目》 in Chinese), the Traditional Chinese Medicine classic by pharmacist Li Shizhen in the Ming Dynasty (1368–1644), stated "Honey can help dispel pathogenic heat, clear away toxins, relieve pain and combat dehydration." With the developing economy and the increasing demand to remain healthy even under high levels of stress, growth in bee product consumption is occurring in the domestic market. For example, the annual consumption of honey per person has grown from 110 g in 2001 to 250 g in 2013 (Chen et al. 2014).

Besides raw consumption, honey, as a sweetener is used in the preparation of some foods like Beijing Roast Duck and also in some health drinks and carbonated drinks. Royal jelly has been introduced into some beverages and cosmetics. However, of all the bee products, beeswax has been, and remains, the most versatile and most widely used material. Its market in China is limited and less than 10% is consumed at home, including the new, yet increasing, use of it as a cosmetic ingredient. The majority is exported to the USA and EU.

Compared with others, propolis is a relatively new bee product. However, judging from the profit to manufacturers and its popularity in China, it has become one of the most valuable products. A large amount of propolis is imported, mainly from South America, to meet the booming demand of the domestic market because of the weakness of the ability of *A. m. ligustica* to produce propolis and the inability of *A. cerana* to collect resin (Zeng 2007).

One of the product uses that has undergone a dramatic increase over the past few years is that of use in cosmetics. Honey, royal jelly, and pollen are included in some cosmetic preparations with claims of rejuvenating and nourishing effects for the skin. Together with its bactericidal and fungicidal characteristics, propolis provides many benefits in various applications in cosmetics.

6.4 Features of Beekeeping in China

6.4.1 *Core of Chinese Beekeeping Techniques: Balancing Between the Number of Frames and Bees*

Colony management with *A. mellifera* in China is extraordinarily laborious because colonies have to be managed elaborately to obtain a high yield of bee products, which is the main goal of Chinese beekeeping. For example, colonies are inspected frequently and, if necessary, fed with sugar or/and a protein diet to build up and maintain a strong population.

One feature of Chinese beekeeping techniques is that beekeepers pay great attention to the balance between the number of frames and number of bees in a hive. In contrast with most European and American beekeepers, who add or remove a hive body, a honey super, or sometimes even a few of the frames, Chinese beekeepers add or remove one or more frames each time. In most times of the year, only the absolute necessary number of frames is provided to the colony to ensure that each comb is covered by a layer of bees. This status is the so-called combs balanced to bees. To avoid extra combs being built at the side of the edge frame and to create a relatively separated space, a wooden board is added at the side of a colony. From the point of view of Chinese beekeepers, in this way, bees are able to take care of all the combs in the hive and carry out hygienic behavior more efficiently to prevent parasites and diseases. This is especially true when keeping *A. cerana* because combs of this honeybee species are more prone to be attacked by wax moths if not all the combs are covered by sufficient bees (Oldroyd and Wongsiri 2006).

This fundamental idea of beekeeping techniques is incorporated in colony management year-round. Honeybee colonies dwindle over winter, resulting in a situation where the number of combs is greater than what the bees can cover. Beekeepers thus remove extra frames before initiating early-season build-up. At this stage, usually only one or a few frames are left in a hive to achieve the status where each comb is covered by several layers of bees. This status is the so-called combs less than bees. The congestion of a colony can stimulate egg-laying by the queen and help the bees maintain appropriate temperature for brood development. Provided with sufficient supplementary feeding of pollen and syrup/honey, and with the necessary thermal insulation, this operation enables a colony to start fostering a brood theoretically any time in winter in a moderate climate. In fact, in Zhejiang province, where the lowest temperature in winter reaches −5 °C and meteorological spring starts in March, beekeepers initiate build-up of colonies as early as the beginning of the year. This provides a colony sufficient time to build up its population before the first blooming season of a year when rape (*Brassica napus*) blooms. Frames of combs are gradually added after new generations of bees emerge. In this process, it is still important to keep the bees crowded. This accelerates the speed of colony build-up, especially when a colony is weak. This process should be ceased before the arrival of swarming season because the crowd situation may stimulate swarming. Frames of combs should be added to provide sufficient space for egg-laying of queens and reach the status "combs balanced to bees" in a colony.

In summer, a few more frames than necessary are kept in a hive, the status of which is the so-called combs more than bees. This is done to increase ventilation and cool down the colony.

The idea of maintaining a balance between the number of frames and bees is regarded as a core of Chinese beekeeping techniques. We believe that this is critical to keeping honeybee colonies healthy and productive. This operation is practical in apiaries where each beekeeper takes care of less than 100 colonies. However, when beekeepers are required to manage large numbers of colonies, it will not be an easy task to manage colonies in this way.

6.4.2 Royal Jelly Production

Royal jelly has been demonstrated to possess numerous functional properties, including activities and properties such as antibacterial, anti-inflammatory, vasodilatory and hypotensive, disinfectant, antioxidant, anti-hypercholesterolaemic, and antitumor (reviewed in Ramadan and Ai-Ghamdi 2012). As a valuable bee product, it has been widely promoted and commercially available as a health food in China. It is used as a general health tonic, for fighting the effects of aging, and for boosting the immune system.

China is the largest producer of royal jelly in the world and satisfies nearly all (95%) of the global demand. Since the 1980s, the production of royal jelly in China has increased from 200 to approximately 3000 tons per year. The increased production of royal jelly in China over the past three decades has only been possible because of the development of a high royal jelly-producing lineage of honeybees (abbreviated as HRJB), which was derived from an Italian honeybee subspecies (*A. m. ligustica*) (reviewed in Cao et al. 2016). Furthermore, the implementation of advanced production techniques have optimized royal jelly production (reviewed in Hu et al. 2017).

Zhejiang Province was one of the first areas to introduce *A. m. ligustica* for honey production. In the 1950s, with the increased demand for royal jelly, beekeepers started production royal jelly using *A. m. ligustica* queen right colonies. At that time, royal jelly production per colony was only approximately 0.2–0.3 kg/year (Chen 2005). After the 1960s, beekeepers in some parts of Zhejiang Province were inclined to select *A. m. ligustica* colonies for improved royal jelly production (CNCAGR 2011). Colonies with the highest royal jelly production were selected and used for rearing daughter queens and drones in each apiary within adjacent regions in the same season year after year. Beekeepers sometimes also used the larvae of high royal jelly-producing colonies from other apiaries for queen rearing. All queens were mated naturally in the open air. After over 20 years of semi-controlled breeding, the royal jelly production per colony in some apiaries had increased to 2.0–3.0 kg/year in the late 1980s (Chen 2005). The great feature of these bees was recognized by the Chinese government, and soon the bees were rapidly introduced into other regions of Zhejiang Province and other parts of China (CNCAGR 2011). Selective breeding continued and the royal jelly production of HRJB colonies in Zhejiang Province increased to 6.0–8.0 kg/colony/year by the 2000s. At present, the yield of royal jelly per HRJB colony averages 200 g in a 72-h production period and the annual yield reaches >10 kg. HRJB, also called the Zhejiang Jiang Feng (*Zhejiang Royal Jelly Bee* in Chinese) was first certified as a new unique honeybee genetic resource by the Chinese government in 2010. Thanks to the wide distribution of HRJB in Zhejiang Province, the honeybee colonies in this province, which account for one eighth of the total number in China, are producing 60% of royal jelly production of the entire country.

The development of related techniques and tools contributed greatly to the high yield. The most involved technique is building up and maintaining a large colony population. As with honey production, a large bee population is the key to high yield of royal jelly production. HRJB builds up population numbers more rapidly if given a sufficient food supply and can maintain a strong population without swarming. The most important tool developed for royal jelly production were plastic queen cups, into which 1-day-old larvae are grafted and nurse bees secrete royal jelly to feed the larvae. The invention of the plastic queen cups, were a small but great change, making it possible to produce royal jelly at a large scale. Previously, wax queen cups had to be made and mounted individually on wooden strips before grafting.

In addition to the genetics of HRJB, the elaborate beekeeping management provides additional support for the maintenance of a strong population (Jin 2008). A standard procedure for royal jelly production in China involves the following steps: (1) a single queen or multiple queens (see Sect. 6.5.5) lay eggs on an empty comb; (2) 1-day-old larvae are grafted into plastic queen cups mounted on bars; (3) the bars are put into the queenless portion of the colony, separated by a queen excluder. Nurse bees then feed the larvae in the queen cups with royal jelly; (4) the bars are removed from the colony after 72 h; (5) larvae are removed from the queen cups using forceps, and (6) the royal jelly is collected with a plastic spatula. An interval of 72 h is normal; however, some beekeepers collect royal jelly at 48 h or even 24 h after grafting (Zheng et al. 2011).

6.5 Recent Development of Beekeeping in China

6.5.1 Honeybee Pollination

Bee products are the main source of income for beekeepers although the demand for pollination is obviously increasing. Over the past decade, the importance of honeybee pollination in agriculture has gained increasing attention. Honeybee colonies are rented or bought for the pollination of watermelons and strawberries in greenhouses. This has been a significant development; just 10 years ago few honeybee colonies were commercially used for pollination in China.

However, compared with the large number of honeybee colonies in China, the colonies used for pollination are a very small percentage. These colonies are mainly from sedentary apiaries, which are near farms or greenhouses. The pollination market is not sufficiently developed to provide sufficient business for migratory apiaries.

6.5.2 Promotion of Beekeeping with Apis cerana

A. cerana is indigenous to China. It is better adapted to the local climate (Oldroyd and Wongsiri 2006) and more resistant to diseases (Page et al. 2016); thus, it does not require as elaborate management as does *A. mellifera*. Moreover, supplementary feeding is usually unnecessary due to the bees' excellent ability to forage from dispersed nectar or pollen sources. Because of these reasons, *A. cerana* is preferred by few professional beekeepers or hobbyists, and by beekeepers of sedentary apiaries in mountainous areas.

Although movable-frame hives allow for inspection and manipulation of colonies, a large percentage of *A. cerana* colonies are still kept in traditional hives, which are cylinder wooden buckets with a few bars at middle height to allow for comb attachment (Fig. 6.2). The traditional cylinder hives are not favorable for colony management, but they are still preferred by some beginners because with these hives, bees can build combs freely and little attention is needed to add or reduce frames.

Compared with *A. mellifera*, these bees are much less bred and selected. They maintain smaller populations and are less productive. Only 5–20 kg of honey can be produced per year by each colony. However, keeping *A. cerana* colonies is still profitable because the price of *A. cerana* honey is several times that of *A. mellifera*. In recent years, governmental projects promoting beekeeping with *A. cerana* have been implemented in many rural areas because beekeeping with *A. cerana* is an ideal source of extra income for rural farmers.

The extension of beekeeping with *A. cerana* and the use of *A. cerana* in pollination service have resulted in a high demand of *A. cerana* colonies. Therefore, another significant development in recent years is that a market supplying established *A. cerana* colonies or packaged bees emerged. However, this has also evoked great concern regarding disease transmission caused by the movement of colonies and ecological effects of the large-distance exchange of honeybee genetics.

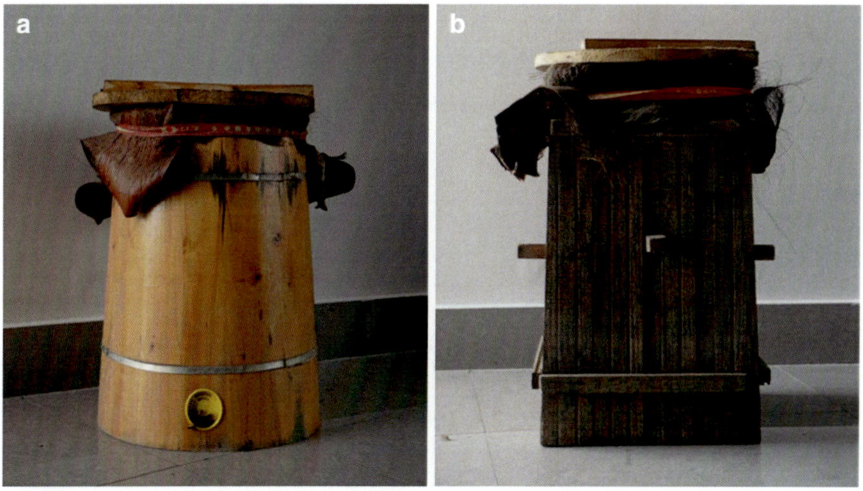

Fig. 6.2 Traditional cylinder hives for keeping *A. cerana* in China. A few bars are installed at middle height of the hives to allow combs to attach. Palm bark is a good material to cover the top of the hives because of its rain-proofing and ventilation functions (**a**) cylindrical shape; (**b**) squared shape

6.5.3 Production of Comb Honey

Most honey in China is produced in the form of extracted honey. Because of the difficulty avoiding the adulteration of honey (see Sect. 6.10) and the rising demand for higher honey quality, more beekeepers are producing comb honey. Instead of cutting comb honey into sections after harvesting, some kinds of frames assembled from plastic or wooden boxes were designed (Fig. 6.3). Foundations are inserted into the boxes and bees are forced to build comb inside each of the boxes. After bees fill the boxes with honey, the comb honey in each box can be packaged directly.

6.5.4 Mechanization of Beekeeping

6.5.4.1 Beekeeping Truck

An estimated 80% of *A. mellifera* beekeepers migrate throughout the country (personal estimate) in pursuit of different honey flows. Basically, colonies are moved from the south to the north, then from the north to the south on a yearly basis. These long-distance migratory apiaries may move more than 3000 km each year. There is also a small percentage of apiaries that only move short distances. They are sedentary, usually near their houses most of the time and only move short distances, i.e., a few hundreds of kilometers, to new locations to make use of other blooming plants. Trains were the first choice among transportation methods 30 years ago. However, this has been replaced by trucks because of the inconvenience of train transportation.

Beekeepers with migratory apiaries spend a lot of time in a truck during transportation and often live in tents after they arrive at a new location. Because forklift

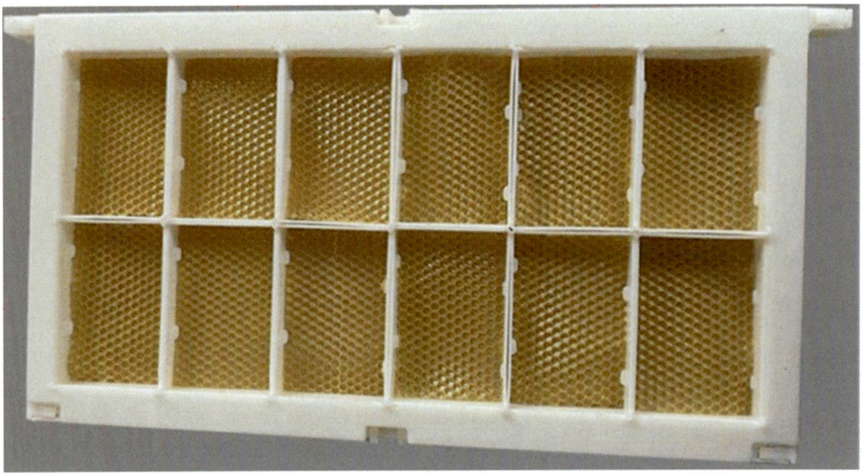

Fig. 6.3 An assembled frame for comb honey production

trucks are not commonly used in the Chinese beekeeping industry, beekeepers have to load and unload hives using manpower. These situations make the life of beekeepers hard and migratory beekeeping is tremendously labor-intensive. To increase the comfort of living and decrease the labor intensity in loading and unloading hives, trucks specialized for beekeeping have recently been made commercially available. A beekeeping truck is modified from a typical truck. Major modifications usually include (1) a carriage added behind the cab for the living space of beekeepers, both during transportation and after arriving; (2) a shelf installed for placement of hives on the truck; (3) a solar panel system installed on the truck to provide electricity to facilitate the comfortable living of beekeepers; and (4) sometimes a small-scale crane installed at the rear end of the truck to load and unload hives.

With a beekeeping truck, loading and unloading hives are not necessary because hives can be left on the shelf after the arrival at a new place. However, it has been noted that drifting of foragers significantly increased because of the high density of hives on the truck. Various methods have been attempted to solve this problem. Apparently, the design of the truck still requires improvement.

6.5.4.2 Mechanization of Royal Jelly Production

Despite advances in breeding and the development of new tools and techniques, royal jelly production is still laborious and time-consuming. This is especially true for the process of grafting larvae. For example, a beekeeper keeping 100 colonies for royal jelly production in a 72 h interval has to graft ~5000 1-day-old larvae every day for 33 royal jelly-producing colonies if one frame of queen cups is used for each colony. Provided with sufficient food and strong populations in the colonies, double frames of queen cups are sometimes used. In this case, grafting of ~10,000 1-day-old larvae are needed every day. A bionic, non-grafting larval depositor has recently been designed, which removes the larval grafting stage and significantly reduces the intensity of labor required in royal jelly production (Zeng et al. 2013; Hu et al. 2017). Machines for withdrawing larvae from queen cells and gathering royal jelly have also been invented (Su et al. 2010; Pan et al. 2013). These inventions show that there are still opportunities to further improve the methodology of royal jelly production, and it can be assumed that large-scale factory-based royal jelly production may be feasible in the near future.

6.5.5 Creation and Maintenance of Honeybee Multiple Queen Colonies

6.5.5.1 Creation of Honeybee, *A. mellifera*, Multiple Queen Colonies

Honeybee, *A. mellifera*, multiple queen colonies (Fig. 6.4) composed of several mated queens able to move around freely were produced by modulating biological factors that evoke fighting and queen elimination within the colony. This was primarily

Fig. 6.4 Four queens peacefully cohabitating on one side of a comb. Provided with close supervision and special care, such multiple queen colonies are sustainable enough for commercial use

achieved by ablating mandibles of queens to avoid inter-queen rivalry and by using young workers to avoid aggressive behavior toward the queens (Zheng et al. 2009a).

The colonies destined to host the multiple queens are prepared as follows: combs of emerging broods are selected and shaken, which triggers flight in the older bees, whereas young bees tend to remain on the comb. The combs are then placed in the hive box with the young bees still clinging to them. The host hives are placed at a distance (5–10 m) from their original location to ensure that all remaining foragers (older bees) do not re-enter them. The number of combs and bees to be used in the multiple queen colony depends on the number of queens to be introduced. Four to six combs are used for three- to six-queen colonies. Additional combs of honey and pollen are added beside the brood combs to provide enough food because the colony is deprived of foragers at the beginning.

Two days after the colonies are prepared, queens are taken out of their original colonies and introduced to different locations in the host hives after a third to half of both mandibles are removed with micro-scissors. The ablation of mandibles reduces their propensity to fight and kill each other (Dietemann et al. 2008). In addition, the large abdomens of the egg-laying queens might further reduce their ability to fight (Spiewok 2006).

The results of a 4-year study showed that the method is viable for the production of sustainable multiple queen honeybee colonies for commercial use. It should be noted that, based on our experience, the success rate is much higher in a blooming season. The majority of these colonies experienced no queen loss for 2 months and most were still stable after 6 months. Of 80 colonies, 55 (68.8%) experienced no queen loss over winter (Zheng et al. 2009a).

6.5.5.2 Maintenance of Honeybee, *A. mellifera*, Multiple Queen Colonies

The rates of egg production of three-queen and five-queen colonies were respectively 199% and 328% greater than that of a single queen colony (Zheng et al. 2009b). It has been demonstrated that keeping multiple queen colonies is a viable way to improve the quantity of broods produced by a colony, despite the fact that the egg-laying rate of the multiple queen colonies was less than directly proportional to the number of queens per colony (Zheng et al. 2009b). This lays a solid foundation for the use of multiple queen colonies in beekeeping practice because the production and maintenance of strong populations is a precondition for maximizing colony productivity. Close supervision and special care of these multiple queen colonies are, however, necessary because they are sensitive to internal and external factors, such as social interactions among individuals and food supply (Zheng et al. 2009b). It is not recommended to keep populous multiple queen colonies to produce hive products. Colonies with several free running queens are more sensitive to external environmental conditions, and therefore require increased care, but perform well when kept as supporting colonies. In this situation, the advantages of keeping multiple queen colonies outweigh the disadvantages involved in the additional work needed for their maintenance. In China, they can contribute to commercial beekeeping by providing larvae for royal jelly production (see Sect. 6.4.2) and capped broods for the rapid build-up of production colonies and workers for package bees.

In addition, this technique will help to increase our understanding of basic questions regarding the evolution of sociality, such as division of reproduction and the evolution of polygyny.

6.6 Bee Forage Resources

Most of the Chinese territory has a temperate climate, with sub-tropical and tropical areas in the south and a cold temperate region in the north. Such variation offers a diverse ecological environment and maintains diverse floral resources for honeybees. Preliminary statistics show that nearly 10,000 kinds of nectar plants are available (Wu et al. 2006).

Rape is widely cultivated in China and is the most important bee plant, providing abundant nectar and pollen. Depending on its genetic varieties and the region in which it occurs, the blooming season spans from December in southern China, e.g., in Yunnan Province, to August in northwestern China, e.g., in Qinghai Province. Thus, a migratory apiary can utilize nectar and pollen sources of rape blossoms to build up the colony population in the early season in southern China and, when it later migrates to the north, it is possible to meet several other blooming seasons to harvest commercial rape honey and pollen during the trip.

Black locust (*Robinia pseudoacacia*) is widely distributed and extensively grown in China. With abundant nectar secretion and excellent nectar quality, it is a major nectar source in early summer. It is famous for producing a fruity and fragrant honey that ranges from water white to lemon yellow to yellowish green. A batch of monofloral black locust honey with little cross-contamination from other flowers can be as clear as a glass jar. The honey is high in fructose, and therefore it can be stored for long periods without crystallizing. Because of the clearness, fragrance, and noncrystallization properties, black locust honey is one of the most popular honey varieties in China.

Loquat (*Eriobotrya japonica*) is indigenous to China. It is distributed mainly in southeastern China, i.e., provinces of Fujian, Taiwan, Zhejiang, Jiangsu, and Anhui. It is a major nectar source in winter. Because of the cold weather in its blooming season, commercial Loquat honey is usually produced by *A. cerana* thanks to the bees' tolerance to cold weather. *A. mellifera* may be able to forage on Loquat flowers, but they usually do not provide surplus honey.

Other major bee plants and their distributions in China are shown in Table 6.1.

Pollen and/or nectar of some bee plants are toxic to bees and even humans. These include *Tripterygium wilfordii, Veratrum nigrum, Tripterygium hypoglaucum, Celastrus angulatus, Macleaya cordata, Aconitum carmichaelii, Rhododendron molle, Alangium chinense, Gelsemium elegans, Datura stramonium, Camellia oleifera, Camptotheca acuminata,* and *Stellera chamaejasme* (Chen et al. 2017).

6.7 Pests and Diseases of Honeybees

6.7.1 Parasites

6.7.1.1 Ectoparasitic Mites *Varroa destructor*

Varroa in China was not confirmed as *V. destructor* until 2004 and was thought to be *V. jacobsoni* (Zhou 2004). *V. destructor* infestation in *A. mellifera* colonies was first reported in 1955 in China (Xia 1957). The infestation has caused devastating losses of *A. mellifera* since 1956, especially in Zhejiang and Jiangsu provinces, southeastern China (Shen 1958; Zhang 1958), and then across China after 1964 (Fan and Huang 1993). From then on, the mite has been the most serious biotic threat to western honeybees in China (Zhou 2004).

A plastic strip containing fluvalinate or flumethrin is the most popular product used for the treatment of *V. destructor* in China since the 1990s because of its high effectiveness, efficiency, and lower toxicity to honeybees (Wei 1991). However, these miticides could accumulate in comb wax and propolis, honey, and other hive products after long-term usage (Tsigouri et al. 2000; Chen et al. 2004). Moreover,

Table 6.1 Major bee plants and their distributions in China

Common name	Scientific name	Distribution in China (provinces)
Rape	*Brassica napus*	Nationwide
Black locust	*Robinia pseudoacacia*	Shandong, Hebei, Henan, Liaoning, Shaanxi, Gansu, Jiangsu, Anhui, Shanxi
Loquat	*Eriobotrya japonica*	Zhejiang, Fujian, Jiangsu, Taiwan, Anhui
Lychee	*Litchi chinensis*	Guangdong, Fujian, Taiwan, Guangxi, Sichuan, Yunnan, Guizhou
Longan	*Dimocarpus longan*	Guangdong, Fujian, Taiwan, Sichuan, Guangxi
Chinese Milk Vetch	*Astragalus sinicus*	Hunan, Jiangxi, Hubei, Anhui, Zhejiang
Vetchleaf Sophora	*Sophora davidii*	Shanxi, Shaanxi, Gansu, Ningxia, Yunnan, Sichuan, Tibet
Chinese Date	*Ziziphus zizyphus*	Hebei, Shandong, Shanxi, Shaanxi, Henan, Gansu
White Sweet Clover	*Melilotus albus*	Northern provinces
Cow vetch	*Vicia cracca*	Jiangsu, Shandong, Shaanxi, Yunnan, Guizhou, Anhui, Sichuan, Hunan, Hubei, Guangxi, Gansu
Linden	*Tilia*	Heilongjiang, Jilin
Chaste tree	*Vitex negundo*	Liaoning, Hebei, Beijing, Shanxi, Inner Mongolia, Shaanxi, Henan, Shandong, Anhui, Gansu
	Eucalyptus	Guangdong, Guangxi, Hainan, Fujian, Taiwan
Sunflower	*Helianthus annuus*	Heilongjiang, Liaoning, Jilin, Inner Mongolia, Xinjiang, Ningxia, Gansu, Hebei, Beijing, Tianjin, Shanxi, Shandong
Sesame	*Sesamum orientale*	Henan, Hubei, Anhui
Komarov Swallowwort	*Cynanchum komarovii*	Inner Mongolia, Ningxia, Shaanxi
Buckwheat	*Polygonum esculentum*	Gansu, Shaanxi, Inner Mongolia
Rugulose Elsholtzia	*Elsholtzia rugulosa*	Yunnan, Guizhou, Sichuan
Ivy Tree	*Schefflera octophylla*	Fujian, Taiwan, Guangdong, Guangxi, Hainan, Yunnan

long-term usage of these products has caused unintended selection for resistance in the parasite. Formic acid and amitraz are also commonly used for the treatment of *V. destructor*. Several essential oils extracted from plants were proven to be effective against *Varroa*. These included methyl salicylate (Anonymous 1963), tobacco extracts (Yang 1965), *Mentha haplocalyx* Briq extracts (Wang et al. 2005), and essential oils of *Truestar Anisetree* (Su et al. 2016). However, their effectiveness is far less than that of miticides like flumethrin, and none of them have been registered.

6.7.1.2 Ectoparasitic Mites *Tropilaelaps mercedesae*

It was confirmed that *Tropilaelaps mercedesae* rather than other *Tropilaelaps* species infested *A. mellifera* in China, after *Tropilaelaps* mites collected from 25 provinces throughout China were analyzed (Luo et al. 2011). *T. mercedesae* mainly parasitizes capped brood cells and normally only survives 2–3 days without a brood because their mouthparts are not suitable to feed on adult honeybees (Woyke 1984; Rinderer et al. 1994). However, it was confirmed that some *T. mercedesae* mites did survive the winter in broodless *A. mellifera* colonies (Wang et al. 1984). Several beekeepers reported a few cases where *T. mercedesae* were found on adult honeybees in wintering colonies (Mo 2003). Because of a rapid reproductive rate, *T. mercedesae* can easily multiply to a threatening threshold and cause severe problems in the colony after spring.

Taking advantage of the broodless period, there are two ways recommended to eliminate *T. mercedesae* from colonies without chemical treatment. One is to remove all brood combs from heavily infested colonies, and the other is to cage the queen for 21 days and clean the rest of the capped brood cells (Woyke 1984). Treatment with a miticide in the broodless condition is more efficient in practice.

Sulfur is the most popular chemical used for the control of *T. mercedesae*. It is dusted in the space between two combs, on the bees, or directly on the surface of capped brood cells. Sulfur has also been mixed with amitraz or other miticides and smeared on the capping of cells to kill mites in the cells (Yu et al. 2011). The dose of sulfur used on the capped brood should not be higher than 0.6 g/comb and no more than 3 g for a colony with 10 frames of bees to avoid adverse effects.

6.7.1.3 *Neocypholaelaps indica*

Neocypholaelaps indica, is a whitish mite that belongs to the family Ameroseiidae (Evans 1963). It was first reported in China in 1964 (Teng and Pang 1964). It lives on more than 30 kinds of flowers and migrates to many flower visitors, including honeybees (Fan and Jiang 2014). In spring, the mite can easily be seen on the thorax of foragers of both *A. mellifera* and *A. cerana* in Fujian, China (Fig. 6.5) (Fan and Jiang 2014). Sometimes there are dozens of mites on the dorsal thorax of foragers. Although foragers work normally, they are obviously uncomfortable with the infestation of the mites, and try to get rid of the mites by sweeping with legs over the thorax and shaking their body in vain (Zhang 1965). Because this mite has not been found to cause any obvious effects on the health of honeybee colonies, no obligate measures are required. However, its effect on the health of honeybee individuals still needs clarification.

Fig. 6.5 A worker of *Apis cerana* infested with dozens of *Neocypholaelaps indica* mites. The infestation is not uncommon in the Fujian Province, but rarely reported in other areas. The effect of the infestation on honeybee health is not clear

6.7.1.4 *Acarapis woodi*

Although the honeybee tracheal mite, *Acarapis woodi,* is widely distributed worldwide, its infestation in honeybee colonies was not scientifically reported in China until 2013 (Yang et al. 2013). This does not necessary mean it was recently transmitted to China. Very little research has been conducted on the mite and little attention has been paid by beekeepers to the mite because it is not visible. No further information is available on its prevalence or impact.

6.7.2 Viruses

6.7.2.1 Honeybee Viruses in China

Recently, several surveys on the prevalence of honeybee viruses have been conducted both on *A. mellifera* and *A. cerana* in China (Ai et al. 2012; Li et al. 2012; Wang et al. 2016; Yañez et al. 2015). Top on the list is deformed wing virus (DWV), the most prevalent virus in *A. mellifera*, and black queen cell virus (BQCV) is the second (Ai et al. 2012). DWV was detected in 94% of *A. mellifera* apiaries from 18 provinces in a survey and sacbrood virus was detected in 86% of *A. cerana* apiaries as the dominant virus (Ai et al. 2012). Furthermore, IAPV and CBPV were detected in *A. cerana*, although with low incidence (Ai et al. 2012). In another survey, BQCV, DWV, KBV, and SBV were widely prevalent in *A. cerana* of Yunnan Province (Wang et al. 2016). It appears that interspecies transmission of viruses may contribute to the wide prevalence (Yañez et al. 2015).

Multiple viruses may simultaneously co-infect a single bee. In one survey, *A. mellifera* workers were infected with both DWV and BQCV at 21% incidence, and SBV + DWV at 43% in *A. cerana* (Ai et al. 2012). Occasionally, some samples were infected by three or four viruses in both honeybee species (Ai et al. 2012). In another report, 240 honeybee samples from diseased colonies collected from 20 provinces were confirmed to be infected with multiple viruses (Wu et al. 2015). Among them, 51.76% were co-infected with five viruses: DWV/BQCV/IAPV/SBV/CBPV or DWV/IAPV/SBV/KBV/CBPV (Wu et al. 2015). Although these investigations only covered a very limited area and limited number of colonies, the results suggested that multiple virus infections in the two honeybee species are common and may be one of the causes of honeybee colony losses in recent years in China.

6.7.2.2 Chinese Sacbrood Virus

Sacbrood disease rarely occurs in *A. mellifera* in China, but it is the most serious disease in *A. cerana* (Gong et al. 2016). The pathogen of sacbrood disease in *A. cerana* is a Chinese sacbrood virus (CSBV), a close strain of SBV. It can wipe out all colonies in an apiary within months. It was first reported in southeastern China, Guandong Province in the spring of 1972, with high virulence and infectiousness the disease quickly swept through the entire country in 1973 (Luo et al. 1998). The typical symptom was uncapped pre-pupae with the pointed end heading outside, which is described as "pin head" disease by beekeepers.

The overall incidence gradually dropped over decades although sporadic reports and local outbreaks in some areas are still common (Liu et al. 2001; Hu et al. 2009). Since 2004, a new outbreak wave erupted in China again, heavy losses were quite common in 2009 (Li et al. 2006; Zhang 2012). In Guangxi Province, approximately 65% of *A. cerana* colonies were seriously infected, especially those located in mountainous areas (Li et al. 2006).

Several methods have been suggested to control these devastating diseases.

1. Breeding anti-CSBV strains: Selection of an anti-CSBV strain has already proved to be an effective measure. Colonies with high resistance to CSBV were achieved by continuous active selection by feeding the colonies a virus solution over 3 years (Gong et al. 1984; Verma et al. 1990; Huang et al. 2008).
2. Re-queening: When a new healthy queen is available, it is suggested to immediately replace the queen of a diseased colony. It may stop the vertical transmission of the virus and help the colony recover.
3. Balancing the ratio of combs to adult workers (see Sect. 4.1): Remove extra combs to have the brood area covered by a sufficient number of adults.
4. Treatment with Chinese herbs: Many traditional herbs have been tried and proved to be helpful for disease prevention. For example, infected larvae decreased, and colonies survived after they were fed with filtrate of boiled *Scutellaria barbata,* whereas all the infected colonies without treatment died within 1 month (Shaanxi Institute of Animal Science and Veterinary Medicine 1977; Gong et al. 1984),

Sophora subprostrata had similar effects (Yang 1978). However, none of these herbs exhibit absolute power to eliminate this disease.

5. Feeding dsRNA: After the colonies were fed with 1 μg/mL dsRNA *VP1* in a field test, larval mortality dropped to 44% compared to 86% of the controls (Zhang et al. 2016). Although a lot of work must be done before field application, feeding dsRNA seemed to be a very promising method to tackle CSBV.

In conclusion, Chinese sacbrood disease is the most devastating disease in *A. cerana cerana*. Although progress has been achieved in virus detection, disease prevention, and control, knowledge regarding CSBV epidemiology is still limited, and a highly efficient treatment is urgently needed.

6.7.3 Bacteria

6.7.3.1 *Melissococcus plutonius*

European foul brood is a common bacterial disease in China both in *A. cerana* and *A. mellifera*. The pathogen is *Melissococcus plutonius*, a Gram-positive bacterium. The disease peaks in spring. The typical symptom on the comb at the early stage is a "scattered brood pattern" resulting from removal of diseased larvae by adult workers. At the late stage, when there are too many diseased larvae to be cleaned, a few dead 2- to 3-day-old larvae that are a light brown color might be left in cells and become dark brown. Occasionally, the diseased colony absconds. *M. plutonius* is a non-spore-forming bacterium, which can be killed by many antibiotics. Oxytetracycline hydrochloride (OTC) is the only drug registered for bacteria treatment in hives. Resistant strains are already ubiquitous in the environment; however, applying doses higher than instructed is not recommended because of the risk of contamination of hive products. Treated colonies should be excluded from hive product production and the extracted honey could be fed back as food after more than 8 weeks storage.

Many herbs have been tried for EFB treatment in China. For example, mixed filtrates of boiled *Rhizoma coptidis*, *Radix scutellariae*, and *Flos lonicerae* showed inhibitive effects on *M. plutonius* with an MIC of 781.2 μg/mL (Zhou et al. 2001). Moreover, the right management to prevent young broods from cold stimulation and to ensure sufficient food storage in early spring is very important to the avoidance of EFB.

6.7.3.2 *Paenibacillus larvae*

American foul brood (AFB) is deleterious disease of the honeybee, *A. mellifera*. It is caused by *Paenibacillus larvae*, a Gram-positive, spore-forming bacterium. The spore is highly persistent in the environment. Contaminated combs, hives, or other appliances could still be infectious after years of storage. To control this contagious disease from spreading, burning of the diseased colonies is suggested in many countries (Elke 2010).

Fortunately, only a few cases of AFB had been reported in *A. mellifera* in China (Jia 2001; Wang 2010). Although no case has been reported about AFB in *A. cerana* in China, *P. larvae* could infect 1- to 2-day-old larva of *A. cerana* by oral inoculation (Chen et al. 2000). It is not thought to be a serious disease in beekeeping in China.

6.7.4 *Fungi*

6.7.4.1 *Ascosphaera apis*

Chalk brood is a refractory fungal disease in *A. mellifera* caused by *Ascosphaera apis*. In the first report, over 50% of colonies and 5–40% of larvae were infected in some apiaries (Fan and Lai 1991). Long-distance transportation of diseased colonies through the country and feeding contaminated pollen were the main two sources of the rapid spread over the entire country after its first discovery in the late 1980s (Lai and Fan 1993). Larvae ingest spores at early ages and die 2 days after capping at the prepupa stage. The fungus quickly grows and covers the whole body with hyphae. The dead larvae are as dry as a mummy, which are easily cleaned. Only when there are too many mummies to clean, the mummies are left in the cells, and produce white or dark green ascocarp (Liu 1992).

Chalk brood disease may persist in some apiaries and cause continuous losses year after year. There is no drug officially allowed to be used in colonies for chalk brood treatment. Proper colony management may help prevent the spread of the disease. For example, colonies that are placed in a dry and ventilated environment, with sufficient food and a strong population are more resistant to the disease (Liu 1997). Combs and hives should be disinfected, e.g., potassium hypermanganate (Li 1993). Pollen fed to colonies should be disinfected before feeding.

6.7.4.2 *Nosema ceranae* and *Nosema apis*

In a study conducted in 2005, 54.5% ± 16.6% of *A. m. ligustica* foragers and 10.6% ± 8.3% of *A. c. cerana* foragers were detected with *Nosema* spores in April (Huang et al. 2006). The spores from *A. c. cerana* were (4.68 ± 0.60) × (2.19 ± 0.35) μm in size, which is significantly different from that of *A. m. ligustica* [(5.35 ± 0.52) × (2.84 ± 0.15) μm] (Huang et al. 2006). However, the difference in the spores from the two honeybee species was not further analyzed with molecular methods. Interestingly, a survey and a molecular analysis were conducted in 2008, which revealed that *N. ceranae* was prevalent in *A. mellifera* colonies, but *N. apis* was rarely found (Liu et al. 2008). Recently, another survey with microscopic and molecular analyses covering 701 colonies in 19 apiaries in Zhejiang Province revealed that 74% of the colonies were infected with *N. ceranae* and none of them with *N. apis* (Chen 2016). Fumagillin is not forbidden in the use of colony treatment. However, it is not available in the Chinese market.

6.7.5 Protozoa

6.7.5.1 *Malphigamoeba mellificae*

Malphigamoeba mellificae was first found in adults of *A. cerana* in 1983 and in *A. mellifera ligustica* in 1985. The diseased bees exhibit "crawling" symptoms and were also found to be infected with *Nosema* (Wang et al. 1985; Li et al. 2009, unsure if it was *N. ceranae* or *N. apis*). One beekeeper from Jiangsu reported a loss of 150 colonies in 2 months with this infection (Wang et al. 1985). However, only a few cases of *M. mellificae* infection have been reported (Wang et al. 1985; Li et al. 2009).

6.7.5.2 *Crithidia bombi* and *Crithidia mellificae*

Crithidia mainly infects the gut of invertebrates. It is a trypanosomatid belonging to Excavata (Euglenozoa: Kinetoplastea: Trypanosomatida: Trypanosomatidae). *C. bombi* was detected by PCR in *A. cerana* workers in 13 provinces, with the highest infection rate from Hubei Province at 97% (Li et al. 2012). Infection by *C. mellificae* was confirmed both in *A. m. ligustica* and *A. c. cerana* by Yang et al. (2013).

6.7.6 Wax Moths

There are two kinds of wax moths in China, the greater wax moth, *Galleria mellonella*, and the lesser wax moth, *Achroia grisella* (Wang 1981). The larvae of the greater wax moth cause unexpected uncapping of immature pupae in *A. cerana*. This will prevent colony growth and can destroy it within a month. It is one of the main reasons for colony absconding in summer. Because uncapped pupae are always found with white heads, beekeepers give it the name "White head pupae" (Wang 1981; Feng 1998). Worker bees uncapped the pupal cell not because of the abnormality of pupae themselves, but they are eager to gain access to drive the wax moth larva out from underneath the pupa. Workers may succeed when there are only a few capped brood on the comb, but they will fail when there are too many capped broods. Because the wax moth larvae can easily bite through the wax cell wall and hide in the adjacent capped cells, it becomes a "catch me if you can" game (Huang and Wang 2001).

Table 6.2 Life history of *Galleria mellonella* at 35 °C and 60–85% RH (Huang and Wang 2001)

Developmental stage	Observed no.	Duration (day)
Egg	1100	8.6
Larva	26	49.4 ± 9.4
Pupa	22	10.8 ± 1.9
Adult-female	30	5.5 ± 2.6
Adult-male	23	9.2 ± 3.3

When wax moth larvae grow too large to hide under pupa in cells, they can make silk tunnels on the comb to protect themselves from attacks of worker bees, and finally form a clot of cocoon on the comb or in the hive corner. Comb frames or hive walls are easily damaged by the larvae.

On the contrary, western honeybee colonies are more resistant to wax moths. However, preserved combs are vulnerable to both greater and lesser wax moths. To protect combs, fumigation is needed, or cold storage is better if facilities are available.

The life history of *G. mellonella* in darkness at 35 °C and 60–85% RH is shown in Table 6.2. The fecundity of the *G. mellonella* moth is 725 ± 148 with 4.3 days of laying duration under the same conditions (Huang and Wang 2001).

6.8 Other Problems of Beekeeping

6.8.1 Colony Losses

In a survey conducted by Liu et al. (2016) covering 100 apiaries from each of 12 provinces, colony losses of *A. mellifera* in winter was estimated to be 8.9% between 2009 and 2013 (before the June). Only two of the provinces had colony losses higher than 15% (Xinjiang and He'nan, 19.0% and 16.2%, respectively). The relatively lower colony losses were suggested to occur because of elaborate beekeeping management and timely treatment of *V. destructor*.

However, massive colony losses of *A. mellifera* were reported in some areas since 2013 (Li and Li 2013; Zheng et al. 2016). This mainly happened in Zhejiang, Jiangsu, and An'hui, which are major beekeeping provinces in China. It peaked in winter, but also occurred in later autumn and early spring (Zheng et al. 2016). In a survey in Jinhua City, Zhejiang Province in 2014, 16 of the 34 apiaries experienced colony losses of more than 50% (Zheng et al. 2016). Eight apiaries had colony losses between 30 and 50%. Acute toxicity of pesticides and pollutant poisoning could be excluded as the major causes of the death. DWV infection rate of diseased colonies was significantly higher than that of healthy colonies; however, the BQCV infection rate had the opposite trend (Zheng et al. 2016). In addition, *N. ceranae* infection rates and spore loads were significantly higher in diseased colonies than in healthy colonies (Zheng et al. 2016). The investigation and analysis suggested that multiple virus infections, high DWV and IAPV titers and *N. ceranae* infection were closely related to the colony losses (Zheng et al. 2016). Considering the role of *V. destructor* in promoting virus proliferation, *Varroa destructor* might also be an important cause of death (Neumann et al. 2012). Moreover, it is worth noting that almost all apparently healthy colonies were infected with a variety of pathogens, which suggested that the health of these colonies were in a seriously impaired condition (Zheng et al. 2016).

6.8.2 Interspecific Robbing

The introduction of *A. mellifera* into China has resulted in an artificial sympatry of *A. mellifera* and *A. cerana*. The physical proximity provides ample opportunity for the interactions of the two honeybee species, including mating interference, food competition, nest site competition, and robbing of stores, among other things (reviewed in Moritz et al. 2005 and Yang 2005). Interspecific robbing between honeybee colonies (Fig. 6.6) has caused great concern in the industry because it invokes conflict of interest between beekeepers keeping *A. mellifera* and those keeping *A. cerana* (Li 2015).

Colonies of the two species can cohabit peacefully within flying distance, even in the same apiary, when foraging sources are abundant. Although in a desert season when nectar flow is not available, robbing between species occurs. It is not uncommon to see *A. cerana* workers entering or leaving *A. mellifera* colonies. But this usually happens for only a few workers. However, when a colony of *A. cerana* is robbed by *A. mellifera* workers, a mass of A. mellifera workers can be recruited within a few hours (personal observation). Because *A. cerana* is smaller in size both for the individual and the colony, they are usually unable to defend their nest in this

Fig. 6.6 An *Apis cerana* colony being robbed by *A. mellifera* workers. (**a**) Entrances are blocked to prevent entry of robbers. (**b**) *A. mellifera* workers at the entrance. (**c**) An *A. mellifera* worker is attacked by *A. cerana* workers on a comb on which honey has been robbed. (**d**) *A. mellifera* workers inside the hive. (**e**) An *A. mellifera* worker stealing a honey store from a comb

case. As a result, stored honey is stolen completely and larvae are removed because of the lack of food. For unknown reason(s), in most cases robbing results in queen loss in *A. cerana* colonies. In other cases, the entire colony absconds.

6.8.3 Wasps

Wasps can cause considerable damage. A persistent attack by wasps weakens the colony and most often the colonies either perish or abscond. Several species of the genus *Vespa* are major predators of honeybees in China, namely *V. mandarinia*, *V. nigrithorax*, *V. bicolor*, *V. basalis*, *V. ducalis*, *V. affinis,* and *V. tropica* (Li et al. 2005). Predation by *Vespa* spp. on commercial apiaries is generally a seasonal problem, with a peak of intensity during September–October. Apiaries situated near the foothills and tropical forests suffer more acutely than those on the plains.

Many measures are applied by the beekeepers to control wasps. These measures include the extermination of queen hornets in early spring to get rid of thousands of would-be enemies in summer and fall, crushing worker hornets flying at the apiaries after these workers are caught by hand nets or beaten by wooden sticks with flat heads, and the destruction of hornet nests by applying insecticides or fire after dusk.

Interestingly, in some areas wasp predation is not as serious as before because of hunting by humans. Wasps are rich with protein (Deng et al. 2013). They are regarded as nutritious and used as materials in dishes, for example, fried wasp pupae (Fig. 6.7) for consumption (Feng et al. 2001). The venom of adult wasp bees is thought to have therapeutic effects on some diseases like rheumatism and painful joints (Guo et al. 2017). Thus, adult wasp bees are immersed in wine for a period and then consumed. Moreover, a wasp nest is a valuable traditional medicine. Therefore, successful hunting of a social wasp nest is very economically beneficial, bringing several hundreds to thousands of RMB income.

6.9 Other Non-*Apis* Species Found in the Region

Bumble bees (*Bombus* spp.) are distributed all over the country, except for the desert area. At least 125 species have been found in China to date (An et al. 2014). Currently, tens of thousands commercially produced *Bombus terrestris* colonies are imported to China for crop pollination in green houses, especially in North China, which has raised great concern regarding the ecological threat of invasive species (He et al. 2013). Chinese research groups are developing techniques to rear native bumble bees, including *Bombus hypocrita, Bombus patagiatus, Bombus ignitus, Bombus lucorum, Bombus pyrosoma,* and *Bombus lantschouensis* (CNCAGR 2011; Williams et al. 2017). Because of the limits of the techniques, the use of native bumble bee species is not common at present.

Stingless bees (*Trigona* spp.) are mainly distributed in southern Yunnan Province, Hainan Island, and Taiwan Province. Despite there being more than 500 species of stingless bees worldwide, only 10 species have been described in China, namely

Fig. 6.7 Fried wasp pupae were prepared as a dish

Trigona ventralis, *Trigona terminata*, *Trigona vidua*, *Trigona lutea*, *Trigona pagdeni*, *Trigona laeviceps*, *Trigona iridipennis*, *Trigona canifrons*, *Trigona smithii*, and *Trigona thoracica* (Wu 2000; Kuang and Kuang 2007). Very few studies have been conducted on stingless bees and artificial rearing of stingless bees is rarely reported in China.

6.10 Future Perspectives

Because the role of honeybee pollination in agriculture has been recognized by government and farmers, beekeeping in China has been undergoing tremendous changes in the past years. In 2008, the beekeeping industry was included in the Modern Agricultural Industry Technology System, initiated by the Chinese Ministry of Agriculture in 2007. Within the system, sustained financial funding was provided to more than 46 teams for the research and promotion of beekeeping. This has apparently accelerated honeybee research and the development of the beekeeping industry in China.

One of the most challenging problems in the bee product industry is the adulteration. Adulteration is a complex problem, which currently has a significant economic impact and undeniable nutritional and organoleptic consequences. Various methods have been utilized for honey adulteration detection by researchers (reviewed by Wu et al. 2017). They are all applicable and provide useful information about each aspect of honey authenticity. However, honey adulterations can occur at each stage of production and processing (Ruiz-Matute et al. 2010), and an overall and accurate result may not be achieved solely relying on one or a few techniques. Moreover, the cost of each of these methods is usually high. It is not practical to the economy to conduct a screen to cover all the possible adulteration for a commercial honey product. As a result, honey adulteration is believed to be more serious than previously thought.

Propolis is a natural remedy that has been employed extensively since ancient times. The high demand and limited availability of authentic propolis motivate adulteration of propolis. Poplar tree gum, which is the artificially brewed extract of Populus buds, leaves, bark, and other tissues, has a color, smell, chemical composition, and antimicrobial activity similar to those of the poplar-type propolis (Vardar-Ünlü et al. 2008), and it has been extensively used for propolis adulteration all over the world since the late 1990s because of its low cost and ready availability. A method based on the detection of an "authenticity factor," namely salicin, was used to distinguish propolis from poplar tree gum (Zhang et al. 2011). Salicin is found in poplar tree gum, but not in propolis because of the hydrolyzation by glucosidase from honeybees during propolis collection and processing. In a survey conducted in 2012, 66% of the samples in the Chinese market was positive for salicin, indicating a high percentage of propolis products were adulterated with poplar tree gum (Zhang et al. 2015).

Another challenge of Chinese beekeeping in the near future will be the colony losses in both *A. cerana* and *A. mellifera*. Besides the fact that significant colony losses in *A. cerana* have continuously occurred primarily because of sacbrood disease, massive colony losses of *A. mellifera* caused by unknown reasons have been reported over the past years. A countrywide survey and comprehensive studies are urgently needed to uncover the reasons and find solutions.

Together with the development of the honeybee pollination market, the importance of other pollinators in agriculture is also being recognized. We believe techniques for rearing native bumble bees will be improved and the use of these bees in pollination services will be successful in the near future. The exploitation of stingless bees is still in the very early initial stage. Research and financial support are needed to make use of the endemic species both for honey production and pollination.

In conclusion, Chinese beekeeping has undergone great changes over the past decade and is entering an era full of opportunities and challenges.

Acknowledgements H. Zheng and F. Hu were supported by the Earmarked Fund for Modern Agro-industry Technology Research System from the Ministry of Agriculture of China (CARS-44-KXJ 22 and CARS-44-KXJ 16). L. Cao was supported by a grant (No. 31602014) from the National Natural Science Foundation of China.

References

Ai H, Yan X, Han R (2012) Occurrence and prevalence of seven bee viruses in *Apis mellifera* and *Apis cerana* apiaries in China. J Invertebr Pathol 109(1):160–164.

An J, Huang J, Shao Y et al (2014) The bumblebees of North China (Apidae, *Bombus Latreille*). Zootaxa 3830(1):001–089

Anonymous (1963) Beijing-bee mite control seminar. China Apic (1):28
Cao LF, Zheng HQ, Hu FL et al (2012a) Genetic structure of Chinese *Apis dorsata* population based on microsatellites. Apidologie 43(6):643–651
Cao LF, Zheng HQ, Hu CY et al (2012b) Phylogeography of the giant honeybee *Apis dorsata* (Hymenoptera: Apidae) from China and neighboring Asian Areas. Ann Entomol Soc Am 105(2):298–304
Cao LF, Zheng HQ, Chen X et al (2012c) Morphometric analysis of the giant honeybees in China. J Apic Res 51(3):245–251
Cao LF, Zheng HQ, Pirk CWW et al (2016) High royal jelly-producing honeybees (*Apis mellifera ligustica*) (Hymenoptera: Apidae) in China. J Econ Entomol 109(2):510–514
Chen MH (2005) Analysis of the current state of royal jelly in China mainland. J Bee 25:17–19
Chen RL (2014) Zhejiang: transformation and upgrading of beekeeping industry based on science and technology. China Anim Ind 24:30–31
Chen XX (2016) Epidemiology and treatment of *Nosema ceranae* (Microsporidian) in Western Honey Bee (*Apis mellifera* L.) in Zhejiang Province. Master degree thesis. Zhejiang University
Chen LH (2017) The Chinese beekeeping industry in the "12th five-year". China Anim Ind 10:45–48
Chen YW, Wang CH, An J et al (2000) Susceptibility of the Asian honey bee, *Apis cerana*, to American foulbrood, *Paenibacillus larvae larvae*. J Apic Res 39(3–4):169–175
Chen L, Zhao J, Zhang Y et al (2004) Monitoring of tau-fluvalinate residue in honey. Pest Sci Admin 25(2):8–9
Chen LH, Wu J, Zhang FX et al (2014) Analysis of domestic bee products market of 2013. China Anim Ind 22:34–35
Chen C, Liu ZG, Pan Q et al (2016) Genomic analyses reveal demographic history and temperate adaptation of the newly discovered honeybee subspecies *Apis mellifera sinisxinyuan* n. sp. Mol Biol Evol 33(5):1337–1348
Chen LH, Wu J, Wang JM et al (2017) A survey on toxic bee plants in China. China Anim Ind 8:53–54
China National Commission of Animal Genetic Resources (CNCAGR) (2011) Animal genetic resources in China-Bees. Chinese Agricultural Press, Beijing
Deng Z, Yang W, Yang C et al (2013) Nutritional analysis and evaluation of *Vespa basalis* Smith. Acta Nutrimenta Sin 35(5):514–516
Dietemann V, Zheng HQ, Hepburn C et al (2008) Self-assessment in insects: honeybee queens know their own strength. PLoS One 3(1):e1412
Elke G (2010) American Foulbrood in honeybees and its causative agent, *Paenibacillus larvae*. J Invertebr Pathol 103:S10–S19
Evans GO (1963) The genus *Neocypholaelaps Vitzthum* (Acari: Mesostigmata). J Nat Hist 6(64): 209–230
Fan ZY, Huang WC (1993) Varroa mite. In: Apiculture Editorial Committee (ed) Encyclopedia of Chinese agriculture. China Agriculture Press, Beijing, pp 266–26
Fan QH, Jiang F (2014) The tiny flower fairies *Neocypholaelaps indica* Evans, 1963 (Acari: Ameroseiidae). Syst Appl Acarol 19(2):248–249
Fan ZY, Lai DZ (1991) First report on honey bee "Chalkbrood disease" in China. Apic China (4):23
Fang BB (2015) The review of Chinese bee products market of 2014 and the forecast of 2015. Apic China 66(5):16–17
Fang BB (2016) The review of Chinese bee products market of 2015 and the forecast of 2016. Apic China 67(5):12–13
Feng F (1998) Biology of wax moth and their control. J Bee 18(8):20
Feng Y, Chen X, Ye S et al (2001) The common edible species of wasps in Yunnan and their value as food. For Res 14(5):578–581
Gong YF, Li SG, Zhang QK (1984) Report on breeding sacbrood resistant in Chinese honeybee. China Apicult (3):1–2
Gong HR, Chen XX, Chen YP et al (2016) Evidence of *Apis cerana* Sacbrood virus infection in *Apis mellifera*. Appl Environ Microbiol 82(8):2256–2262

Guo YJ, Li ZY, Tao SB (2017) The current status and prospects of industrialization of wasp resources. China Agric Inf (2):94–96
He WZ, Wu KM, Shao YQ et al (2013) Exotic bumblebee may be a rist for *Bombus* resources in China. Apic China 64(3):38–40
Hu JW, Xi JP, Li XT (2009) Investigation of sacbrood virus in Gangsu province. Apic China 60(2):23–24
Hu FL, Bíliková K, Casabianca H, Daniele G, Espindola FS, Feng M, Guan C, Han B, Kraková TK, Li J-K, Li L, Li X-A, Šimúth J, Wu L-M, Wu Y-Q, Xue X-F, Xue Y-B, Yamaguchi K, Zeng Z-J, Zheng H-Q & Zhou J-H (2017) Standard methods for *Apis mellifera* royal jelly research. J Apic Res. https://doi.org/10.1080/00218839.2017.1286003
Huang SK, Wang JD (2001) Studies on the biology of the greater wax moth (*Galleria melonella* L.) Apic China 52(5):8–11
Huang SK, Ye SZ, Dong J (2006) The infectivity of *Nosema* from *Apis mellifera ligustica* to different aged adult workers of *Apis cerana cerana*. Apic China 58(8):8–10
Huang K, Zheng ZJ, Yang WY (2008) CSBV resistance breeding in Chinese honeybee. J Jiangxi Agric Univ 30(5):883–887
Jia SC (2001) Separating colony for treatment of American foulbrood. Apic China 52(4):20
Jin SH (2008) Studies on Pinghu Italian Bee. Zhejiang University Press, Hangzhou, p 69
Kuang BY, Kuang HO (2007) Protection and utilization of bee genetic diversity. J Bee S1:86–88
Kuang BY, Li YQ, Wang HR et al (1983) Studies on *Apis andreniformis*. Apic China 34(6):10–11
Lai DZ, Fan ZY (1993) Controlling Chalk brood disease by using "Bai-erJin No.2" and "Feng Kang" additives. Apic China 115(2):22–23
Li XT (1993) Honey bee chalk brood. Apicult Technol (4):20–22
Li YG (2015) Solving the dispute caused by interspecific robbing between the Western honeybees and the Eastern honeybees for the sustainable development of beekeeping industry. J Bee 35(4):44–45
Li JZ, Li Z (2013) The hurt to beekeepers: several cases of heavy colony losses in Shaanxi province in 2013. Bee J 33(12):39–41
Li JJ, He SY, Tan K (2005) Advances in research on honey bee's natural enemy-wasp. Bee J 25(8):26–28
Li ZL, Ou X, Hu JJ (2006) Investigation and control of Chinese sacbrood disease. Guangxi J Anim Husb Vet Med 22(6):276–277
Li H, Zhang ZY, Luo LJ et al (2009) Morphology of Yunnan *Apis dorsat*a and *Apis laboriosa*. Apic China 60(4):11–13
Li JL, Qin HR, Wu J et al (2012) The prevalence of parasites and pathogens in Asian honeybees *Apis cerana* in China. PLoS One 7:e47955
Liu ZH (1992) The pathogen of chalk brood disease and its transmission route. Beekeeping Technol (1):47–48
Liu X (1997) Comprehensive methods for Chalkbrood control. Yunnang Agric Technol (1):33–34
Liu CT, Xu ZY, Zhang JS (2001) Investigation and analysis of outbreak of Chinese sacbrood disease. J Bee 21(2):22–23
Liu F, Wang Q, Dai PL et al (2008) Natural stripe of microsporidia of honeybee in China. Chin Bull Entomol 45(6):963–966
Liu ZG, Chen C, Guo HK et al (2016) Mortality in *Apis mellifera* L. colonies in China from 2009 to 2013. Acta Ecol Sin 36(10):3005–3012
Luo YX, Chen LH (2014) The current status of beekeeping with *Apis cerana cerana* in China and the suggestions for the future development. China Anim Ind 24:22–23
Luo QH, Zhou T, Wang Q et al (2011) Identification of *Tropilaelaps* mites (Acari, Laelapidae) infesting *Apis mellifera* in China. Apidologie 42(4):485–498
Luo YX, Zhang XF, Chen HS (1998) Chinese sacbrood disease and its control. Apic China 16(4):13–14
Lv LP, Diao QY, Shi W et al (2014) The apiculture and west district development. J Anhui Agricult Sci 42(16):5217–5219
Mo YG (2003) *Tropilaelaps* mite can survive the winter without the secondary host. J Bee 23(7):26

Moritz R, Härtel S, Neumann P (2005) Global invasions of the western honeybee (*Apis mellifera*) and the consequences for biodiversity. Ecosci 12(3):289–301
Neumann P, Yañez O, Fries I et al (2012) Varroa invasion and virus adaptation. Trends Parasitol 28:353–354
Oldroyd BP, Wongsiri S (2006) Asian honey bees: biology, conservation, and human interaction. Harvard University Press, Cambridge, p 199
Page P, Lin Z, Buawangpong N et al (2016) Social apoptosis in honeybee superorganisms. Sci Rep 6:27210
Pan QZ, Lin JL, Wu XB et al (2013) Research and application of key technique for mechanized production of royal jelly (IV)-design and application of a machine for gathering royal jelly. Acta Agricult Univ Jiangxiensis 35:1266–1271
Peel MC, Finlayson BL, Mcmahon TA (2007) Updated world map of the Köppen-Geiger climate classification. Hydrol Earth Syst Sci 11:1633–1644
Ramadan MF, Ai-Ghamdi A (2012) Bioactive compounds and health-promoting properties of royal jelly: a review. J Funct Foods 4(1):39–52
Rinderer TE, Oldroyd BP, Lekprayoon C et al (1994) Extended survival of the parasitic mite *Tropilaelaps clareae* on adult workers of *Apis mellifera* and *Apis dorsata*. J Apic Res 33:171–174
Ruiz-Matute AI, Weiss M, Sammatora D et al (2010) Carbohydrate composition of high-fructose corn syrups (HFCS) used for bee feeding: effect on honey composition. J Agric Food Chem 58:7317–7322
Shaanxi Institute of Animal Husbandry and Veterinary Medicine (1977) Effectiveness of Sculellaria barbatain curing Chinese sacbrood disease. Shaanxi Agric Sci (11):30
Shen ZZ (1958) Treatment of bee mite. Apic China (1):13–14
Spiewok S (2006) When killers grow old: decrease of fighting ability in honeybee queen ontogeny. In: The IUSSI congress
Su SK, Wang YX, Pan SF et al (2010) A new machine for gathering royal jelly-new impetus for mechanise production of royal jelly. Apic China 61:30
Su XL, Hua QY, Zhao DX (2016) The effect of essential oils of *Truestar Anisetree* against mites in honeybee colonies. J Environ Entomol 38(6):1251–1255
Tan K, Meixner MD, Fuchs S et al (2006) Geographic distribution of the eastern honeybee, *Apis cerana* (Hymenoptera: Apidae), across ecological zones in China: morphological and molecular analyses. Syst Biodivers 4:473–482
Teng GP, Pang ZW (1964) An Ameroseiid mite (*Neocypholaelap sindica* Evans,1963) (Acarina: Ameroseiidae). Acta Entomol Sin 13(5):771–772
Tian SH, Zhao ZL, Zhao HT et al (2014) Study of the population structure of *Apis cerana* based on mitochondrial gene variation. Chin J Appl Entomol 51(2):448–459
Tsigouri A, Menkissogluspiroudi U, Thrasyvoulou AT et al (2000) Determination of fluvalinate residues in beeswax by gas chromatography with electron-capture detection. J AOAC Int 83(5):1225–1228
Vardar-Ünlü G, Silici S, Ünlü M (2008) Composition and in vitro antimicrobial activity of *Populus* buds and poplar-type propolis. World J Microbiol Biotechnol 24:1011–1017
Verma LR, Rana BS, Verma S (1990) Observations on *Apis cerana* colonies surviving from Thai sacbrood virus infestation. Apidologie 21(3):169–174
Wang JD (1981) Biology of Achroiagrisella and control study. Apic China 6:7–12
Wang DC (2010) Erythromycin can treat American foulbrood. J Bee 30(2):34
Wang JD, Zhang WH, Hong M et al (1984) A primary study on the overwintering of small mite. J Fujian Agric Coll 13(4):299–303
Wang JD, Liang Q, Zhang WH, et al (1985) First report of *Malpigamoeba mellificae* on *Apis cerana cerana* and its control. Apic China (4):1–2
Wang DS, Zhang JY, Li XT et al (2005) Study of effect of Chinese herb medicines as bee acricides. J Tradit Chin Vet Med 24(3):14–16
Wang M, Bi J, Wang L et al (2016) Prevalence of four common bee RNA viruses in Eastern bee populations in Yunnan Province, China. J Vet Sci Technol 7(1):284
Wei HZ (1991) "MangPu"—a new miticide. Bull Agric Sci Technol (8):24
Williams P, Huang JX, An JD (2017) Bear wasps of the middle kingdom: a decade of discovering China's bumblebees. Antenna 41(1):21–24

Woyke J (1984) Survival and prophylactic control of *Tropilaelaps clareae* infesting *Apis mellifera* colonies in Afghanistan. Apidologie 15(4):421–434
Wu YR (2000) Fauna sinica. Science Press, Beijing
Wu J, Li J, Li JK (2006) Major honey plants and their utilization in China. Am Bee J 2:153–157
Wu YY, Jia HR, Wang Q et al (2015) Multiple virus infections and the characteristics of chronic bee paralysis virus in diseased honey bees (*Apis mellifera* L.) in China. J Apic Sci 59(2):95–106
Wu L, Du B, Heyden YV et al (2017) Recent advancements in detecting sugar-based adulterants in honey—a challenge. Trends Anal Chem 86:25–38
Xia ZY (1957) Mites found in the Southern of Jiangsu and the western of Zhejiang. Apic China 11:12–15
Yañez O, Zheng HQ, Su XL et al (2015) Potential for virus transfer between the honey bees *Apis mellifera* and *A. cerana*. J Apic Res 54(3):179–191
Yang QH (1965) Biological characteristics of *Varroa jacobsoni* and control in colony. J Appl Entomol 1:42–43
Yang XM (1978) Herbs used for treatment of Chinese sacbrood disease. Apic China (5):23
Yang GH (2005) Harm of introducing the western honeybee *Apis mellifera L.* to the Chinese honeybee *Apiscerana F.* and its ecological impact. Acta Entomol Sin 48(3):401–406
Yang B, Peng GD, Li TB et al (2013) Molecular and phylogenetic characterization of honeybee viruses, *Nosema* microsporidia, protozoan parasites, and parasitic mites in China. Ecol Evol 3(2):298–311
Yu YS, Zhang ZY, Lu HX et al (2011) Control effect of sublimed sulfur paired with acaricide on bee mites. J Anhui Agric Sci 39(13):7676–7677, 7684
Zeng ZJ (2007) Honeybee biology. China Agriculture Press, Beijing, p 47
Zeng ZJ, Wu XB, Zhang F et al (2013) Research and application of key technique for mechanized production of royal jelly (I)-a design of bionic non-grafting larvae ovipositor. Acta Agric Univ Jiangxiensis 35:842–847
Zhang ML (1958) Treatment of bee mite. Apic China (1):15
Zhang W (1965) Some supplement about *Neocypholaelaps indica*. China Beekeeping (4):103
Zhang DL (2012) Investigation of prevalence of sacbrood virus disease in Liaoning province. Apic China 63(4):19–20
Zhang ZY, Li H, Luo LJ et al (2009) Morphology of Xishuangbanna honeybees *Apis florea* and *Apis andreniformis*. Apic China 60(1):8–9, 12
Zhang CP, Zheng HQ, Liu G et al (2011) Development and validation of HPLC method for determination of salicin in poplar buds: application for screening of counterfeit propolis. Food Chem 127:345–350
Zhang CP, Ping S, Wang K et al (2015) A survey of the incidence of poplar tree gum in propolis products on the Chinese retail market. J Apic Res 54(1):30–35
Zhang J, Zhang Y, Han R (2016) The high-throughput production of dsRNA against sacbrood virus for use in the honeybee *Apis cerana*, (Hymenoptera: Apidae). Virus Genes 52(5):698–705
Zhao WZ, Tan K, Zhou DY et al (2014) Phylogeographic analysis of *Apis cerana* populations on Hainan Island and southern mainland China, based on mitochondrial DNA sequences. Apidologie 45:21–33
Zheng HQ, Jin SH, Hu FL et al (2009a) Sustainable multiple queen colonies of honeybees, *Apismellifera ligustica*. J Apic Res 48(4):284–289
Zheng HQ, Jin SH, Hu FL et al (2009b) Maintenance and application of multiple queen colonies in commercial beekeeping. J Apic Res 48(4):290–295
Zheng HQ, Hu FL, Dietemann V (2011) Changes in composition of royal jelly harvested at different times: consequences for quality standards. Apidologie 42:39–47
Zheng H, Gong H, Su X et al (2016) Investigation and analysis of large scale colony losses of *Apis mellifera* in winter in Jinhua, Zhejiang. Apic China 67:20–24
Zhou T (2004) Identification of *Varroa* mites (Acari: Varroidae) infesting *Apis cerana* and *Apis mellifera* in China. Apidologie 35(6):645–654
Zhou T, Feng F, Dong B et al (2001) Studies on the sensitivity of honey bee EFB pathogen to drugs. Chin J Anim Vet Sci 32(3):283–288

Chapter 7
Beekeeping on Taiwan Island

Mei-Chun Lu

Abstract Taiwan Island possesses a very diverse plant fauna that has enriched its nectar sources. Beekeeping started 300 years ago for native eastern honeybee (*Apis ceranae*) cultivation, and the large-scale rearing did not develop until the introduction of western honeybee (*Apis mellifera*) in 1910. The main nectar sources in Taiwan Island are longan (*Dimocarpus longan*), lychee (*Litchi chinensis*), and tea tree (*Camellia sinensis*). Most of the beekeepers have performed migratory beekeeping in pursuance of the nectar sources from south to north of Taiwan for honey collection. Currently, approximately 184,000 hives are kept, that most common honeybee species of the honeybee species is *Apis mellifera*. The annual yield of honey is ranged from 8000–12,000 tons, royal jelly 370–460 tons, beewax 290–464 tons and bee pollen 500 tons. Small-scale beekeeping (<300 hives) is dominant operation, which stands for 90% of beekeepers. Honeybee has provided approximately 80 million USD income for more than 940 families annually as well as ecosystem services. To ensure sustainable beekeeping development, a series of policies has been promulgated to enhance the quality of bee products, such as Chinese National Standard (CNS) for honey and royal jelly, certified and traceable code for bee products and the longan honey championship activity. However, bee diseases and pests have greatly hindered development of apiculture. Parasites and diseases such as *Varroa* mite, American foulbrood, European foulbrood, chalkbrood disease, Nosema disease, and bee viruses are frequently occurred in the apiaries throughout Taiwan Island. Beekeeping industry is also affected by global climate change and the use of pesticides in the region.

Keywords Beekeeping · *Apis mellifera* · Nectar source · Pest · Disease · Bee product · Taiwan

---

M.-C. Lu
Miaoli District Agricultural Research and Extension Station, Council of Agriculture, Executive Yuan, Gongguan, Miaoli, Taiwan
e-mail: lumj@mdais.gov.tw

P. Chantawannakul et al. (eds.), *Asian Beekeeping in the 21st Century*,
https://doi.org/10.1007/978-981-10-8222-1_7

7.1 Introduction

Beekeeping plays an important role in sustainable agriculture and the production of bee products. Due to Taiwan's warm and humid climate, multiple and diverse nectar sources have provided a suitable environment for honeybee habitats to widely propagated. In addition to native eastern honeybee (*Apis cerana*), the western honeybee (*Apis mellifera*) was introduced and cultivated in 1910 (Inamura 1914). Since *A. mellifera* produce high honey yield and it can be easily kept, the honeybee has been widely spread all over the island for beekeeping industry. The main goal of beekeeping is the production of honey, royal jelly, beewax, and pollen. The application of honeybee as pollinator is a subsidiary utilization. Beekeepers propagate the bee colonies in early spring for honey collection. The honey harvest season is from February to June. The royal jelly production, new queen breeding and grafting are performed during September to October. In the early autumn, the beekeepers perform pollen collection, the dominant pollen source is tea tree (*Camellia sinensis)*. Beekeepers feed their bee colonies with syrup and pollen cakes as a subsidiary nutrition when the nectar sources are in shortage (Lu 2016a, b).

7.2 Historical Record of Beekeeping

According to the record of Inamura (1912), the exploitation of wild bees in Taiwan Island has been traced back to around 300 years ago. The immigrants, from mainland China to Taiwan Island, have brought the beekeeping techniques to this island. They caught the wild bees (*A. cerana*) from rock or tree holes and reared them in simple beehives made of bamboo slices, mud, and cement (An et al. 2004). The wild bees had a black-brown appearance, clustered with thousands of workers in each colony, and they often swarmed (An et al. 2004). Since that time, the modern beekeeping techniques have gradually been established and expanded throughout the island (An et al. 2004; Sung et al. 2006). Modern beekeeping in Taiwan started in 1910, for which the western honeybee (*Apis mellifera ligustica*) was introduced from Japan (Inamura 1912, 1914). The introduction of this bee species progressed for the next 20 years (An et al. 2004). Due to favored geographic locations and climates, *A. mellifera* beekeeping operation in Taiwan Island has been rapidly expanded. In order to conduct a breeding program, honeybee subspecies were introduced by Miaoli District Agricultural Research and Extension Station (MDARES, former Sericulture and Apiculture Improvement Station). According to the record, the Caucasian honeybees were introduced from Hawaii in 1991. Besides, Italian gold subspecies was introduced from Japan in 1993, Italy subspecies, Carnica subspecies, and Cape species were also introduced from Italy in 1997 (Wu 2002; Sung et al. 2006).

The professional and large-scale beekeeping was developed in 1975 due to the exportation market of royal jelly (An 1990). Fresh royal jelly cost 120 USD per kilogram, and the total product value was over ten million USD (An 1990). The high profits encouraged the investment and extended the scales of apiaries. The beekeepers reached to a maximum of 2,000 families, and there were a total 260,000 beehives in 1976 (An 1990). Royal jelly quickly became the main product made from bees instead of honey, and the business model of apiculture was established at that time (An 1990). However, the competition of royal jelly exported from mainland China occurred, due to the cheaper manpower. From 1978 to 1987, the price and productivity of royal jelly fell dramatically (An 1990). Consequently, beekeepers and beehives decreased. In 1988, there were only 711 beekeepers and 120,000 beehives in Taiwan Island (An 1990). Some of the beekeepers went abroad, such as Thailand, to perform beekeeping and then sold the bee products back to Taiwan. Recently, apiculture industry has been expanded. Beekeeper families of 943 produced 8000–12,000 tons of honey, and 370–460 tons of royal jelly, 290–464 tons of beewax, and bee pollen about 500 tons (personal communication with Taiwan Beekeepers' Association) (Council of Agriculture, 2017a). Furthermore, the safety of bee products has been paid more attention by government authority and consumers. More strict regulations were enforced to prevent adulteration and pesticide residues of honey. The growing demand for honeybee as pollinators, for greenhouse crops, has increased. The application of honeybees to pollinate cucumbers, melons, and fruits in greenhouse has also opened a new era for crop production.

7.3 Beekeeping Industry on Taiwan Island

Taiwan is the island located in the southeast coast of the Asia continent. There is a central mountain pass across in the middle of the island. This creates diverse climate: tropical, subtropical, and temperate climates. The main nectar sources are longan (*Dimocarpus longan*) and lychee (*Litchi chinensis*), which are distributed in the South and central part of Taiwan (Fig. 7.1). Due to the climate differences, flower blooming of longan and lychee occurs from south to north area, in February to May annually. Consequently, the beekeepers, like nomads, moved their beehives from southern to central Taiwan for honey collection during the blooming season. A family is the traditional basic unit of beekeepers because of labor costs and technical demand. The age of beekeepers was between 40 and 70 years old, which represented 70% of the population (Fig. 7.2). Twelve percent of beekeepers were more than 70 years old. This reflects an aging problem among beekeepers (Lu 2016a). However, the aging problem in traditional agricultural industry was common, and an alternative solution, such as developing low labor-cost automatic devices was necessary. The total amount of beehives in 2016 was 184,254 (Fig. 7.3) (Council of Agriculture, 2017a, b). About 65.7% of beekeepers owned 100–199 beehives, 3% owned 1–100 beehives, and 22% owned 200–299 beehives (Fig. 7.4). Beekeepers who owned more than 300 beehives were only 9.3% (Fig. 7.4). The result reflected that small-scale management of apiaries was performed in Taiwan Island (Lu 2016b). On the other hand, a total value of bee products was calculated about 80 million USD (Fig. 7.5). However, the pollinator demand, especially for cultivated

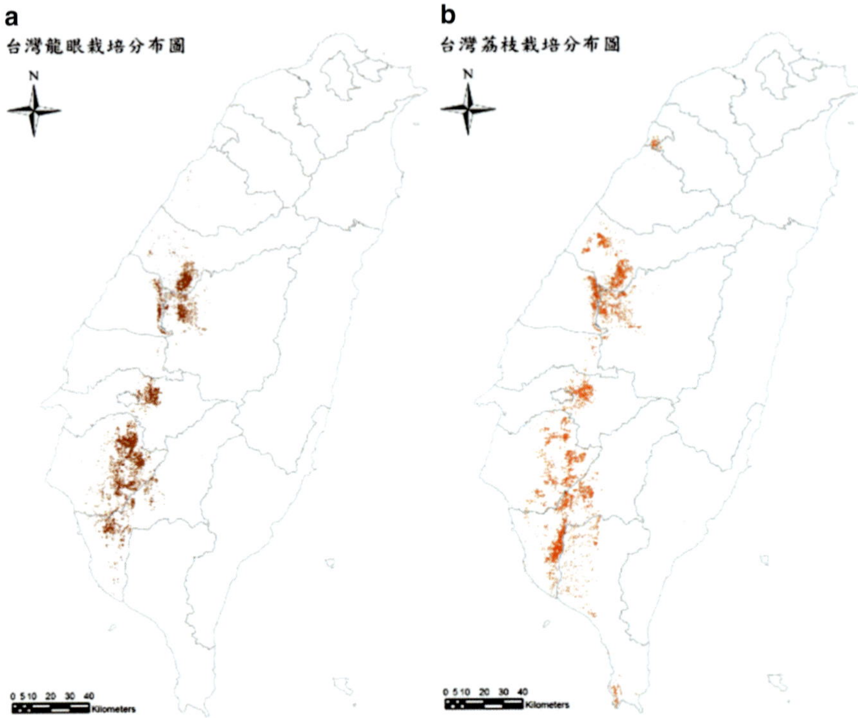

Fig. 7.1 Distribution (showed as the red spots) of longan and lychee cultivation in Taiwan Island. (**a**) Longan, (**b**) Lychee

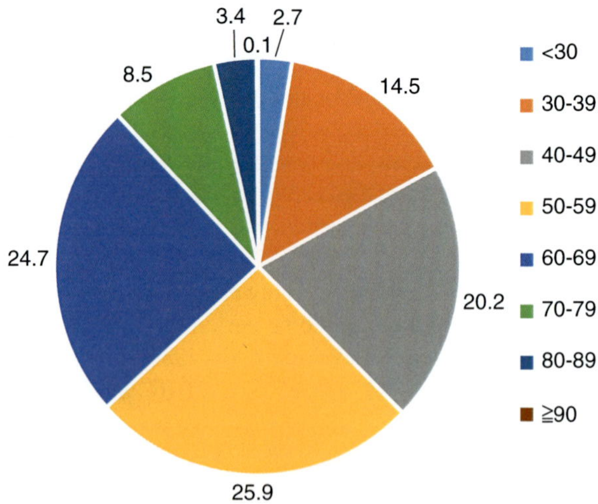

Fig. 7.2 The age distribution for beekeepers in Taiwan Island. Data was statistically investigated by Council of Agriculture (2017a, b) (http://agrpmg.afa.gov.tw/agr-Sed/agrJsp/login.jsp)

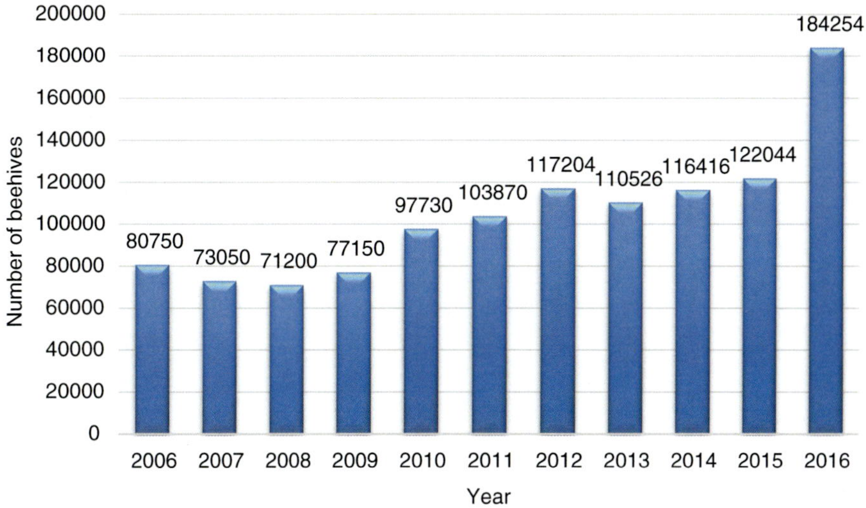

Fig. 7.3 Number of beehives during 2006–2016. Data were from the Agriculture Statistics Yearbook, 2017

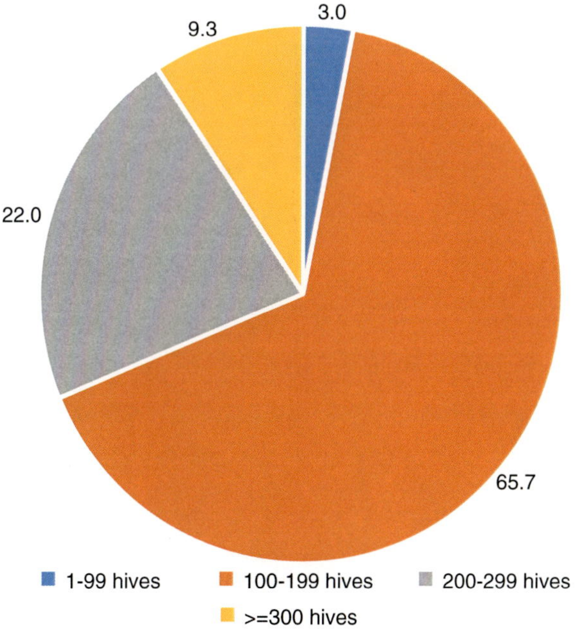

Fig. 7.4 Percentage of beehives per beekeeper in Taiwan Island, 2017 (http://agrpmg.afa.gov.tw/agr-Sed/agrJsp/login.jsp)

crops (bitter melons, cucumbers, strawberries, and melons) in the greenhouse (Fig. 7.6), was increased due to the high cost of labor and high quality products for market. Honey export as well as import ranged from 2,500 to 5,000 tons per year (Council of Agriculture, 2017b). The imported honey was mainly from Thailand,

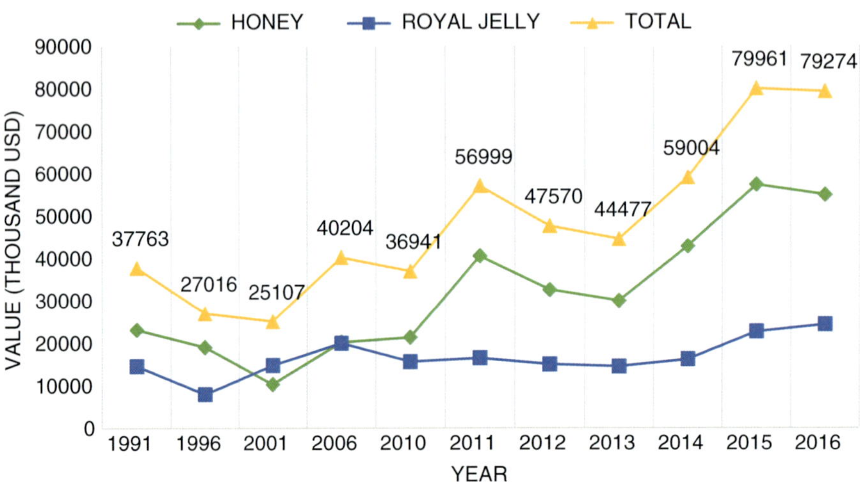

Fig. 7.5 Value of bee products. Data were from the Agriculture Statistics Yearbook, 2017

Fig. 7.6 Plants depending on bee pollination. (**a**) Bitter melon; (**b**) Strawberry; (**c**) Tomato; (**d**) High productivity of fruits of bitter melon after bee pollination

and exported honey was transported to the United States. The exportation of royal jelly has recently declined, and only 11–16 tons in recent 5 years (Council of Agriculture, 2017b).

7.3.1 Regulation and Management of Bee Products

In order to enhance the bee product quality, a series of regulation policies for bee products were introduced in Taiwan Island. The Chinese National Standard (CNS) for honey, which states the minimal requirement for honey, in which the content comprises water, sucrose, reducing sugar, insoluble material, diastase activity, and hydroxymethylfurfural (HMF) (Table 7.1) (CNS 1305, http://www.cnsonline.com.tw/). The Chinese National Standard (CNS) for royal jelly established in 2014 in which the content comprises water, ash, acid, starch, crude protein, and 10-hydroxy-20-decenoic acid (10-HDA) (Table7.2) (CNS15474, http://www.cnsonline.com.tw/). Furthermore, in order to enhance the Taiwan's honey quality the "Logan Honey Competition" was held annually during the harvesting season (in May). It is the most important honey festival for all beekeepers in Taiwan Island. The award-winning longan honey should meet the Chinese National Standard (CNS1305) of longan honey and should be antibiotic and pesticide free with a good aroma, taste, and color. To date, more than 400 farmers per year have participated in the activity and their profits have increased more than 40% for honey sales. The honey and royal jelly that are certified possessed the traceable codes (Fig. 7.7). Consumers who demand to "know where

Table 7.1 Chinese National Standard for honey and longan honey

Constituents	Honey	Longan honey
Water (%)	<20	
Sucrose (%)	<5	<2
Sugar (%)	>60	>70
Insoluble (%)	<0.1	
Acid (meq H^+/1000 g)	<50	<30
Diastase (Schade unit)	>8	
HMG (mg/kg)	<40	<30

http://www.cnsonline.com.tw/

Table 7.2 Chinese National Standard for royal jelly

Constituents	Amount
Water (%)	63–68
Ash (%)	>1.5
Acid(meq H^+/1000 g)	32–53
Starch	0
Crude protein (%)	11–15.5
10-Hydroxy-20-decenoic acid (10-HAD) (%)	>1.6

http://www.cnsonline.com.tw/

Fig. 7.7 The certification mark for the bee product. "H" means honey following the serial number. The product information could be found in website of Taiwan Beekeepers' Association (http://www.bee.org.tw/) by entering the serial number

Table 7.3 Amount of certified honey and royal jelly during 2009–2016

	Honey		Royal jelly	
Year	Weight (kg)	Certified honey tag (no.)	Weight (kg)	Certified royal jelly tag (no.)
2009	100,298	98,900	11,660	11,750
2010	197,923	176,040	11,145	22,470
2011	361,329	300,712	7680	15,360
2012	359,885	334,190	9223	18,410
2013	223,168	301,620	8515	17,980
2014	739,606	763,356	6187	11,040
2015	635,678	638,556	5808	11,600
2016	551,578	599,798	4836	6670

Data from Taiwan Beekeeper's Association, 2017

their products come from" could be satisfied by these codes. In addition, producers would be asserted for excellent quality control of bee products. On the website of Taiwan Beekeeper's Association (http://www.bee.org.tw/), information about producers could be found by entering the traceable code. The certified honey and royal jelly from 2009 to 2016 are shown in Table 7.3. Since 2015, the bee products (honey and royal jelly) have been integrated into the "Taiwan Agricultural Products Production Traceable System" (https://qrc.afa.gov.tw/), which provided the QR code for each agricultural product.

Furthermore, in order to provide the re-education chances to the new farmers, the "Farmer's Academy" (http://academy.coa.gov.tw/list.php?id=28) for beekeepers has been held every year in Miaoli District Agricultural Research and Extension Station. The re-education system delivered new knowledge and provided basic techniques for young beekeepers, and average 250–300 well-trained beekeepers were achieved per year.

7.4 Nectar Resources

Geographically, Taiwan Island is located in the oriental region. Together with the influence of seasonal wind and sea currents, the island is warm and humid all year around. Furthermore, the interaction between the Philippine sea plate and Eurasia plate has built up high and steep mountain in the center of the island. Combining all the above conditions, Taiwan possesses a very diverse plant fauna including tropical, subtropical, and temperate alpine areas. Also, it possesses an enriched nectar sources. There are 54 families, 134 species of plants that produce sources of nectar in Taiwan Island. Among them, there are 8 monocotyledones for 17 species, and 46 dicotyledones for 117 species (An et al. 2004). The most important ones are longan, lychee, and tea. There were also wild or sporadic species, such as *Bidens pilosa*, *Schefflera octophylla*, *Brassica campestris*, *Cucumis* sp., *Sapium sebiferum*, *Astragalus* sinicus, *Citrus* sp., *Vitex negundo*, and *Ternstroemia gymnanthera* (An et al. 2004). The following is the distribution and blooming for each of the main nectar sources in Taiwan Island.

7.4.1 Longan

Longan (*Dimocarpus longan* Lour) is mainly located in the central and southern regions, including Taichung, Kaohsiung, Nantou, and Chiayi. The cultivation area is about 11,000 ha (Council of Agriculture, 2017a). The blooming season for longan is from March to April. Therefore, the beekeepers move their bee hives along the longan blooming route for the collection of longan honey. The longan honey is the highest quality honey with an amber color and aromatic aroma. It is also the most popular honey in Taiwan.

7.4.2 Lychee

Lychee (*Litchi chinensis*) is the Sapindaceae family, litchi genus. The distribution of lychee is mainly in the South and central Taiwan, including Kaohsiung, Taichung, and Nantou. Lychee blooms from mid-January to March. Lychee honey has a slightly fragrant aroma and has a pale yellow color.

7.4.3 Citrus

Citrus trees are widely distributed across Taiwan Island. This variety of species contains citrus, tankan, oranges, pomelo, lemon, grapefruit, kumquat, and lime. The production area was around 25,000 ha (Council of Agriculture, 2017a). However, the citrus honey served as the regional bee product. Citrus honey tends to crystallize and looks slightly yellow in color.

7.4.4 *Bidens*

Biden (*Biden pilosa* var. radiata), is a wild plant. It was introduced as nectar plants from Ryukyu, Japan in 1976 (Huang et al. 2012), and later distributed all over the Taiwan island. The Biden honey has a light fragrant and a pale yellow color.

7.4.5 *Tea Tree*

The tea tree (*Camellia sinensis*) is widely cultivated in central Taiwan in the mountainous regions. It is the most important pollen source for honeybee (Lu 2016b). However, risk of pesticide contamination in pollens derived from tea tree cultivation is being concern.

7.5 Bee Pests

Pests and diseases of bees weaken the colonies, decrease the productivity of bee products, and can cause colony loss. Bee diseases and pests also decrease the quality of bee products and the pollination capacity of honeybees. In Taiwan, bee colonies also suffer the bee pests and diseases, such as *Varroa* mite, American foulbrood, European foulbrood, chalkbrood disease, and Nosema disease (Chen and Chen 2007; Lu 2016b). Among them, *Varroa* mite, chalkbrood disease, and American foulbrood frequently occur and are difficult to be controlled.

7.5.1 Varroa *Mite*

Varroa mite (*Varroa destructor*) fed by body fluid of bees is an ectoparasitic pest of honeybee (*A. mellifera*) (FAO 2006; Nazzi and Le Conte 2016). This is the most serious pest in the apiculture industry. Occurrences of *Varroa* mite, result in a seriously damaged bee population (Nazzi and Le Conte 2016; Rosenkranz et al. 2010). The symptoms of *Varroa* mite parasitism include weight loss, deformed wings or abdomen, mobility (flight, feeding) deteriorates, short life, and larva death (Nazzi and Le Conte 2016). The mites also can decrease the production of royal jelly (Rosenkranz et al. 2010). Furthermore, *Varroa* mites are proven to be the transmitters of bee viruses (McMenamin and Genersch 2015) and show synergistic interaction with nosema, chalkbrood, and American foulbrood diseases to cause colony loss (DeRyche et al. 2002; Evison 2015; Bahreini and Currie 2015).

The control methods of mites are classified roughly into chemical and non-chemical methods (FAO 2006; Rosenkranz et al. 2010; Lu et al. 2015). The chemical method is to apply synthetic acaricides, organic acids, and essential oils. The non-chemical methods would be operated by beehive equipped with the wire screen bottom boards or the removal of useless capped drone combs (Chen and Chen 2007; Lu et al. 2015). Tau-fluvalinate is the only legal acaricide for beekeeping in Taiwan (TACRI 2016). However, pesticide resistance in some apiaries decreases its effect. The organic acids, such as formic acid and oxalic acid, are the alternative choices. Formic acid is dripped on the rock wool. Beekeepers apply formic acid to the top of beehives for three consecutive days per 6 day's interval in three cycles. The chemical method using formic acid displays 65–75% efficacy against *Varroa* control (Lu et al. 2015). Alternatively, spraying the oxalic acid syrup (3–4% v/v) to avoid the larva area was recommended by Chen and Chen (2008). Because mites preferentially reproduce in drone brood (Wantuch and Tarpy 2009), a "drone-brood trapping" by inserting drone comb to beehives were recommended during the non-mating season. In daily management of apiaries, the removal of parasitic drone combs will effectively decrease the mite population. To sum up, the integrated control strategy to control *Varroa* mites was applied during beekeeping, and the less chemical treatments were advocated.

7.5.2 Hornets

Hornets are natural predator of the honeybees. There are seven species of hornets in Taiwan, including *Vespa affinis, Vespa analis, Vespa basalis, Vespa mandarinia nobilis, Vespa ducalis pseudosoror, Vespa wileman, and Vespa velutina flavitarsus* (An 2015). A nesting hornet builds as a bowl-like structure with 6–10 central hexagons. Hornets create their nests from April to June. Eggs are hatched 40 days after being laid (An 2015). During September to November, the population gradually grows and reaches its maximum (An 2015). Hornets cause great damages to the colonies of honeybees. The elimination method was frequently adopted by using capture nets. The captured hornet was then immersed in alcohol to be used as a traditional Chinese medicinal liquor. Alternatively, the trapping method is also recommended and is prepared as the following: a plastic bottle with cross-cut of 1.2–1.5 cm in diameter in the lower portion as a trapping unit. A bait mixture is composed of pineapple (150 g), sugar (10 g), yeast extract (1 g), detergent (1 g), and water (53 g) (An 2005). The trapping units were then placed nearby the beehives in the apiaries.

7.6 Bee Diseases

7.6.1 American Foulbrood

American foulbrood (AFB) is caused by the spore forming bacteria *Paenibacillus larvae* (FAO 2006; Ebelling et al. 2016; Chen and Chen, 2007). The infected dead larvae produce foul odor. They die in capped cells, and the small holes could be observed on the surfaces of cells (FAO 2006). In Taiwan, American foulbrood outbreaks often occur since the pathogenic bacteria can spread quickly via contaminated food (e.g., pollen) across the beehives and the bacterial spores are difficult to eradicate (Chen and Chen 2007). Furthermore, beekeeping practice also promotes the disease transmission among honeybee colonies. No bactericide is approved in Taiwan to control AFB. The disease control measure including renewing the old comb, sanitation of beehives, removal of infected hives, and clean bee food, should be done to prevent the disease transmission and outbreak of AFB.

7.6.2 European Foulbrood

European foulbrood (EFB) is caused by *Melissococcus pluton*, which infected the larva of honeybee (FAO 2006; Forsgren 2010). However, it is not considered as a serious bacterial disease in Taiwan. It occurs frequently in spring and autumn (Chen and Chen 2007). They also can spread via contaminated foods. The infected larvae are frequently removed by worker bees. Sometimes, the disease weakens the colonies but does not destroy the entire colony. The recommended method for controlling EFB is focusing on management and sanitation practices.

7.6.3 Nosema Disease

Nosema disease is caused by *Nosema ceranae* (FAO 2006; Higes et al. 2013). The infected bees crawl on the ground and showing bloated abdomen with disconnected wings. Nosema disease is also transmitted by contamination bee food (Higes et al. 2013). Nosema shortens the bee longevity and lower the colony productivity. The disease often occurs in spring and autumn in Taiwan. The management practices, such as food supply, sanitation, and removal of old combs, are recommended (FAO 2006; Chen and Chen 2007).

7.6.4 Chalkbrood Disease

Chalkbrood disease is caused by the fungus *Ascosphaera apis* (FAO 2006). In Taiwan, the chalkbrood disease season is from May to June and September to November (Chen and Chen 2007). The infection weakens the bee population, however rarely causes the colony loss. The fungal pathogen can be spread by adult bees, *Varroa* mites, and contaminated beekeeping instruments (FAO 2006). It is advised to disinfect the combs by burning and the hives are sterilized by 500 ppm of NaOCl (Chen and Chen 2007). In addition, maintaining the strong bee population is important for combating chalkbrood disease.

7.6.5 Bee Virus

According to the survey, there were at least 6 bee viruses prevalence in Taiwan (Wu et al. 2002; Lu et al. 2010, 2014). Deformed wing virus was the most prevalent virus found in adult *A. mellifera* (22.6%). The existence of *Varroa destructor* virus 1 (VDV1) (18.3%), Kashmir bee virus (KBV) (7.8%), Kakugo virus (7.8%), black queen cell virus (BQCV) (4.9%), and sacbrood virus (SBV) (3.7%) were also detected in bee colonies (Lu et al. 2014). In some cases, the Nosema disease and *Varroa* mites were simultaneously detected in the same colonies. In this case, the infected bee colonies should be isolated and destroyed. The beehives could be sterilized and fumigated by sterilization agent after the bee removal. Replacing a new queen is also recommended (Lu et al. 2010).

7.6.6 Management Strategies

Proper disinfection is necessary when a disease occurs in a beehive. Keeping the apiaries and water supply clean is essential. Dead bees or discarded combs should be removed or buried to prevent disease reoccurrence. Bee colonies are best located in dry, airy, and sunny areas. This would decrease disease transmission. The chosen bee food should be fresh and clean.

7.7 Future Perspective

Beekeeping has been developed for over 300 years in Taiwan Island. The beekeeping industry is closely linked to crop production. Honeybees are praised as the "Wing of Agriculture." Due to abundant nectar sources, beekeeping is prosperous all around the island and creates profit. In order to enhance the quality and quantity of bee products, effective policy and regulation such as certified mark and traceability codes should be put in place. Over the past 5 years, honey production, especially lychee and longan honeys, has been greatly influenced by climate change including flooding, drought, low temperature during flower initiation, and late blooming of the main nectar sources. Bee breeding program for stress-tolerance traits of honeybee will be necessary. *V. destructor* is the most threatening pest in apiculture industry and often synchronized or is transmitted with other bee diseases. New acaricides or non-chemical approaches should be explored and applied because the tau-fluvalinate-resistance mites have been found in Taiwan apiaries. Furthermore, the pesticide threat during blooming seasons poses great risk for bee colonies.

References

An JK (1990) Development of the beekeeping industry in Taiwan in early twenty centuries. Chung-Hsing Univ Entomol Soc 23:63–70. (in Chinese)
An JK (2005) Prevention the damage of hornet. Agric Ext MDARES 30:13–16. (in Chinese)
An JK (2015) Dance with hornets. Independent Author Press, Taiwan, p 234. (in Chinese)
An JK, Ho KK, Chen YW (2004) Apiology. Huasianyuan, Taipei, p 524. (in Chinese)
Bahreini R, Currie RW (2015) The influence of *Nosema* (Microspora: Nosematidae) infection on honeybee (Hymenoptera: Apidae) defense against *Varroa destructor* (Mesostigmata: Varroidae). J Invertebr Pathol 132:57–65
Chen YW, Chen PL (2007) Bee pests and diseases in Taiwan. Bureau of Animal and Plant Health Inspection and Quarantine Press, Taiwan, p 43. (in Chinese)
Chen YW, Chen PL (2008) The control of *Varroa destructor* using oxalic acid syrup in brood-right honeybee colonies. Formosan Entomol 28:31–41. (in Chinese)
Council of Agriculture (2017a) Agricultural statistics yearbook. Council of Agriculture, Taipei, p 393. (in Chinese)
Council of Agriculture (2017b) Agricultural trade statistics of the Republic of China. Council of Agriculture, Taipei, p 120. (in Chinese)
De Ryche PH, Joubert JJ, Hoseinian SH et al (2002) The possible role of *Varroa destructor* in the spreading of American foulbrood among apiaries. Exp Appl Acarol 27:313–318
Ebelling J, Knispel H, Hertlein G et al (2016) Biology of *Paenibacillus larvae*, a deadly pathogen of honey bee larvae. Appl Microbiol Biotechnol 100(17):7387–7395. https://doi.org/10.1007/s00253-016-7716-0
Evison SEF (2015) Chalkbrood: epidemiological perspectives from the host-parasite relationship. Curr Opin Insect Sci 10:65–70. https://doi.org/10.1016/j.cois.2015.04.015
FAO (2006) Honeybee diseases and pests: a practical guide. Food and Agriculture Organization of the United Nations, Rome, 42 pp
Forsgren E (2010) European foulbrood in honey bees. J Invertebr Pathol 103(Suppl 1):S5–S9. https://doi.org/10.1016/j.jip.2009.06.016
Higes M, Meana A, Bartolome C et al (2013) *Nosema ceranae* (Microsporidia), a controversial 21st century honeybee pathogen. Environ Microbiol Rep 5(1):17–29. https://doi.org/10.1111/1758-2229.12024
Huang YL, Chen SJ, Wen-Yuan K (2012) Floral biology of *Bidens pilosa* var. *radiata*, an invasive plant in Taiwan. Bot Stud 53:501–507
Inamura S (1912) Import of honeybee. Taiwan Agric Newspaper 68:645–647
Inamura S (1914) Beekeeping business of Taiwan. Taiwan Agric Newspaper 89:397–399
Lu MC (2016a) Current status and future perspective of apiculture industry in Taiwan. In: Abstracts of Apimondia Philippines 2016 symposium and workshop on indigenous bees, Tagaytay City, p 56
Lu MC (2016b) Introduction of the development and challenge of apiculture industry in Taiwan. Agric Policy Rev 290:72–75. (in Chinese)
Lu MC, Wu HH, How FW (2010) Detection of bee viruses in Taiwan. Agric Policy Rev 217:56–58. (in Chinese)
Lu MC, Wu HH, Lo YC (2014) Distribution and prevention strategy of honeybee (*Apis mellifera* L.) viruses in Taiwan. In: Proceedings of the symposium on the agriculture technology, p 151. (in Chinese)
Lu MC, Chen CC, Liao CH, Yan CC (2015) Evaluation of the fluvalinate, amitraz, and formic acid efficacy in control of *Varroa destructor*. In: The 44th Apimondia international apicultural congress, Daejeon, 15–20 Sept 2015
McMenamin JA, Genersch E (2015) Honeybee colony losses and associated viruses. Curr Opin Insect Sci 8:121–129. https://doi.org/10.1016/j.cois.2015.01.015
Nazzi F, Le Conte Y (2016) Ecology of *Varroa destructor*, the major ectoparasite of the western honeybee, *Apis mellifera*. Annu Rev Entomol 61:417–432. https://doi.org/10.1146/annurev-ento-010715-023731

Rosenkranz P, Aummeier P, Ziegelmann B (2010) Biology and control of *Varroa destructor*. J Invertebr Pathol Suppl 1:S96–119. https://doi.org/10.1016/j.jip.2009.07.016

Sung IH, Chiang CH, Ho KK et al (2006) A review on history of beekeeping in Taiwan and the development of its research works (Hymenopteera, Apidai, Apinae). In: Proceedings of the symposium on the honeybees and bee products, Taiwan Insect Association Press, Taiwan, p 230. (in Chinese)

Taiwan Agricultural Chemicals and Toxic Substances Research Institute (TACRI) (2016) Plant protection manual, p 93 (in Chinese)

Wantuch HA, Tarpy DR (2009) Removal of drone brood from *Apis mellifera* (Hymenoptera: Apidae) colonies to control *Varroa destructor* (Acari: Varroidae) and retain adult drones. J Econ Entomol 102(6):2033–2040

Wu DJ (2002) Apiculture research in Taiwan: past and future. Sci Agric 50:189–193. (in Chinese)

Wu CY, Chen YY, Jiang JH et al (2002) Honeybee picorna-like virus in Taiwan. In: Proceedings of the symposium of Taiwan insect, vol 4, pp 79–92. (in Chinese)

Chapter 8
Beekeeping in Korea: Past, Present, and Future Challenges

Chuleui Jung and Myeong-lyeol Lee

Abstract The history of beekeeping in Korea dates back 2000 years to the earlier kingdom of Gogurye. In the present day, the native honeybee *Apis cerana* Fabr. is mostly kept in log hives in the deep mountains, whereas the introduced *A. mellifera* L. occupies habitats at the interface of mountains and open land. The beekeeping industry has drastically increased to two million hives in the early 2000s, and honey production is now 30,000 tons per year. Even with diverse floral composition, honey production relies highly on the black locust *Robinia pseudoacacia*. The mites *Varroa* (Korean haplotype) and *Tropilaelaps* provide challenges in mite pest management for *Apis mellifera.* A recent outbreak of Sacbrood Virus has threatened *A. cerana* beekeeping. Challenges for the beekeeping industry in Korea include invasion of new pests, such as the yellow legged hornet (*Vespa velutina*) and small hive beetle (*Aethina tumida*), and degradation of black locust viability.

Keywords Rapid development · Korean haplotype varroa · *Tropilaelaps* · IT technology · Quality control

8.1 Introduction

Korea is located in the northeast of Asia at a latitude of N 33–43 and longitude of E 124–132, neighboring China, Russia, and Japan. Two species of honeybees are currently found in Korea: the native Asiatic honeybee, *Apis cerana* (native honeybee = *tobong or jaerae-ggulbeol*), and the European honeybee, *A. mellifera,* which is introduced. The native honeybees are kept in log hives or smaller tetragon woodbox system, whereas the European honeybee are mostly housed in standard Langstrogh hives (Fig. 8.1). Most of the hive boxes are made of wood, with a few

C. Jung (✉) • M.-l. Lee
Department of Sericulture and Apiculture, National Institute of Agricultural Science, RDA, Wanju, JB, Korea
e-mail: cjung@andong.ac.kr

P. Chantawannakul et al. (eds.), *Asian Beekeeping in the 21st Century*,
https://doi.org/10.1007/978-981-10-8222-1_8

Fig. 8.1 Typical hive arrangement of *Apis cerana* (upper) and *Apis mellifera* (lower)

being made of styrofoam. Beehives of *A. cerana* are usually kept deep in the mountains, whereas those of *A. mellifera* are usually kept on the outskirts of the mountainous regions, including agricultural fields. The Korean population of *A. cerana* shares characteristics with the Chinese population of *A. cerana cerana* (Lee and Choi 1986). Many subspecies of the European honeybee, *A. mellifera*, have been introduced in the modern apiculture history of Korea. In the early 1900s, black strain European honeybees were introduced, followed by yellow strain European honeybees, *A. m. ligustica,* in the 1960s and 1970s. During the 1980s and 1990s, *Apis mellifera mellifera and A. m. canica* were introduced.

8.2 Historical Records and Recent Development of Beekeeping

8.2.1 Early History of Beekeeping

The early history of beekeeping in Korea goes back 2000 years to the three-kingdom period. Historical records show that the first honeybees were transferred from India through China to Goguryeo, when Great King Dongmyeongsung established the country between 37 BC and 19 BC (Lew 1988). We consider this to be the time of some technology transfer or some honeybee strain movement along with the propagation of Buddhism from India through China into old Korean kingdoms, but not the transfer of the honeybee species itself (Lew 1988; Ryu et al. 2007; Jung 2014). This record is much earlier than the published record of Crane (1995). According to Crane, the earliest record of beekeeping in box hives was from China in the Han dynasty, about 25–150 AD. Beekeeping knowledge and technology was transferred from the Baekjae Kingdom of Korea to Japan in the mid-seventh century (Japanese chronicle, Seogi). The various history records illustrated that beekeeping was widely practiced and honey was considered to be one of the three miraculous medicines along with ginseng and deer antlers (Dongeui bogam; traditional, SamKuk Sagi; the chronicle of three kingdoms). During the Goryeo (Korea = *Corea*, named after Goryeo by the ginseng merchandizers) and Chosun dynasties, beekeeping was widely adopted in the country, which is reflected by diverse records in productions, stories, and literature (Lew 1988, 1991). The basic production system and management practices for *A. cerana* have remained unchanged. In 1685, there was a historic record of Korean beekeeping by Ik Lee in his book "Sungho-Saseol," using traditional Chinese characters. He recognized that one bee king, who was not a working bee, ruled the colony. He also observed polymorphism and polyethism in the honeybee colony. He also described 15 kinds of honeybee pests as well as the importance of humidity, wind, and other environmental conditions for honeybee health. In the late eighteenth century, another Korean publication on beekeeping by Heonanjeon Lee was produced; she was from the Gyoungbuk area. This implies that beekeeping of *A. cerana* was widely practiced and the knowledge and technology were easily available to ordinary people, even women, who did not have easy access to publications with Chinese characters.

8.2.2 Beginning of Modern Beekeeping

Modern beekeeping began with the invention of three components: a movable frame, beewax comb, and centrifugal extraction (Crane 1992). Lorenzo L. Langstroth found that a distance of 8 mm between the hive and comb was essential. In 1851, he invented in the USA the movable frame comb, which is easy to lift from the top of the hive. Johannes Mehring in Germany produced the first matrix of beewax comb foundation in 1857. Franz von Hrsuchka in Austria developed the honey extractor by rotating centrifugal force in 1865. Modern beekeeping was established in the Western world in the second half of the nineteenth century. However, in Korea, full adoption of the Western science and technology innovation was delayed by political and social issues. Modern Korean beekeeping only started in the early 1900s. There is debate on the starting time of modern beekeeping and the first introduction of the European honeybee, *A. mellifera*, to Korea. Germany was involved with the beginning of modern beekeeping in Korea. During the 1910s, there were increased activities of beekeeping development, education, and publications. The first beekeeping textbook, Experimental Beekeeping (*Sil-heom-yangbong*), was published by Sin-Young Yoon in 1917 (Fig. 8.2). He had received higher education in Germany from the 1890s to early 1900s. When he returned to Korea, he brought honeybees and

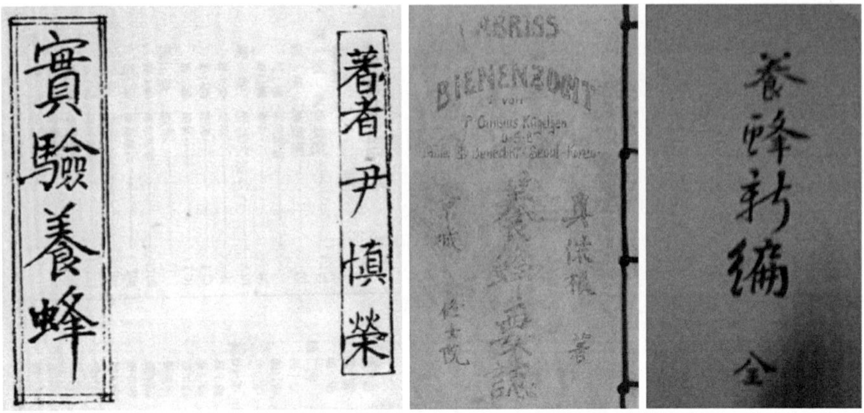

Fig. 8.2 The first modern beekeeping textbook (*Sil-heom-yangbong*) written by Korean Sin-Young Yoon in 1917 from Jung-Ang-Seo-Gwan, with six chapters in 145 pages (left), and the other two early Korean publications by P. C. Kugelgen, a German missionary (middle), and *Yangbong-sinpyeon* by Keunyoung Lee in 1918 from Daegu Korea (right, from Lew 1988, Jung 2014)

knowledge of apiculture technology. In 1918, the booklet titled "Abriss Bienenzucht" (양봉요지 = Abstract of Beekeeping) was published by Von P. Canisius Kugelgen underlined by O.S.B. Abbey Benedict (Jung 2014). While working on religious missionary jobs, Von P. Canisius Kugelgen reared honeybees and taught beekeeping to his followers. In the booklet, there is no mention of the honeybee species, but it is assumed to be *A. mellifera*.

8.2.3 Current Status of Beekeeping in Korea

As noted earlier, there are two species of honeybees in Korea. Until 2013, around 20,000 beekeepers were spread all around the country but more predominantly in the mountainous provinces such as Gyoungbuk, Gyoungnam, Jeonnam, and Jeonbuk for *A. mellifera* and Gangwon for *A. cerana*. About half of the beekeepers keep native honeybees and the other half keep the European honeybees. In 2009, the average number of hives per household was 22 for *A. cerana* and 89 for *A. mellifera*. Most of the beekeepers have fewer hives than the average, with a few beekeepers keeping honeybees in very large numbers. Some of the largescale beekeepers also use their bees to pollinate crops. Renting honeybees for pollination of crops such as strawberry, oriental melon, watermelon, and other fruit and vegetables, has become more important for the beekeeper's income (Jung and Cho 2015). The number of Korean beekeepers peaked in the 1960s and then again in the 1980s. In the first peak, it gradually increased from the 1950s to 1960s and then decreased in the early 1970s. This can be explained by the fact that people were leaving farms and moving to the city to seek higher earning jobs. In the second peak, the number of beekeepers increased from the mid-1970s to 1980s. The number of beekeeping householders peaked in 1983–1986 at more than 50,000, then gradually declined and stabilized between 40,000 and 45,000 for most of the 1990s. However, the beekeeper numbers decreased again to 20,000 in the early 2010s. Until 2000, the proportion of beekeepers of *A. mellifera*:*A. cerana* was 1:2. However, after 2000, beekeepers with *A. mellifera* outnumbered beekeepers of *A. cerana* in a ratio of 2:1. The number of bee hives of *A. cerana* and *A. mellifera* did not change significantly until the 1980s. Due to government investment in beekeeping, as well as scientific/societal support, it began to increase dramatically. The number of colonies of *A. mellifera* increased from 50,000 to 150,000 over 10 years, reaching the highest number of 170,000 in 2004, and then stabilized at 150,000 thereafter. The number of *A. cerana* colonies has been stable at around 30,000–40,000 for 20 years but it rapidly decreased in the 2010s due to the epidemic of the Sacbrood Virus. Changes in honey production directly reflect the changes of colony numbers. Until the early 1980s, national honey production was less than 5000 tons per year. During the 1980s, it ranged between 5000 and 10,000 tons. By 2000, the honey production began to increase to 20,000 tons, then fluctuated between 20,000 and 27,000 tons after that. The pattern of honey production is slightly variable and it is predominantly influenced by the increase in *A. mellifera* colonies.

8.2.4 Beekeeping Economy

Annual production from the beekeeping industry is approximately 400 billion Korean won (US$400 million). Honey production is the main income generator of Korean beekeepers. Royal jelly and propolis come next, followed by the pollen production market. Selling the colonies spliced and with a pollination service could make up 10–20% of the apicultural market (Jung and Cho 2015). Higher demand and production of royal jelly, propolis, and honeybee venom is also a unique aspect in Korean apiculture. Total yearly honey production is about 25,000–30,000 tons, depending on weather, honey flora, and the timing of flower blossoms. Among the various sources of honey, approximately 70% is from the black locust (*Robinia pseudoaccaci*), and is called Acacia honey, followed by Castania Chestnut honey (Jung and Chon 2016). Recently, there has been an increase in the production of honey from plants in the genera *Styrax*, *Tilia,* and *Hovenia*. People prefer Acacia honey to the other types because it is clear (white) and has a honey smell with a mild sweet taste. Of the 30,000 tons of yearly honey production, 99% is used for domestic consumption. Very few exports were made but imports of honey and other bee products are increasing. In 2009, honey (approximately 500–600 tons) was imported mostely from New Zealand, Australia, and the USA. Most of the imports target consumers with interest in honey for health benefits, e.g., Manuka honey possesses high antibacterial activity.

8.3 Bee Forager Resources

8.3.1 Nectar/Pollen Plants

In Korea, 555 plant species are recorded as nectar/pollen sources for honeybees (Jung and Ryui 2007). Important tree species are Cherry, Styrax, Acacia, Chestnut, Bee Bee Tree, Raisin Tree, Kalopanax, and Vitex (Fig. 8.3). The distribution of these nectar/pollen sources depends on the local conditions. It must be noted that Korea is a mountainous country, with mountains occupying 65% of the total area. Table 8.1 describes some of the major honey plants in Korea. For the commercial production of honey, only a few tree species are utilized. Amongst these, the black locust (*Robinia pseudoacacia*) accounts for 70% of the total honey production in Korea. However, the apiculture industry is seeking alternative honey plants because the black locust species has been in decline for various reasons (e.g., natural aging, dieback, and damage from the black locust Gall Midge, *Obolodiplosis robinae*). In particular edible and medicinal plants were studied as alternatives as honey plants. *Hovenia dulcis* and *Acanthopanax senticosus* have been used in liver and arthritis treatments, and they flower for 17 and 22 days, respectively. Those flowering periods are longer than that of the black locust. The number of flowers per individual tree of *H. dulcis* is much higher than that of the black locust. In addition, nectar secretion per flower was estimated at 4.15 ± 1.11 μL and 3.51 ± 2.12 μL for *H. dulcis* and *A. senticosu,* respectively, which is higher than for black locust (2.20 ± 1.18 μL) (Han and Kim 2008). Also, based on analysis of the antioxidant activity using the DPPH method, the honeys of the two new plant species have shown twice as much antioxidant activity as that of Acacia honey (Kim et al. 2015). Therefore there are efforts to utilize *H. dulcis* var. *koreana* Nakai and *A. senticosus* Harms as potential alternative honey plants.

8.3.2 Agricultural Crops and Pollination Services

In Korea, rice is the most important crop and yields are relatively high. Rice represents about 90% of total grain production. Barley is the second most important crop. The other crops include millet, corn, sorghum, buckwheat, soybeans, and potatoes. Recently, horticultural crop production has become more important, and includes fruit and vegetables such as apples, pears, grapes, mandarin oranges, peaches, Welsh onions, Chinese cabbages, red peppers, persimmons, cabbages, and radishes. Relatively smaller plantations of rape seed and buckwheat provide nectar sources for the honeybee. In the agricultural landscape, honeybees are the most common flower visitor in various fruit tree crops (Kim et al. 2009). About 60–70% of flower-visiting hymenopterans are the European honeybee, *A. mellifera* (Table 8.2) (Michener et al. 1994). In Korea, the economic contribution of honeybee pollinations was estimated to be US$5.9 billion while fruit and horticultural crop productions were about US$12 billion (Jung 2008). Apart from this, Lee et al. (2014a, b) reported that 11% of beekeepers produce honeybee colonies for a pollination

Fig. 8.3 Major nectar plants in Korea (from the top left, Cherry, Styrax, Acacia, Chestnut, Bee Bee Tree, Raisin Tree, Kalopanax, and Vitex)

Table 8.1 Characteristics and distribution of some important honey/nectar plants in Korea

Tree name	Scientific name	Source[a]	Month of flowering	Distribution
Cherry	*Prunus serrulata*	N	March	Southern
Styrax	*Styrax japonica*	N	April	Southern
Oak Tree	*Quercus* spp.	P	April	All
Black Locust (Acacia)	*Robinia pseudoacacia*	N	May	All
Liriodendron	*Liriodendron tulipifera*	N, P	May	Middle
Chinese Toon Tree	*Cedrela sinensis*	N	May	Middle
Lacker Tree	*Rhus vermiciflua*	N	June	Middle
Chestnut	*Castanea crenata*	N	June	Southern
Linden Tree	*Tilia amurensis*	N	July	Northern
Raisin Tree	*Hovenia dulcis*	N	July	Middle
Kalopanax	*Kalopanax septemlobus*	N	August	Middle
Bee Bee Tree	*Evodia daniellii*	N	August	Middle to north
Acanthopanax	*Acanthopanax sessiliflorus*	N	September	Middle to south
Vitex	*Vitex negundo*	P	September	All

[a]Source for nectar (N) or pollen (P)

Table 8.2 Abundance of hymenopteran flower-visiting insects from selected fruit trees in 2007 (data from Kim et al. 2009)

Pollinators	Apple	Pear	Peach	Persimmon
Total Hymenopterans	308.6 ± 136.4	23	80	42 ± 7.26
Symphyta	5.3 ± 1.3	5	6	0
Apocrita	303.4 ± 135.3	18	74	41 ± 7.25
Apidae	288.8 ± 133.9	12	69	36 ± 8.7
Apis mellifera	213.4 ± 100.1	12	68	29 ± 6.1
P (%)	69.1	52.2	85.0	69.6

Numbers of orchards studied are 8, 1, 1, and 4 for apple, pear, peach, and persimmon, respectively
P is the percentage abundance of honeybee among pollinators

service. The number of honeybee colonies used for pollination services was estimated to be 226,000 colonies per year, which can be valued at about US$15.8–22.6 million. The target crops are mostly fruit-bearing crops such as strawberry (48%), watermelon (21%), oriental melon (16%), pepper (6.7%), and field fruit crops such as apple, pear, peach, and berries.

8.4 Pests and Diseases of Honeybees

Honeybees suffer from diseases caused by pathogens (such as viruses, bacteria, and fungi) or pests (such as parasites), and predators. In most cases, antibiotics, chemicals, and plant products are used to reduce losses to the colony and beekeepers. The pressure from diseases and pests seems to be increasing as evidenced from the

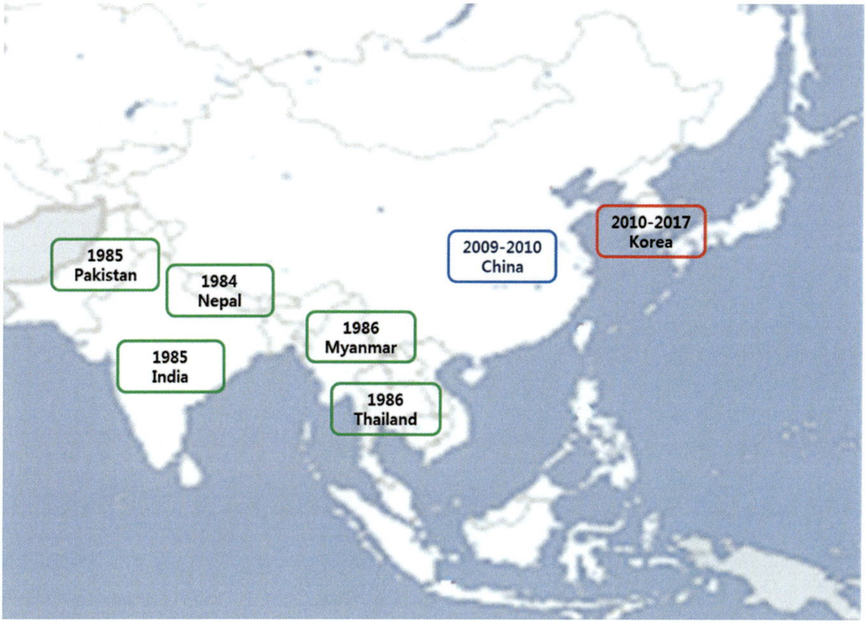

Fig. 8.4 Recent outbreaks of Sacbrood Disease in *Apis cerana* in Asian countries (data from Kang et al. 2012)

Colony collapse disorder (CCD) in the West or the recent outbreak of Sacbrood Virus in Asia (including Korea) (Fig. 8.4; Kim et al. 2008; Choi et al. 2010). A list of important pests and diseases occurring in the Korean beekeeping industry is presented in Table 8.3 and their prevalence in Fig. 8.5.

8.4.1 Diseases

Viral, bacterial, and fungal pathogens attack honeybee larvae or adults, and cause serious diseases in some cases. Among the bacterial diseases, American Foulbrood often impacts *A. mellifera* colonies but rarely *A. cerana*. Chalkbrood disease occurs during wet seasons where colony strength is weak in *A. mellifera*. Stonebrood disease often occurs in both species of honeybees. Nosema (Deformed Wing Virus) and Black Queen Cell Virus are more problematic to *A. mellifera* than to *A. cerana*.

Nosema Disease (*Nosemosis*) is generally regarded as one of the most destructive diseases of adult honeybees, affecting workers, queens, and drones alike. Seriously affected worker bees are unable to fly and may crawl about at the hive entrance or stand trembling on top of the frames. The bees appear to age physiologically: their lifespan is much shortened and their hypopharyngeal glands deteriorate;

Table 8.3 Diseases and pests of Korean honeybees

Group	Disease-pathogen and Pest	Severity[a]		Reference
		Am	Ac	
Fungi	Chalkbrood—*Ascosphaera apis*	**		Lee et al. (2005)
	Stonebrood—*Aspergillus flavus*	**	*	
	Nosema apis			Hong et al. (2013)
	Nosema ceranae	**	*	
Bacteria	American Foulbrood—*Paenibacillus larvae*	**		
	European Foulbrood—*Melissococcus plutonius*			Ha et al. (2005)
Virus	Black Queen Cell Virus	*	*	Choi et al. (2008)
	Chronic Bee Paralysis Virus			
	Cloudy Wing Virus			
	Deformed Wing Virus	*	*	
	Israeli Acute Paralysis Virus	*		
	Kashmir Bee Virus	*		
	Sacbrood Virus		***	
Mites	*Varroa destructor*	***		Woo and Lee (1993)
	Tropilaelaps mercedesae	**		Woo and Lee (1993), Jung et al. (2014), since 1993
	Varroa underwoodi			Woo and Lee (1993)
Predators	*Vespa analis*			
	Vespa crabro flavofasciata			
	Vespa ducalis			
	Vespa dybowskii			
	Vespa mandarinia	***	**	
	Vespa simillima simillima			
	Vespa simillima xanthoptera			
	Vespa velutina nigrithorax	***	*	Jung et al. (2007a, b), since 2003
	Aethina tumida			Kim et al. (2017), Lee et al. (2017) since 2017
	Formica spp.			
	Achroia grisella			Kang et al. (2015)
	Galleria mellonella		**	

[a]Severity expressed as * weak, ** modest, and *** serious on *Apis mellifera* (Am) or *A. cerana* (Ac). Blank means no concrete information or severity

the result is a rapid dwindling of colony strength. Other important effects are abnormally high rates of winter losses and queen supersedures. In climates with long periods of flight restrictions (i.e., no flight opportunities, even for a day), the infection easily reaches a severe stage that visibly affects the strength of the colony. The damage caused by Nosema Disease should not be judged by its effect on individual colonies alone as collectively it can cause great losses in apiary productivity. The disease is caused by the fungus *Nosema apis* or *Nosema ceranae*, whose 5- to 7-mm

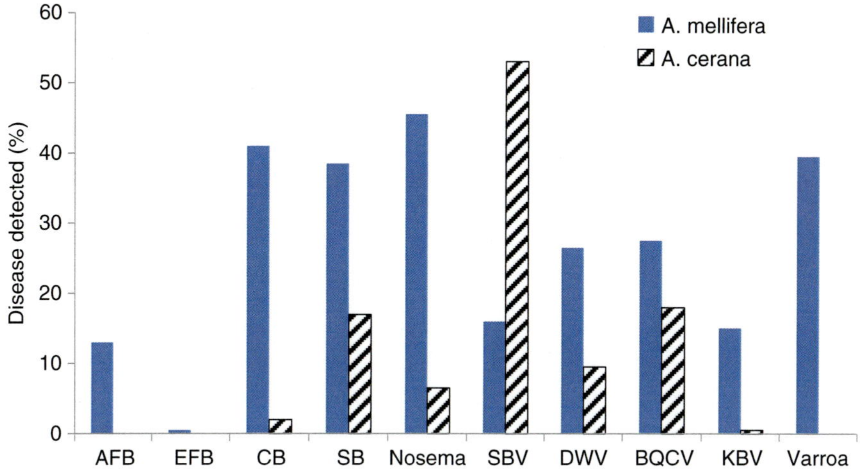

Fig. 8.5 Relative importance of honeybee diseases and *V. destructor* mite on *A. mellifera* and *A. cerana* revealed by molecular markers during 2010 and 2011 winter seasons. Data are from Kang et al. (2012). Abbreviations: American Foulbrood (AFB), European Foulbrood (EFB), Chalkbrood (CB), Stonebrood (SB), Nosema, Sacbrood Virus (SBV), Deformed Wing Virus (DWV), Black Queen Cell Virus (BQCV), Kashmir Bee Virus (KBV), and *V. destructor* (Varroa)

spores infect the bees, are absorbed with the bee's food, and germinate in the bee's midgut. After penetration into the gut wall, the cells multiply, forming new spores that infect new gut cells or can be defecated. The nutrition of the bees is impaired, particularly protein metabolism. A notable symptom of the disease is the bee defecates on the wall of the hive or bess tumble in front of the hive entrance, especially in the cooler spring, e.g., in April in Korea (Hong et al. 2013). However, diagnosis should be made in the laboratory by separating the intestine from the abdomen or macerating to check for spores under a microscope.

8.4.1.1 Viral Diseases

There are at least 18 virus types and strains that have been recorded as disease pathogens of adult bees and bee offspring. Nearly all of them are RNA viruses. The damage caused to colonies by viral infection varies considerably according to a number of factors, which include the type and strain of virus involved, the strength of the colony, weather conditions, the season, and food availability. The recent outbreak of Sacbrood Disease in *A. cerana* decimated the native honeybee population to near extinction (Fig. 8.4). The causal agent was identified as the Korean Sac Brood Virus (KSBV), which is different to the closely related Thai Sac Brood Virus (TSBV), or Chinese Sac Brood Virus (CSBV), or even the virus strain infesting *A. mellifera* (Choi et al. 2010). The disease symptoms in *A. cerana* appeared to be different to those in *A. mellifera.* Once infected, workers of *A. cerana* deliberately pulled the diseased young larvae to outside of the hive entrance, even before capping and swollen sac development (Kang et al. 2012; Choi et al. 2010). *A. cerana* populations suffered from sacbrood diseases in south and south-eastern Asia including Nepal, Pakistan, India, Thailand, and Myanmar during the 1980s to north-eastern Asia (China and recently Korea) (Kang et al. 2012). In 2010–2015, about 90% of *A. cerana* colonies in Korea were lost during an outbreak of KSBV (Jung and Cho 2015).

8.4.2 Parasitic Mites

Parasitic mites are the worst enemies of honeybees. The success or failure of beekeeping operations with *A. mellifera* depends largely on mite control and management. Several species of mites have been reported to cause devastation to both *A. mellifera* and *A. cerana* beekeeping. Several species of pollen-feeding mites are occasionally found in hives or attached to foragers. These phoretic mites are mostly innocuous to beekeeping. Notably pestiferous mites are *Varroa destructor* and *Tropilaelaps mercedesae* (formerly known as *T. clarea*, see Jung et al. 2014).

8.4.2.1 *Varroa* Mite

Among the management concerns, *Varroa* mite is the most important. *Varroa destructor* (previously known as *Varroa jacobsonii* but now separate species) is quite large compared to other mite species, and it can be seen with the naked eye (Anderson and Trueman 2000). This species was first recorded in Java (Indonesia) in southern Asia and first recorded in Korea in the 1970s (Choi and Woo 1974; Choi et al. 1986). Currently the majority of the *Varroa* populations in the world are considered to be the Korean haplotype (Anderson and Trueman 2000). The *Varroa* mites seemed to have occurred earlier but they were noticeable and widespread from 1968. From 1980, this mite became a chronic problem (Choi and Woo 1974). The *Varroa* mites cause varroasis symptoms in honeybees, which decreases their weight and shortens their lifespan. It also hampers the pollen-collecting activities of the workers (Sammataro et al. 2000). The *Varroa* mites also transmit viruses (ABPV and DWV) while feeding from honeybees (Bailey and Ball 1991). From a local survey, 88% of beekeepers recorded that their most serious pest was the *Varroa* mite (Jeong et al. 2016). In Korea, the government strictly regulates the application of synthetic miticides that may leave chemical residues in honeybee products (KFDA, 2014). Most beekeepers still rely on synthetic chemical miticides for mite control (Choi et al. 1986; Woo et al. 1994). Only a few well-known miticides have been used for 20 years; however, there has been no information on mite resistance in Korea (Jung et al. 2000). Kim and Jung (2010) tested their efficacy in mite control. They found that tau-fluvalinate was the most effective miticide followed by amitraz and flumethrin. Coumaphos and flavonoid+citric acid showed lower efficacies for mite treatments during summer and fall. Among the less hazardous materials, formic acid and thymol showed higher efficacies. Additionally, queen exclusion or drone brood removal are practiced by some beekeepers to avoid using chemical methods to control mites in their apiaries (Lee et al. 2004). Population modeling for the *Varroa* mite was used to gain a systematic understanding of the population dynamics and control simulation or future climate-related simulation (Jung 2012c, 2015). Simulation results indicated that the efficiency of miticide and the timing of the miticide application are important factors determining the long-term mite population projection. Under future climatic conditions where the honeybee activity window is enlarged, mite pressure could increase. Also, action threshold infestation level for miticide treatment has been estimated as 10% infestation on adult bees from a sugar-shaking method. At this level and if not managed properly, most honeybee colonies tend to collapse (Kim and Jung 2010).

8.4.2.2 *Tropilaelaps* Mite

Tropilaelaps mite is a native parasite of the Giant Honeybee (*A. dorsata*). The mite is widely distributed throughout tropical Asia but has host-expanded to *A. mellifera*. Invasion of *Tropilaelaps mercedesae* (formerly known as *T. clarea*) was from China in 1993, named the Chinese Spiny Mite in Korean (Woo and Lee 1993). When surveyed in 2005, the percentage of mite infestations by *Varroa* mite was much higher than that of *Tropilaelaps* (91% versus 25.7% infestation rate). Mite density was recorded as 18.7 *Varroa mites* and 1.7 *Tropilaelaps* mites per 100 bees (Fig. 8.6, Lee et al. 2005). Recently there are increasing patterns of occurrence of the *Tropilaelaps* mites in honeybees (Jung et al. 2014). Dual parasitism of *A. mellifera* colonies by both *Varroa mites* and *Tropilaelaps* mites can result in more damage to honeybee colonies. When this happens, the population of *Tropilaelaps* is often greater than that of *Varroa*, as the *Tropilaelaps* mite can almost completely prevent multiplication of the *Varroa* mite (Kim 2016).

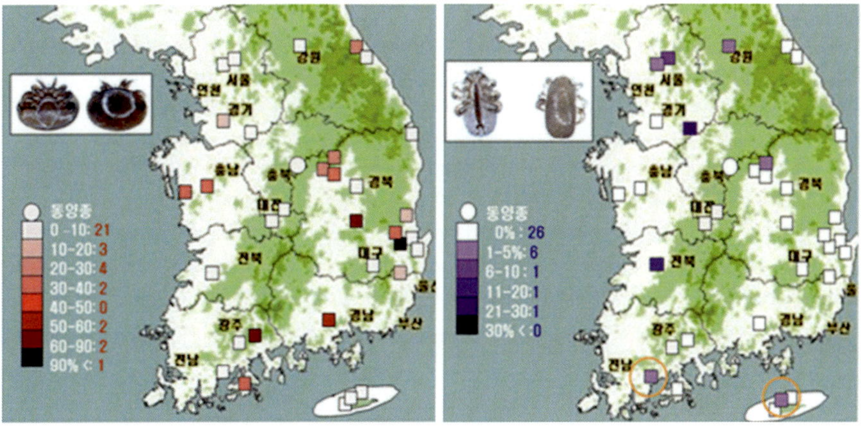

Fig. 8.6 Infestation levels of *Varroa* mite (left) and *Tropilaelaps* mite (right) surveyed over Korea in 2005 (figure from Lee et al. 2005)

8.4.2.3 Tracheal Mites

Tracheal mites (*Acarapsis* spp.) have not been found infesting honeybees in Korea.

8.4.3 Predatory Hornets

Hornets from the family Vespidae of Hymenoptera are the major natural enemies of honeybees, *Apis mellifera* and *A. cerana,* in Korea as well as in the world (Ono et al. 1995; Jung 2012a, b). From August to November, the worker hornets appear in the vicinity of the honeybee hive entrance, and hunt the honeybees to feed to the hornet larvae (Jung et al. 2007a). The annual economic impact of the hornets to Korean beekeeping was estimated to be 100–170 billion Korean won (US$100–170 million) (Jung 2012b). In Korea, *Vespa* diversity is very high, with nine species present. This is higher than in European countries where only one to two species can be found (Villemant et al. 2008). There are six hornet species that frequently visit apiaries: *Vespa ducalis, V. mandarinia, V. crabro flavofasciata, V. similima simillima,* and *V. dyvoski.* (Fig. 8.7). Other genera of Vespidae are small in size and do not harm honeybees. Recently there was an invasion of the *Vespa velutina* and its distribution is expanding rapidly and threatening Korean beekeeping (Kim et al. 2006; Choi et al. 2012; Jung et al. 2009; Jung 2012a). The rate of spread has been estimated as 12.4 km/year, which is 5.6 times slower than that from France (67.3 km/year). However, the diffusion coefficient (D) is still increasing, implying a greater risk of spread. The community structure of *Vespa* was different before and after the invasion. After the *V. velutina* invasion, the survey in 2010 from Gyeongnam province showed that *V. velutina* comprised 67% of total *Vespa* hornets caught to the trap, and showed reduced proportion of the smaller sized hornets such as *V. analis, V. crabro,* and *V. simillima simillima* compared to the bigger sized hornets such as *V. mandarinia* or *V. ducalis*. The native honeybee *A. cerana* uses defensive behavior toward the hornets such as balling and colony defense (Ono et al. 1995; Tan et al. 2007). However, *A. mellifera* lacks an effective way of defending the colony. For this reason, protection from hornet predation is important during the fall season. In search of a hornet nest site and their behavioral boundary, spatial distribution range was estimated via variogram model (Park et al. 2014). They estimated the independent sampling distance to be 70–90 m with slight species-specificity. Also, Park and Jung (2016) explained the past spread pattern by climate-related factors and provided the risk prediction map based on the CLIMEX model (Fig. 8.8). Research on the suppression of hornet population size and spread as well as the method to

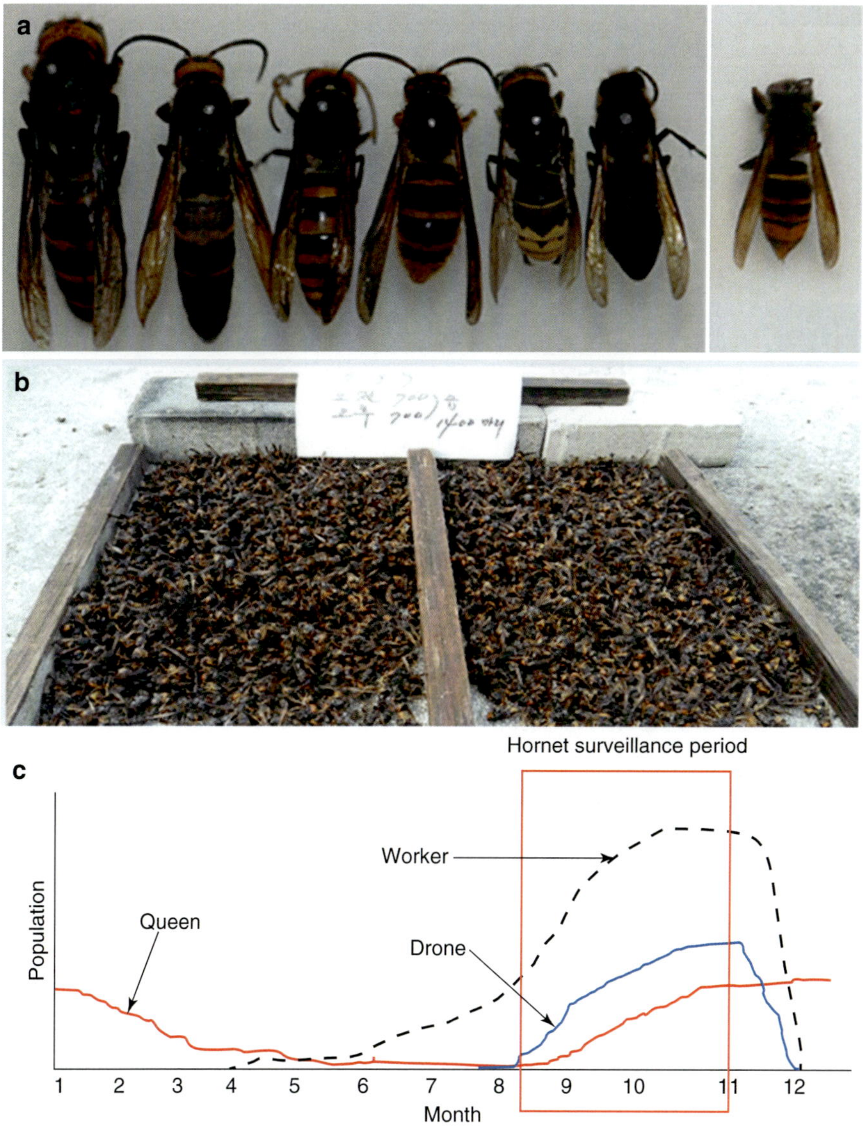

Fig. 8.7 Common *Vespa* hornets found in apiaries in Korea. From left, *V. mandarinia, V. ducalis,, V. analis parallela, V. similima similima, V. crabro flavofasciata, V. dyvoski,* and *V. velutina nigrothorax* (**a**), photograph showing number of hornets trapped in 1 day at a seriously damaged apiary in Masan, Gyeongnam province (**b**), and generalized population structure and size showing the critical hornet surveillance period (**c**)

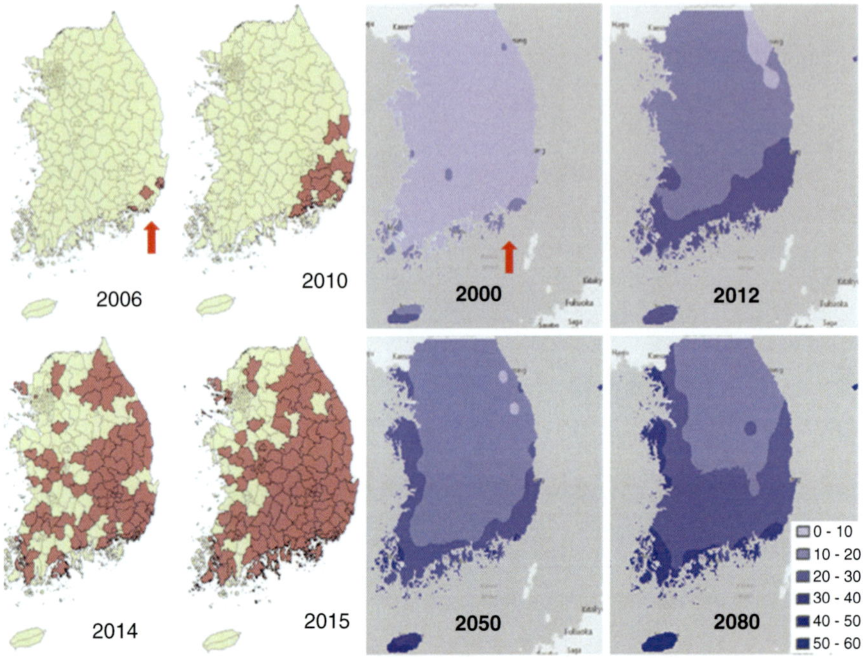

Fig. 8.8 Spread of invasive *Vespa velutina* in Korea after its first detection in Busan Port (arrow) in 2003. Figures from Park and Jung (2016)

protect the ecosystem health and apicultural industry is urgently needed (Jeong et al. 2016).

8.5 Other Problems of Beekeeping (Pesticides and Other Abiotic Factors)

Habitat deterioration (including fragmentation and degradation of natural landscapes), pest and pathogen prevalence, climate change, land use intensification (including agriculture and urbanization), and species invasion (Goulson 2003; Furst et al. 2014) have led to lowered plant diversity and increased usage of chemical inputs such as fertilizers and pesticides(Kluser and Peduzzi 2007). The consequences of pesticides are mostly negative and these chemicals pose risks to pollinators (including honeybees) (Cresswell 2011; Cressey 2013). The non-target effects of neonicotinoid insecticides have been studied and this work has resulted in the banning of these chemicals for commercial use in Europe (EC 2013). Not only foragers are exposed to pesticides, but also hive broods and young adults feeding on pollen and nectar in the combs. Therefore, different life stages of honeybees (larvae and adults of various ages or involved in different duties) or even genotypes could be exposed inside or outside of their hives and respond differently (Laurino 2013). Also, the concentration and duration of the pesticide exposure can vary: the quantity of chemicals, the persistence of residues, and the frequency of spraying on the crop (Kluser and Peduzzi 2007). In Korea, beekeepers have historically experienced honeybee poisoning from agricultural pesticides (Kim and Jung 2013b). There have been at least three cases of bee poisoning that attracted public attention. First, fipronil poisoning near a paddy field in ChungNam in 1999. Fipronil is a broad-spectrum insecticide in the phenylpyrazole chemical family that disrupts the insect's central nervous system by blocking GABA-gated chloride channels and glutamate-gated chloride (GluCl) channels, resulting in central nervous system toxicity. This causes hyperexcitation of contaminated insects' nerves and muscles. GluCl channels do not exist in mammals. Widescale bee poisoning occurred when the chemical was sprayed by helicopter to control the Rice Water Weevil. The causation was confirmed and appropriate compensation was made but it provoked public attention to the non-target effect of the pesticide.

The second case involved acephate misuse in the apple-growing area of Gyoungbuk in 2006. Acephate is an organophosphate foliar insecticide of moderate persistence with the residual systemic activity of about 10–15 days at the recommended rate. It targets aphid, miners, caterpillars, and thrips. But the problem began with the mis-labeling of the chemical's safety guidelines. These guidelines stated that this product could be used "before or after 20 days post-bloom". Thus farmers understood that this could be used even 1 day before the bloom and they consequently sprayed acephate. The issue was not bee poisoning but the failure of the trees to bear fruit because of the absence of pollinating insects in the sprayed apple orchards. Later this problem was partly resolved, but the causation was not scientifically solved.

The third case involved bee poisoning during apple blooming. Many apple growers use cabaryl as a flower thinner; in fact, this is registered as an insecticide and also a fruit-thinning agent. Carbaryl, a carbamate insecticide, is widely known as "Sevin" with a cholinesterase inhibitor. This was favored because it is of low toxicity and degrades quickly in vertebrates, but is highly toxic to beneficial insects such as the honeybee, and also harmful to humans (and is classified as carcinogenic). This chemical is related to the Bhopal disaster (1984 India), the largest industrial accident when carbaryl was produced using methyl isocyanate.

Choi and Lee (1986) reported that most of the beekeepers had experienced bee poisoning from pesticides and damage ranged from 10–50%. Most damage to honeybees was associated with apples and field peppers. Kang and Jung (2010) reported that most damage to honeybees is associated with insecticides rather than fungicides or miticide. Also, even with the same pesticide, toxicities were different between the bumblebee and honeybee. A carbaryl residual study on *B. terrestris* and *A. melliferea* showed no toxic effect on the bumblebee but a high toxicity effect on the honeybee (Kim and Jung 2013a). A recent questionnaire study on honeybee poisoning incidents from beekeepers and apple growers revealed interesting insights (Kim and Jung 2013b). As was expected, most of the honeybee poisoning incidents were from apple orchards, where the chemical fruit-thinning agent carbaryl was used. Most of the incidents occurred during May. Mitigation is possible if there is mutual communication between beekeepers and apple growers. However, even though apple growers understood the value of honeybee pollination on their orchards, they were reluctant to pay attention to the honeybees' health or invest more into protecting populations of insect pollinators and their health. Also, neonicotinoid insecticides have been tightly linked to bee toxicity and pollinator losses (Lee et al. 2016), and further careful monitoring and risk assessment in the Korean agricultural environment are needed (Jeong 2017).

8.6 Other Non-*Apis* Species Found in Korea

Twenty-four species of *Bombus* have been imported since 1995 (including *Bombus terrestris*). *Bombus ignites* is the most dominant bumblebee in Korea (Table 8.4) and is being considered for commercialization. So far, *B. terrestris* is mostly used in cherry tomato pollination in greenhouses and is being expanded for use in other greenhouse crops (Yoon et al. 2012). There are 31 species of Megachilidae reported

Table 8.4 Distribution of field collection of overwintered queens among *Bombus* species (partial data from Yoon et al. 2012)

Year	No site	*Ba*	*Bi*	*Bhs*	*Bu*	*Bcw*	*Bt*	Total
2010	22	436	816	155	101	16	4	1528
2011	14	286	361	182	7	5	2	843
2012	19	285	293	345	155	9	12	1099
Total	55	1007	1470	682	263	30	18	3470
Percentage	1.6	29.0	42.4	19.7	7.6	0.9	0.5	

Species abbreviations: *Ba, Bombus ardens; Bi, B. ignitus; Bhs, B. hypocrita spporoensis; Bu, B. ussurensis; Bcw., B. consobrinus wittenburgi; Bt, B. terrestris*

Table 8.5 Species distribution and sex ratio of Osmia Bees collected from released bamboo traps (Kim and Jung, unpublished data)

	Oc		Op		Ob		Ot		Os		
Site	f	m	f	m	f	m	f	m	f	m	Total
YW1	43	137	2	42	25	27	0	0	1	5	282
YW2	53	111	10	74	9	24	8	10	0	0	299
YW3	62	143	7	33	6	16	13	21	10	4	315
YW4	49	82	6	229	26	31	0	0	7	5	435
DY1	106	193	2	7	18	21	13	17	0	0	377
DY2	57	107	6	27	24	33	4	16	6	10	290
Total	699	1444	45	484	175	224	41	72	24	24	1998
Female ratio	0.32		0.07		0.42		0.37		0.50		

Species abbreviations: *Oc; Osmia cornifrons, Op; O. pedicornis, Ob; O. benefica, Ot; O. Taurus, Os; O. satoi* in Korea. Among them, *Osmia* is of particular interest for managing pollinators. From this genus, nine species of Osmia are known in Korea. Table 8.5 lists common Osmia species and their biological characteristics such as sex ratio. Currently, *Osmia cornifrons* is being used for apple pollination in some places. *Osmia cornifrons* has been recognized as an economically important pollinator of apple, blueberry, and lettuce in China, Japan, Korea, and e thUSA (Lee et al. 2007). *O. cornifrons* is also known as the Mason Bee because it uses mud to construct cell partitions and build linear series of cells in a nest. Females participate in nest construction and pollination (Jeong and Jung 2011).

8.7 Future Perspectives

Even though beekeeping is favored by not only the beekeepers but also the public, recent threats have decreased the size of the industry. Major threats include outbreaks of pests and diseases, and changes of nectar floral composition and limited nectar sources related to recent climate changes.

8.7.1 New Pests and Diseases

From 2005, the number of beekeepers declined by about 50% and the number of bee colonies declined by 25%. The decline is largely caused by the epidemics of Sacbrood Virus in the native honeybee *A. cerana*. Since the beginning of the Korean Sacbrood Virus epidemic in 2010, 80–90% of native honeybee colonies died, almost causing extinction (Jung and Cho 2015). The disappearance of income from native honeybee keeping resulted in many native beekeepers leaving the industry, and the number of colonies dropped to 10%. Effective management tactics against the Sacbrood Virus have not been developed. Selecting out the resistance stocks and enhanced nourishment are being trialed. There has been a new alien pest of honeybees reported in September 2016. From apiaries in Miryang City, the south-eastern part of Korea, heavy infestations of the small hive beetle, *Aethinia tumida,* were detected (Lee et al. 2017). Even though further infestation or spread was not reported, the invasion of the new pest brought another concern about whether it would further damage *A. mellifera* or attack *A. cerana,* which is already in a threatened state. Also, another pest invasion such as tracheal mite is highly likely since it has been found in the neighboring country of Japan since 2010 (Kojima et al. 2011; Maeda 2015). The mites feed on bee hemolymph in the trachea and air sacs of adult honeybees. A heavy infestation could cause serious damage to bee colonies.

8.7.2 Change of Climate and Bee Flora

As mentioned above, there are 555 species of wild and domesticated plants that provide nectar or pollen for bees. However, the majority of flowers bloom in spring (May). Honeybees and beekeepers in Korea are highly dependent on one floral source, black locust (*Robinia pseudoaccaci*). The dependency of honey production on the plant is about 70%. Recent climate change brought some changes in the beekeeping industry. First, simultaneous blooming is a recent phenomenon (Lee et al. 2014a, b). Most honey plants bloom in spring. In the case of Acacia (black locust), it takes about a month for blooming from the southern part of Korea to the north, so beekeepers could migrate from the south to the north, even though the country is small. Recently black locust flower blooms within a 1-week or 2-week period from the southern part to the northern part of South Korea (simultaneous blooming). This makes the beekeepers' migration difficult because one has to spend at least 7–10 days in the place where one has moved one's bee colony to collect and harvest honey. Second, degradation of the black locust trees in the mountain area is a problem (Lee et al. 2009). The black locust is a leguminous plant with a nitrogen-fixing capability that allows it to survive in the bare, nutrient-poor soil condition. Through the national campaign of transplanting and protecting forest areas, forests in Korea are now in a stable, semi-mature state where soil nutrients are conserved in the humus layer. Thus, the natural succession of the vegetation is not favorable for the black locust survival. Aged trees are dying and also new pest infestations such as yellowing or leaf mining fly exacerbates the problem (Ryu and Jang 2008). Furthermore, the planting area of black locust is decreasing. Alternatives to the black locust tree are sought (Han and Kim 2008). Candidate species such as the Raisin Tree, Liriodendron Tree, and Bee Bee Tree are under investigation and seedling cultivation stage. Some of these species are already in the plantation process with the hope that they can compensate for the black locust dieback. However, the honeybees and beekeeping industry will suffer food shortages for a considerable period until the new plants can replace the black locust plants.

Acknowledgements Authors express gratitude to the Journal of Apiculture for the permission to use published information.

References

Anderson DL, Trueman JWH (2000) *Varroa jacobsoni* is more than one species. Exp Appl Acarol 24:165–189

Bailey L, Ball BV (1991) Honey bee pathology. Academic Press, London

Choi SY, Lee ML (1986) A questionary survey on the injury to honeybees by pesticide poisonings in Korea. Korean J Apic 1:76–89

Choi SY, Woo KS (1974) Studies on the bionomics of the bee mite (Varroa jacobsoni Oudemans) and its chemical control (II). Res Rep RDA 16 (Livestock 69–76). Suwon, Korea

Choi SY, Woo KS, Kim YS (1986) Questionary survey on the varroa mites in Korean beekeeping. Korean J Apic 1:62–75

Choi YS, Lee ML, Lee MY et al (2008) Occurrence of seven Honeybee viruses and research of disease occurrence in Korean apiaries. Korean J Apic 23:153–159

Choi YS, Lee MY, Hong IP et al (2010) Occurrence of Sacbrood virus in Korean apiaries from *Apis cerana* (Hymenoptera: Apidae). J Apic 25:187–191

Choi MB, Martin SJ, Lee JW (2012) Distribution, spread and impact of the invasive hornet *Vespa velutina* in South Korea. J Asia Pac Entomol 15:473–477. https://doi.org/10.1016/j.aspen.2011.11.004

Crane E (1992) The world's beekeeping-past and present. In: Graham JM (ed) The hive and the honey bee. Dadant & Sons, Hamilton, pp 1–22

Crane E (1995) History of beekeeping with *Apis cerana* in Asia. In: Kevan PG (ed) This Asiatic hive bee. Apciulture, biology, and role in sustainable development in tropical and substropical Asia. Environquest. Ltd, Cambridge, ON, pp 3–18

Cressey D (2013) Europe debates risk to bees. Nature 496(7446):408. https://doi.org/10.1038/496408a

Cresswell JE (2011) A meta-analysis of experiments testing the effects of a neonicotinoid insecticide (imidacloprid) on honey bees. Ecotoxicology 20(1):149–157. https://doi.org/10.1007/s10646-010-0566-0

EC (European Commision) (2013) Commission implementing Regulation (EU) No 485/2013 of 24 May 2013 amending Implementing Regulation (EU) No 540/2011, as regards the condition of approval of the active substances clothianidin, thiamethoxam and imidacloprid, and prohibiting the use and sale of seeds treated with plant protection products containing those active substances. Off J Eur Union 139:12–14

Furst MA, McMahon DP, Osborne JL et al (2014) Disease associations between honeybees and bumblebees as a threat to wild pollinators. Nature 506:364–366. https://doi.org/10.1038/nature12977

Goulson D (2003) Effects of introduced bees on native ecosystems. Annu Rev Ecol Evol Syst 34:1–26

Ha JS, Lee HM, Kim DS et al (2005) A PCR detection method of *Melissococcus pluton* for rapid identification of European foulbrood. Korean J Apic 20:9–18

Han JG, Kim SH (2008) Flowering and nectar secretion characteristics of honey plant Hovenia dulcis var. koreana Nakai. J Apic 23:199–205

Hong IP, Lee MY, Woo SO et al (2013) Prevalence of Honeybee Nosema Disease and Black Queen cell virus on flowering periods of *Robinia pseudoacacia* in 2013. Korean J Apic 28:199–203

Jeong SM (2017) Relation between the honey bee mortality and the pesticide residue detected during the pear and apple blooming season. Dissertation, Andong National University

Jeong SA, Jung C (2011) Effect of temperature and relative humidity on the emergence of overwintered *Osmia cormifrons* (Hymenoptera: Megachilidae). Korean J Apic 26:261–266

Jeong SM, Lee CY, Kim DW et al (2016) Questionnaire study on the overwintering success and pest management of honeybee and damage assessment of *Vespa* hornets in Korea. J Apic 31:201–210

Jung C (2008) Economic value of honeybee pollination on major fruit and vegetable crops. Korean J Apic 23:147–152

Jung C, Kim DW, Lee HS, Baek H (2009) Some biological characteristics of a new honeybee pest, Vespa velutina nigrithorax Buysson, 1905 (Hymenoptera: Vespidae). Kor J Apic 24:61–65

Jung C (2012a) Spatial expansion of an invasive Hornet, *Vespa velutina nigrithorax* Buysson (Hymenoptera: Vespidae) in Korea. Korean J Apic 27:87–93

Jung C (2012b) Initial stage risk assessment of an invasive hornet, *Vespa velutina nigrithorax* Byusson (hymenoptera: Vespidae) in Korea. Korean J Apic 27:95–104

Jung C (2012c) Incorporation of damage function of *Varroa* mites into honeybee-varroa mite interaction model. Korean J Apic 27:215–221

Jung C (2014) A note on the early publication of beekeeping of western honeybee, *Apis mellifera* in Korea. Yangbong Yoji (Abriss Bienenzucht) by P. Canisius Kugelgen. Korean J Apic 29: 73–77

Jung C (2015) Simulation study of *Varroa* population under the future climate conditions. J Apic 30:349–358
Jung C, Cho SK (2015) Relationship between honeybee population and honey production in Korea: a historical trend analysis. J Apic 30:7–12
Jung C, Chon JW (2016) Quality assessment of honey from different floral origin in Korea. J Apic 31:103–111
Jung HG, Ryui JB (2007) Korean nectar/pollen plants. Gyeongnam publisher, Gyeongnam
Jung JK, Lee MY, Mah YI (2000) Infestation of *Varroa jacobsoni* and *Troilaelas clareae* in some apiaries during spring and fall seasons, 1999~2000 in South Korea. Korean J Apic 15:141–145
Jung C, Kang MS, Kim D et al (2007a) Vespid wasps (Hymenoptera) occurring around apiaries in Andong, Korea I. Taxonomy and life history. Korean J Apic 22:53–62
Jung C, Kang MS, Kim DW (2007b) Vespid wasps (Hymenoptera) occurring around apiaries in Andong, Korea: II. Trap catches and seasonal dynamics. Korean J Apic 22:63–70
Jung C, Kim DW, Kim JW (2014) Redefined species of *Tropilaelaps mercedesea* (Acari: Laelapidae) parasitic on *Apis mellifera*. Korean J Apic 29:217–221
Kang MS, Jung C (2010) Ecotoxicology of several acaricides used in apple orchards to the honeybee, *Apis mellifera*: from the laboratory to the field study. Korean J Apic 25:155–161
Kang SW, Yoo MS, Noh JH et al (2012) Occurrence and prevalence of Honeybee Disease in *Apis mellifera* and *Apis cerana*. Korean J Apic 27:187–195
Kang AR, Lee ML, Lee MY et al (2015) Biological control of wax moth, *Galleria mellonella* L. (Lepidoptera:Pyralidae) by *Bacillus thuringiensis*. J Apic 30:275–280
KFDA (2014) Korean food standards codex. Korea Food & Drug Administration, p 5297
Kim DW (2016) A study of competitive interaction of two ectoparasitic mites, *Varroa destrctor* (Mesostigmata: Varroidae) and *Tropilaelaps mercedesae* (Mesostigmata: Laelapidae) of honeybee, Apis mellifera (Hymenoptera: Apidae). Dissertation, Andong National University
Kim DW, Jung C (2010) Comparison of population monitoring techniques for the ectoparasitic mite *Varroa destructor* in honeybee (*Apis mellifera* L.) Korean J Apic 25:201–209
Kim DW, Jung C (2013a) Comparative analysis of the perception on the honeybee damage by the pesticides between beekeepers and apple growers. Korean J Apic 28:33–38
Kim DW, Jung C (2013b) Oral acute toxicity of chemical compounds used for flower/fruit thinner of apple against buff-tailed bumblebee, *Bombus terrestris* (Hymenoptera: Apidae). Korean J Apic 28:25–32
Kim JK, Choi MB, Moon TY (2006) Occurrence of *Vespa velutina Lepeletier* from Korea, and a revised key for Korean *Vespa* species (Hymenoptera: Vespidae). Entomol Res 36:112–115
Kim HK, Choi YS, Lee ML et al (2008) Detection of Sacbrood virus (SBV) from the honeybee in Korea. J Apic 23:103–109
Kim DW, Lee HS, Jung C (2009) Comparison of flower-visiting hymenopteran communities from apple, pear, peach and persimmons blossoms. Korean J Apic 24:227–235
Kim SG, Hong IP, Woo SO et al (2015) Comparison of anti-oxidant activity of ethyl acetate extracts of different floral honeys from Korea. J Apic 30:67–70
Kim JM, Lim SJ, Tai TA et al (2017) Development of rapid detection system for small hive beetle (*Aethina tumida*) by using ultra-rapid PCR. J Apic 32:119–131
Kluser S, Peduzzi P (2007) Global pollinator decline: a literature review United Nations Environment Programme, p 10
Kojima Y, Toki T, Morimoto T et al (2011) Infestation of Japanese native honey bees by tracheal mite and virus from non-native European honey bees in Japan. Microb Ecol 62:895–906
Laurino D (2013) Toxicity of neonicotinoid insecticides on different honey bee genotypes. Bull Insectol 66:119–126
Lee ML, Choi SY (1986) Biometrical studies on the variation of some morphometrical characters in *Apis cerna* and *A. mellifera* in Korea. Korean J Apic 1:5–23
Lee ML, Park YM, Li BR (2004) Effect of drone brood trap to control *Varroa* mites (*Varroa destructor*) in honeybee, *Apis mellifera* L., colonies in spring season. Korean J Apic 19:133–138
Lee HM, Lee DB, Han SH et al (2005) Rapid identification of *Ascosphera apis* causing chalkbrood disease in honeybee by Real-Time PCR. Korean J Apic 20:109–116

Lee SB, Seo DK, Kim SJ et al (2007) The pear flower-visiting insects, and the characteristics on pollinating activity of honeybee and bumblebee at pear orchard. Korean J Apic 22:125–132
Lee SW, Byun JK, Ji DH et al (2009) Soil physiochemical properties in leaf-yellowing black locust (*Robina pseudoacacia* L.) stands. Korean J Soil Sci Fertil 42:409–414
Lee H-S, Kim J-H, Yun JI (2014a) Recent trends in blooming dates of spring flowers and the observed disturbance in 2014. Korean Soc Agr Forest Meteorol 16:396–402. https://doi.org/10.5532/KJAFM.2014.16.4.396
Lee KY, Lee SG, Lee YB et al (2014b) Current status of honeybee production for pollination service in 2013. J Apic 29:245–256
Lee CY, Jeong SM, Jung C et al (2016) Acute oral toxicity of neonicotinoid insecticides to four species of honey bee, *Apis florea, A. cerana, A. mellifera*, and *A. dorsata*. J Apic 31(1):51–58. https://doi.org/10.17519/apiculture.2016.04.31.1.51
Lee SH, Cho KJ, Choi YS et al (2017) Review of the subgenus *Aethina* Erichson s. str. (Coleoptera: Nitidulidae: Nitidulinae) in Korea, reporting recent invasion of small hive beetle, *Aethina tumida*. J Asia Pac Entomol 20:553–558
Lew YS (1988) Study of modern beekeeping in Korea. Korea Beekeeping Association, Seoul
Lew YS (1991) The study on beekeeping in the late Chosun dynasty. Korean J Apic 6:106–140
Maeda T (2015) Infestation of honey bees by tracheal mites, *Acarapis woodi*, in Japan. J Acarol Soc Japan 24:9–17
Michener CD, McGinley RJ, Danforth BN (1994) The bee genera of north and central America (Hymenoptera: Apoidea). Smithsonian Institution Press, Washinton, DC
Ono M, Igarashi T, Ohno E et al (1995) Unusual thermal defense by a honeybee against mass attack by hornets. Nature 377:334–336
Park JJ, Jung C (2016) Risk prediction of the distribution of invasive hornet, *Vespa velutina nigrothorax* in Korea using CLIMEX model. J Apic 31:293–303
Park JJ, Kim DW, Kim CY et al (2014) Effective sampling ranges for monitoring vespid hornets (Hymenoptera: Vespidae), using Variogram Model. Korean J Apic 29:9–14
Ryu JB, Jang JW (2008) Newly found honeyplants in Korea. Korean J Apic 23:221–228
Ryu JB, Bae JS, Jang JW (2007) Honey producing counties in 4 major geographic books during Chosun dynasty. Korean J Apic 22:93–100
Sammataro D, Gerson U, Needham G (2000) Parasitic mites of honey bee: life history, implications, and impact. Annu Rev Entomol 45:519–548
Tan K, Radloff SE, Li JJ et al (2007) Bee-hawking by the wasp, *Vespa velutina*, on the honeybees *Apis cerana* and *A. mellifera*. Naturwissenschaften 94:469–472
Villemant, C, Perrard A, Rome Q et al (2008). A new enemy of honeybee in Europe: the invasive Asian hornet *Vespa velutina*. In: Proceeding of the 10th international Congress of Zoology, Paris
Woo KS, Lee JH (1993) The studies on the mites inhabiting the bee-hives. Korean J Apic 8:140–156
Woo KS, Cho KS, Lew YS (1994) Analysis on the level of damages caused by honeybee mites. Korean J Apic 9:33–39
Yoon HJ, Lee KY, Kim MA et al (2012) Breeding of the Korean Native Bumblebee, *Bombus ignites*. Korean J Apic 27:179–186

Chapter 9
Beekeeping in Mongolia

Khaliunaa Tsevegmid, Selenge Dooshin, Samuel Ramsey, and Panuwan Chantawannakul

Abstract Notwithstanding harsh climate conditions, the country of Mongolia is rich in insect fauna and is welcoming for managed honey bees. Since 1959, several different races or subspecies of managed honey bees have been introduced and subsequently maintained in Mongolia. These include *Apis mellifera mellifera*, *Apis mellifera caucasia*, *Apis mellifera carpatica*, (the Russian Far East bee), and the local Haliun bee line. These honey bees are well adapted to the climate condition of the country. A survey of honey bee species was conducted and revealed *Varroa destructor, Nosema ceranae,* and four viral infections distributed throughout Mongolia. Despite pathogens here the main stressors are climate variability. Cold conditions exist as the primary stress with the confluence of predators forming a secondary stressor for honey bees (bear attacks to hives, wasps, snakes, and ants). Aside from honey bees, nearly 30 species of bumble bees, in addition to other non-*Apis* bees, have been recorded in Mongolia. Little is known in the country about their biology or their importance for local pollination but it is assumed that they contribute essential pollination services as non-*Apis* bees do in other regions. Based on all given factors, it can be concluded that Mongolia is a great place for beekeeping and honey bee research.

Keywords Honey bees · Mongolia · Pathosphere · Disease · Pollination · Agriculture

K. Tsevegmid (✉)
Mongolian University of Life Sciences, Ulaanbaatar, Mongolia

S. Dooshin
Batsumber Beekeeping Research Station, Mandal, Tov Province, Mongolia

S. Ramsey
Department of Entomology, University of Maryland, College Park, MD, USA

P. Chantawannakul
Department of Biology, Faculty of Science, Chiang Mai University, Chiang Mai, Thailand

International College of Digital Innovation, Chiang Mai University, Chiang Mai, Thailand

P. Chantawannakul et al. (eds.), *Asian Beekeeping in the 21st Century*,
https://doi.org/10.1007/978-981-10-8222-1_9

9.1 Introduction

9.1.1 Features of Mongolian Geography and Climate in Relation to Beekeeping

9.1.1.1 Geography

Mongolia is a landlocked country in Central Asia, with terrain on average 1580 m above sea level. The country is geographically divided into six different natural zones, namely the high mountain forest, taiga forest, forest steppe, steppe, steppe desert, and Gobi desert regions. Of these zones, only the mountain forest, forest steppe zone (particularly in the river bank depression), and partially taiga forest zones are suitable for beekeeping. These zones are naturally rich in nectar (often called wild honey) and pollen producing plants compared to other zones (Shagdar 1974; Selenge 1989). Vegetation cover is suitable for honey bee foraging, even though these regions experience relatively lower temperatures and a shorter daytime compared to the steppe desert and Gobi desert zones of the country (Badarch 1969). In the mountain forest and forest steppe zones, which belong to the northern and central regions of the country, beekeepers actively keep honey bee colonies. These regions are also a key location for important cultivated plants in Mongolia, such as fagopyrum, rape, sunflower, wheat, and vegetables. Some of these plants also serve partly as forage for honey bees. The steppe zone of the northern and eastern regions is in a depression of 600–900 m above sea level (Chadraa 2000). This zone too is rich with nectar and pollen producing plants for honey bees, and has relatively warm weather throughout the year. Therefore, some places, particularly in the low depression areas of river banks, are highly productive for beekeeping. In other zones, with the exception of the steppe desert and Gobi desert where there are good reserves for nectar and pollen producing plants (Ochirbat 2011), honey beekeeping can also occur, but with less productivity.

Over 80% of plant species worldwide are pollinated by honey bees and other species for human food production and medicine (Potts et al. 2010). However, most crops in Mongolia do not rely on honey bee pollination.

9.1.1.2 Climate

Mongolia is known for its extremely cold and dry continental climate—cold, long continuing winters, and hot, short summers. Even with hot temperatures, the short continuous nature of the summers does not allow for the effective development of honey bee colonies.

The coldest period in the country is usually in January. Each region sustains a different range of cold temperatures: between −30 and −34 °C in the valleys of the mountains, −25 to −30 °C in the high mountainous areas, and −20 to −25 °C in steppe zones. The western part of the country presents the coldest area, with low temperatures estimated at −22.6 to 32.8 °C in the day time, and the coldest recorded temperatures estimated at between −45 and −55 °C in river basin areas (Batima et al. 2005). The fluctuation of air temperatures in the northern and central regions is quite stable in January, with temperatures ranging between −22.6 and −23.2 °C (Jambaajamts 1986; AIACC 2006). During the cold winter period in all regions, honey bee colonies are contained live in winter houses to overwinter as the cold season does not allow for them to remain outside.

The warmest period in the country is in July and the beginning of August, when honey bee colonies in Mongolia reach their highest population size (Jaltsav 1974; Selenge 1994). The average air temperature in July in the central and northern regions is 17.7 °C (Mongolian statistical year book 2000–2014). Consequently, in July and in the beginning of August, honey bees develop more actively compared to other periods of the warm season. However, the strength of local honey bee colonies depends on the number of sunny days, amount of precipitation, soil humidity, and air temperature that ultimately affect nectar-producing plants. For example, in low depression areas such as in the eastern and western parts of the country, the best period for honey bee development occurs from May until late September; in the central and northern regions, it can continue from May to the middle or end of September, though it may end earlier.

For honey bees, the most appropriate places in Mongolia are found in the mountain forest and forest steppe zones, with the latter being most appropriate (Selenge 1994; Ochirbat 2011). This zone covers about 238,108.0 km^2, equivalent to approximately 15.2% of the territory of Mongolia (Dash et al. 2003). Both of these zones are excellent for honey bee foraging. However, the warm summer is shorter in these areas compared to the southern and eastern parts of the country. Long-lasting summer regions occur in the south, south-west, and south-east parts of the country. Despite the long, warm season, there is scarce vegetation cover, particularly for honey bee foraging.

9.1.2 Diversity

To Mongolia, since 1959, several different races or subspecies of managed honey bees have been introduced, including *Apis mellifera mellifera, Apis mellifera caucasia, Apis mellifera carpatica,* and the *Apis mellifera vostoca* [referred to as Russian Far East bee in the works of Jaltsav (1974) and Selenge (1989)].

9.1.2.1 ***Apis mellifera mellifera***

The European black or dark bee (Middle Russian race) is thought to have to come to Mongolia toward the end of the 1950s. National scientists carried out research on these honey bees, which revealed that it adapted well to the conditions of Mongolia.

This subspecies appears to display some characteristics that differ from other races of honey bees in Mongolia, such as aggressiveness and an increased tendency to swarm (Selenge 1974–1995).

9.1.2.2 *Apis mellifera carpatica*

This honey bee is originally from the Carpathian mountain region, and it is closely related to *A. m. carnica* (Ruttner 1988). *A.m. carpatica* has inhabited Mongolia since 1974, with many studies considering its management in the country. For example, *A.m. carpatica* appears to be acclimated to the country, and exhibits a calm temperament (Selenge 1974–1995).

9.1.2.3 *Apis mellifera caucasia*

Originally from Central Caucasia (Ruttner 1988), this subspecies is commonly known as the grey mountain bee. They are referred as the Gruzinian bees in Mongolia. These bees were mainly kept in the northern region of the country until about 1990. They are good at obtaining nectar from plants because of their relatively long proboscis, though they do not overwinter as successfully as other honey bees found in the same climate conditions (Jamts 1974).

9.1.2.4 *A.m. vostoca*

According to Ganaev (1965a, b), this subspecies inhabited the southern portion of the Far Eastern region of the Russian Federation. Early settlers to the Far Eastern region of Russia delivered *A. m. mellifera, A. m. caucasia,* and *A. m. ligustica* in the 1870s. Their mixed breed is named as *A.m. vostoca.* This honey bee is distinguished from others by its high ability to forage, even at extreme temperatures. It also displays good hygienic behavior, resistance to foulbrood disease, low aggression, and high overwintering success (Jaltsav 1975). Based on its positive traits, it has been frequently used for breeding purposes, including long-term crossbreeding with other races of honey bees such as *A.m. mellifera* and *A.m. carpatica.*

9.1.2.5 Haliun Bee

As a result of crossbreeding of three geographically distinct races, researchers obtained the mixed honey bee line named "Haliun bee" in Mongolia. This bee line differs from other honey bees by demonstrating a high resistance to cold temperatures, good foraging ability, low aggressiveness and swarming tendency, and stable egg production. This honey bee line (Fig. 9.1) is best adapted for regions of Mongolia

Fig. 9.1 Haliun bee Queen with retinue (*Photo by Yondonjamts G. Central region of Mongolia. July, 2012*)

despite the country's difficult climate condition and geographical altitude, as ~630 to 1500 m above sea level (Selenge 1994).

Since 1990, economic change in the country has had a large effect on beekeeping and the scientific study of apiculture. Only recently has beekeeping become revitalized. In recent years, the importation of honey bee colonies from abroad, as well as breeding efforts, has expanded the industry in Mongolia. The current lack of regulation on bee breeding and honey bee importation makes it difficult to define exact subspecies of bees in apiaries, and may have resulted in the introduction of pests from abroad (Tsevegmid et al. 2014). Based on these observations, the current level of diversity of honey bees in Mongolia is a recent phenomenon, and it may be dynamic.

9.2 Historical Record

Western honey bees, *Apis mellifera*, are not native to Mongolia. Therefore, beekeeping does not have a long history in the country, despite some historical records that suggest beekeeping occurred in the region during the Chinggis Khaan (Genghis Khan) period. During the Mongolian empire dynasty, honey was used as food, but not widely. There are several documented records that Chinggis Khaan's golden tribes, and their successors, consumed beekeeping products such as honey and mead. For example, during the Ogodei Khaan (Chinggis Khaan's second successor) period, it was noted that the center of Khara Khorum (Qara-Qorum) city contained a large silver fountain shaped like a tree with five tubes which distributed special drinks, one of which was a honey drink.

The tree was built by Parisian master Guillaume Bouchier (Tudev 2015), and was used during special ceremonies or holidays (Smith 2000; Tatar 2003). The Silk Road, running through Mongolia, played an important trading role for many products, including imported honey from Europe (Dorjgotov 2006). It is likely that honey was a favorite food of the Khaan's since honey bread or cake was made for the Great Khaan's meals (Buell 1999). To our knowledge, no records were kept in Mongolia about honey bees and beekeeping products after the decline of the Mongolian empire.

At the beginning of the nineteenth century, the northern border of Mongolia developed as a main trading route, in particular with Russia (Sandag 1971; Ochir 2003). Records suggest that some beehives were kept by local people in the Buryatia border area neighboring the former USSR (Jaltsav 1964).

In the middle of the twentieth century, Mongolia started a broad strategy of intensification of its national agrarian industry by prioritizing fruit and beekeeping sciences. Therefore, the 1950s and 1960s was a fundamental period for beekeeping in Mongolia. In 1957 and 1958, seven Mongolian beekeeping trainers were sent to Buryatia and formed the first modern apiary in Mongolia in 1959. It contained 20 honey bee colonies. At the time, those honey bees were believed to be *A. m. mellifera* or *A. m. caucasia.* The imported Russian bees were well acclimated to the harsh climate of Mongolia, likely because they were also adapted to conditions of Siberia. Comprehensive characterization and analysis of the DNA of managed honey bees in Siberia showed that 64% belonged to *A. m. mellifera* (Middle Russian race), 28% originated from a southern subspecies, and 8% were mixed honey bee colonies (Ostroverkhova et al. 2016). Unfortunately, the authors did not clearly define the southern Russian subspecies. However, Avetisyan (1982) suggests that the southern honey bee race in Russia is *A.m. caucasia*. Sometime after 1960, merchants imported local Russian bee colonies into the Northern part of Mongolia (Jaltsav, pers. comm.), who was one of the beekeeping trainers in 1957/1958. The subspecies or races' defined records of these honey bees were not found.

In 1974, the Mongolian government decided to develop local beekeeping with a scientific basis, and to support the bee product industry. This resulted in the formation of the Bee Breeding Research Unit (BBRU), which started to conduct bee studies in Mongolia. In the same year, the first Mongolian bee genetic-specialist graduated from the Russian State Agrarian University—Moscow Timiryazev Agriculture Academy (*previously known Moscow K.A. Timiryazev Agriculture Academy*). She officially brought 30 mated *A. m. carpatica* queens from Russia. Thereafter, researchers intensively and effectively bred honey bees suitable to the local Mongolian ecological conditions. This work employed *A. m. mellifera, A. m. caucasia,* and *A. m. carpatica*. While studying these subspecies, researchers revealed the morphological phenotype traits of *A. m. vostoca*, which were distinguished by their body color (Jaltsav 1975). As a result of the selection work of the BBRU, a new honey bee line (Haliun bee) was developed by selecting the best phenotypes from *A. m. mellifera, A. m. carpathica*, and *A. m. vostoca.* Alongside these breeding efforts, Haliun bee colonies were distributed at 18 locations (Fig. 9.2) of mountain forest, forest steppe and steppe zones across the country together with other races (BMS 2010).

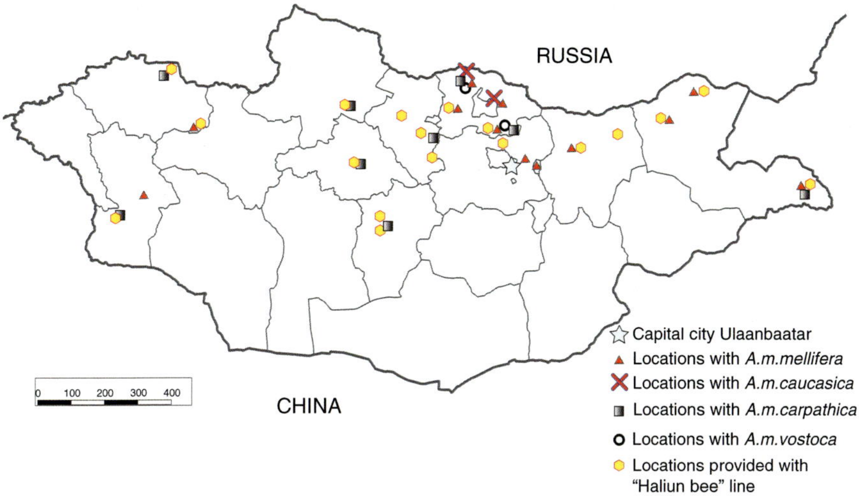

Fig. 9.2 Distribution of honey bees in Mongolia by subspecies (Selenge 1974–1995)

After the dramatic economic changes in the early 1990s, beekeeping became inactive. Recently, there has been renewed interest in beekeeping. Honey bee colonies have been recently imported from abroad, with most being located in the northern part of the country where the most appropriate areas for bee foraging exist. This has resulted in local people from the northern region becoming experienced at beekeeping. In 2005, approximately 800 honey bee colonies were managed in Mongolia (MBM report 2013). This number had increased to 1676 by 2008, 2707 by 2012, and 9276 by 2016 (AHS 2016). This increase may be partly due to a World Vision Organisation two-stage project aimed at improving the livelihoods of communities in the Khentii, Selenge, and Tuv provinces of Mongolia. As part of this work, ~800 honey bee colonies were imported into Mongolia and distributed to rural people (WVIO 2011).

As beekeeping is a growing industry in the country, it is estimated that ~800 beekeepers presently manage the country's 14,000 (Tuya 2017) honey bee colonies; most of which being farmers and small beekeeping enterprises. Roughly, 2000 honey bee colonies are located in the Western and Eastern regions, with the remaining located in the northern and central parts of the country (Selenge 2013). In the northern region, honey bees are considered to be mostly *A. m. mellifera* and *A. m. carpathica,* although *A. m. caucasia* may also occur. The "Haliun bee" line, and a hybrid of *A. m. mellifera, A. m. carpathica,* and *A. m. caucasia* are maintained in the Eastern, Western, and Central regions of Mongolia.

9.3 Bee Forage Resources

The main forage resources for honey bees in Mongolia are nectar and pollen producing plants. Feral vascular plant families such as *Fabaceae, Rosacea, Asteraceae, and Onagraceae* play a significant role for bee forage. They are the largest plant families in Mongolia (Ganbold 2000), and can grow in the places where good soil composition occurs such as in the central, northern, and eastern parts of the country (Tuul 2004).

Scientists of the BBRU have determined that 545 species of plants utilized by honey bees occur in Mongolia; 23 of these are tree species and 72 are bushes (Ochirbat 1994). Important beekeeping forage plants primarily bloom starting from the second or third week of April to the middle of August. Main plants in the North and Central regions providing nectar and pollen are at the end of April and all of May include *Salix caprea, Salix viminalis, Pulsatilla turczaninovii, Potentilla acaulis,* and *Padus asiatica.* From the end of May through June, *Ribes nigrum, Caragana arborescens, Crataegus sanguinea, Thermopsis lanceolata, Trollius asiaticus, Oxytropis microphylla,* and *Robinia pseudoacacia* are available. Afterwards in July, abundant blossoms of nectar and pollen producing plants can be found, such as *Potentilla anserina, Fragaria orientalis, Rosa acicularis, Ranunculus acer* (*borealis*), *Vicia amoena, Vicia costata, Vicia cracca* and *Valeriana alternifolia, Polemonium caeruleum, Onobrychis sibirica, Melilotus* spp., *Trifolium* spp., *Chamaenerion angustifolium, Geranium pratense, Geranium sibiricum,* and *Geranium pseudosibiricum* (Table 9.1) (Selenge 1989; Ochirbat 1994; Otgonbileg 1997). The aforementioned plants grow in the mountain forest and forest steppe zones, which are occupied primarily in the main three mountain ranges of Mongolia—the Altai Mountains in the west, and the Khangai and Khentii mountains in the north. The mountains slopes and their depression areas are rich in natural flowers (Ganbold 2000). Consequently, the mountain forest and forest steppe

Table 9.1 Main honey and pollen plants and their blooming period in Mongolia (Ochirbat 1975; Selenge 1989; Otgonbileg 1997)

Plants' bloom period of the year	Plant species
April	*Salix viminalis* L., *Pulsatilla turczaninovii* Kryl et Serg., *Potentilla acaulis* L.
May	*Gagee pauciflora* Ldb., *Armeniaca sibirica* L., *Spiraea media* Franz Schmidt., *Taraxacum officinale* Wigg., *Padus asiatica* L., *Padus avium* (Mill.), *Malus baccate* L., *Caragana arborescens* Lam.
June	*Thalictrum petaloideum* L., *Papaver rubro-aurantiacum* (Fisch. Ex. DC), *Lamium album* L., *Vicia cracca* L., *Thymus gobicus* Tschern., *Polemonium racemosum* (Regel) Kitam., *Valeriana alternifolia* Ldb.
July	*Vicia cracca* L., *Thymus gobicus* Tschern., *Polemonium racemosum* (Regel) Kitam., *Valeniana alternifolia* Ldb., *Phacelia tanacetifolia* Bent., *Chamerion angustifolium* (L.) Scop., *Allium schoeneprasum* L., *Lilium pumilum* Delile., *Medicago falcata* L., *Phlomis tuberosa* L. *Veronica longifolia* L., *Onobrychis arenaria* (Kit.) DC. *Trifolium lupinaster* L., *Vicia amoena* Fisch., *Geranium pratense* L., *Melilotus officinalis* Desr., *Allium odorum* L.
August	*Sonchus arvensis* L., *Cirsium esculentum* (Sievers.) C.A.Mey., *Echinops latifolius* Taush., *Orostachys spinosa* (L.) C.A.Mey, *Carduus crispus* L., *Lomatogonium rotatum* (L.) Fries ex Fern., *Helianthus annuus* L., *Fagopyrum esculentum* Moench

zones are well suited for beekeeping in Mongolia. Surveys carried out by researchers of the BBRU showed the Khangai and Khentii mountain areas are more appropriate for beekeeping. The research defines a reserve of honey at several sites. According to Banzragch's (1964) study of the Khangai mountain area, it is relatively rich in flora, particularly nectar and pollen producing plants. Further study by Ochirbat (1975) affirmed that there are good nectar reserves in these areas.

9.4 Local Knowledge and Practices for Beekeeping

Beekeeping does not have a long tradition in Mongolia, unlike other livestock such as horses, camels, cattle, sheep, and goats which have been kept since ancient times. Only since the 1960s have honey bees become familiar to Mongolians. Since then, beekeeping in the country has evolved over nearly 60 years and nowadays beekeeper numbers are estimated at 800. Most of these are hobbyists or beginners, with only 5% considered professionals working in the field with 20+ years' experience (Tsevegmid 2015).

9.4.1 Historical Knowledge About Bees

Due to aspects of Mongolian culture, there is very little direct use of insects. However, some mysteries and proverbs have accumulated over the years related to honey bees and can be found among oral literature of Mongols that is passed between generations. There is scarce information and a lack of well-documented evidence about honey bees, to our knowledge but apparently, Mongols had knowledge related to Apidae, which is expressed in lyrics and proverbs. For example "… To love my darling is a honey taste …" (Danzanravjaa mid-1800s), "Not easy to stop an eager bear addicted to honey," and "The person who ate the honey, licks his hand or fingers" (Chuluunjav 2000).

The nomadic lifestyle of Mongols includes accommodation in a "ger"—a traditional round tent covered with skins, sheep wool and used as a dwelling by nomads in the steppes of Central Asia (also known by the Russian word "Yurta"). The structure comprises an angled assembly or latticework of pieces of wood for walls, a door frame, ribs, and a crown compression ring (Chadraa 2000). In the summer, the ger's door is open as is the opening frame on top of the ger. If buzzing bees or bumblebees visit gers through these openings, Mongols happily predict that the year will bring a good harvest and the cattle will be strong. Then they may say to the bees in Mongolian language "Go, go, fly high, here is not place that you should be" (Dulam pers. comm.). It highlights that Mongols knew quite early on the importance of bees.

Nowadays, beekeeping is of interest to a wide range of people such as businessmen and various specialists. Since the first establishment of the scientific beekeeping field in Mongolia in the 1970s, the BBRU expanded to become the Bee Research

Station. Its scientists have since focused on developing a number of documents about bee forage plants (Ochirbat 1994), subspecies of bees, bee breeding and rearing technology (Selenge 1974–1995), and bee forage intensification (Otgonbileg 1997). Furthermore, technological advances have allowed Mongolian beekeepers to be at the forefront. For example, the queen larvae transferring method was conducted for first time in the conditions of Mongolia (Selenge 1985). Work to develop well-adapted queen bees, which could bear the long and cold winters of Mongolia, enhanced our ability to develop colonies (queen's egg laying potential, her life activity and longevity).

9.4.2 Features of Local Beekeeping (Equipment, Activities of Bee Colonies)

Honey bees are maintained in Prokopovych hives (Lukoyanov 1974; Avetisyan and Cherevko 2001), which suits the conditions of the country well. Hive parts are almost the same as in other worldwide bee practices. The use of Prokopovych and Langstroth hives are particularly similar to Russian beekeeping which was used as a model for Mongolians in the early days. Mongolian hive body and compartments with frames are made from dried, high quality wood like pine or cedar. The inside of the hive contains 10 wooden 435 × 230 cm frames with wax foundation sheet for the comb. For good and intensive development of bees for summer time, growth to four hive bodies may occur. Each has a hole for ventilation in front that is 25 mm in diameter. Inside the hive, the top of the frames are covered by a bright cotton cloth to separate frame tops from the warming cushion (Fig. 9.3A–H). These factors help to protect *A. mellifera* from the cold and heat, as well as provide good ventilation and warmth. The cushion is preferably made from natural materials to allow for ventilation, such as sheep wool or cotton.

In most cases, the hive lid is covered by a metal sheet on the top, and the outside surface of the hive is painted with good adhesion oil-based binders to insulate hives from the harsh features of the region, such as strong wind and precipitation. All compartments of the hive are detachable, which is ideal for the nomadic migration of Mongolian beekeepers with their honey bees. This type of hive gives the beekeepers opportunity to work easily with combs of each section and to separate the brood from honey combs. Several modified types also exist in the beekeeping community of Mongolia, for instance, the central and western regions of the country use hive and bottom board with thick wall made with two layers of wood. This is meant to protect bees from the cold, wind, and heat, although the basic construction design principle remains the same.

The busiest workload for beekeepers is the spring and autumn just after and just before housing the bees for the winter. The most enjoyable work to be found is collecting the honey harvest, which usually occurs at the end of June, July, and August. Honeycombs prepared by beekeepers are collected where equipment for honey

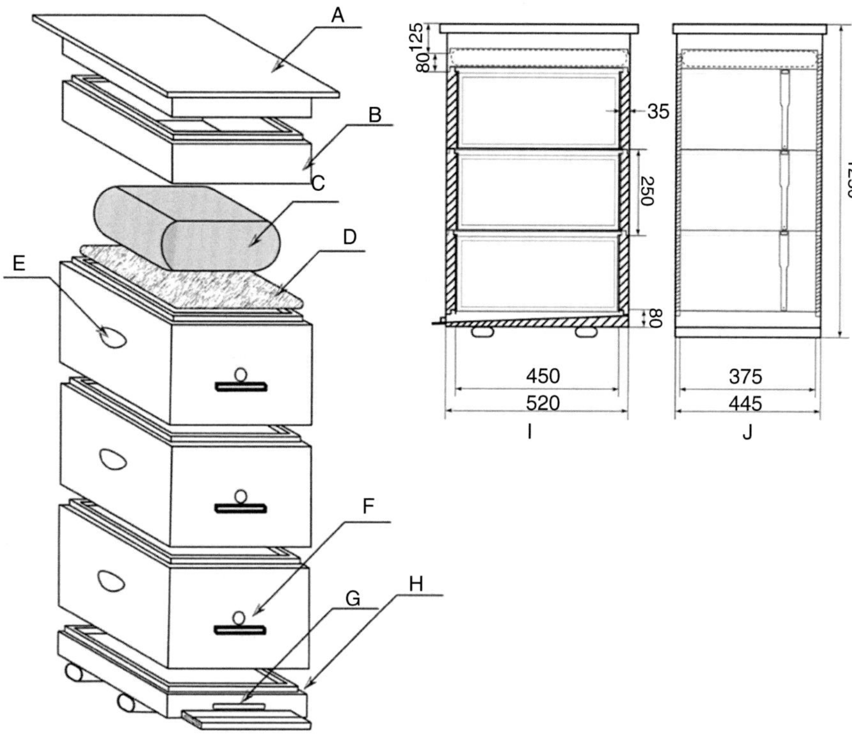

Fig. 9.3 Honey bee hive compartment (*A* hive lid with metal shield on the top, *B* neck for body hive, *C* cushion, *D* cotton cloth for cover top frames, *E* hive body, *F* ventilation hole with 25 mm in diameter, *G* bee visiting entrance slit on the bottom of the hive, *H* hive bottom board, *I* side view, *J* front view)

extraction is available, mostly near apiaries. In Mongolia, two and four frame stainless steel honey extractors are mostly used (Fig. 9.4). Spinning the extractor is easy with no wind and an outside air temperature of 28–33 °C. Before sending the combs with honey to the extractor, they are uncapped manually because apiaries of nomads are not generally supplied with electricity. Traditional dung smoke is used by beekeepers to discourage bees from visiting the centrifuge during the honey extraction process in the open air. Other hive tools and equipment used by Mongolian beekeepers are common among beekeepers around the world.

9.4.2.1 Wintering

Overwinter survival or "cold-hardiness" is a complex strategy for all kinds of organisms (Pullin 1996; Sjursen and Somme 2000), and especially so for insects in the extreme continental climate (Lee 1989; Eskov 2002). In Mongolia, honey bee colonies typically move to winter houses between the third week of October and the first

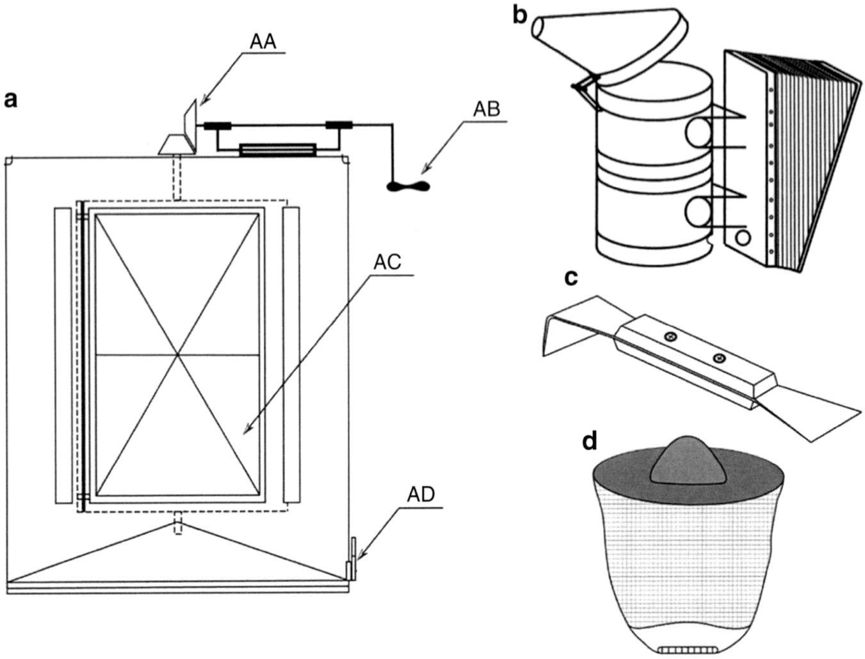

Fig. 9.4 Beekeeping tools (**a**) Honey collector (**AA** Shaft gear, **AB** Handle bar for rotation, **AC** Frame box and **AD** honey drain) (**b**) Smoker, (**c**) Hive tool, (**d**) Beekeeper hat in Mongolia

week of November. During this period, the nectar and pollen plant flowering period ends and outside air temperature drops below 5–8 °C during the daytime while temperatures at night are substantially below zero. In Mongolia, honey bees spend about 153–171 days in winter houses per year (Jaltsav 1976; Selenge 1974–1995). In temperate countries, many different types of winter houses are used for honey bees, and they may be located above or below ground (Avetisyan 1982). Because of the lengthy, frigid winter in Mongolia, most bee houses are underground to take advantage of the improved insulation.

Underground winter houses in Mongolia are built with good quality wooden material, including bee hive shelves (Fig. 9.5). The winter house wall outside is typically embedded under dry soil (sand), moss plants, or other soft plastic materials, which can separate the wall from the soil. One ventilation tube in the roof of the winter house exists per 8–10 square meters. On the floor, there is usually dry sand or pebbles. Properly designed winter houses built with good materials can maintain stable temperature and humidity of −1 °C to +1 °C and 65–70% relative air humidity, respectively, despite substantially lower outdoor temperatures. Experiments conducted since 1975 have shown that honey bees are able to overwinter well within a range of −2 °C to +2 °C and 65–75% relative air humidity in winter houses (Table 9.2). According to surveys, the lowest temperature in winter house was −6 °C, which resulted in decreased survivorship (Table 9.2). In Mongolia, hive lids

Fig. 9.5 Underground winter house for bees in condition of Mongolia. Circles show ventilation tubes. (**a**) Outside view, (**b**) inside view (*Photo by Yondonjamts G. Central region of Mongolia, November, 2014*)

Table 9.2 Regional comparisons of some key indicators for Mongolian beekeeping (average year temperature, colony strength, product (honey) collection and vegetation honey and pollen plant period. Study covers 1974–1995)

	Regions		
Indicators	Central	Northern	Eastern
Average year air temperature, Celsius degree	−2.0 (−4.0)[a]	−1.0[a]	0.2–0.3[b] −0.4[c]
Bee colony strength by number of frames with bees (July to August)	11–13[d]	14.8[e]	16.01[f]
Average honey collection, kg, 15 July to 10 August	15.8[j]	16.5[k] ~15–20[m]	33.8[f]
Honey and pollen plant bloom period	2nd week of May to 3rd decade of Aug. and mid Sept.[l]	3rd week of April to end Sept.[l]	2nd week of April to September[f]

[a]Jambaajamts (1986)
[b]Otgonbileg (1997)
[c]Altantsetseg et al. (2007)
[d]An av. data (1975–1985) Selenge (1974–1995)
[e]Jamts (1974)
[f]Baljinnyam (1974)
[g]An av. data Selenge (1974–1995)
[h]Jaltsav (1974)
[i]Ochirbat (1994)
[j]Shagdar 1974

are removed to improve ventilation during winter house storage. Beekeepers also leave the top cotton cloth open up ¼ under the warming cushion.

The overwinter survival of honey bees in underground winter houses ranges from 85% to 100% in Mongolia, depending on the honey bee subspecies, geographical location, and weather features (Selenge 1974–1995; Jaltsav 1976). Additionally, beekeeper experience appears to play an important role for successful overwintering (Baljinnyam 1974). For example, beekeepers with 10+ years of experience saw far greater overwintering success rates in Mongolia.

While bees are overwintering, beekeepers are able to work other primary jobs. The long dearth period also allows beekeepers to conduct necessary preparation work for the upcoming spring. This may include fixing and repairing the outer cover or body hives, cleaning and sterilizing old equipment, and preparing new hive bodies. For sterilization, beekeepers use acetic acid vapor to clean comb and warming cushions. Flame torches are used to sterilize equipment made from wood and metal.

9.4.2.2 Bees After Winter House

With spring comes inspection of food reserves. When food stores are insufficient, beekeepers use sugar syrup with a concentration of 1:1; they may also add natural honey, pollen, vitamins and microelements like Vitamin B12 and magnesium (Klochko and Luganski 2009), and sometimes concoctions or juice from medical herbs. Depending on the weather and the needs of the colony, additional feed may be provided once or twice. Furthermore, beekeepers must clean their hives, winter house of dead bees, and feces that accumulated over winter.

Wind in Mongolia is a major factor affecting colony strength, in particular for spring time. The highest wind speed occurs in Mongolia in the spring (Natsagdorj et al. 2003), with an average of 7.4 and 14.5 m/s, sometimes reaching 24.0 m/s in April and May, respectively, in the central region (Selenge 1974–1984). To protect colonies from the wind, beekeepers use greenhouses (Fig. 9.6).

With the end of April comes stable air. At the temperature above 10 °C honey bees are able to fly and seek forage. To benefit from early spring flowers, beekeepers start moving their colonies to the mountains or to where there is a good source of upcoming floral blooms.

9.4.2.3 Active Period

An intensive period of development of honey bee colonies in the country starts from June to the beginning of August. This period is the warm season, and there is good vegetation coverage including nectar and pollen producing plants. In particular, honey bee forage plants bloom everywhere. In the northern, western, and eastern regions, the colony developing period could occur a little earlier than the central region. Following warming air temperature, proper precipitation, and blooming plants, colony strength increases though it varies by region. For instance, due to

Fig. 9.6 (**a**) Moving bee colonies to mountain on the vehicles, (**b**) Greenhouses are used during the cold spring season for protection from strong wind and cold temperatures (*Photo by Yondonjamts G. Central region, April, 2013*)

Table 9.3 Bee products given by different subspecies of honey bees in Mongolia (3-year rate)

Years	Year 1 (1976)		Year 2 (1977)		Year 3 (1978)	
Bee species	Honey per hive, kg	Comb, number	Honey per hive, kg	Comb, number	Honey per hive, kg	Comb, number
A.m.mellifera[a]	14.0 ± 0.214	7.46 ± 0.58	8.9	7.1	29.1 ± 3.382	8.0
A.m.caucasia[a]	23 ± 0.321	9.41 ± 0.52	6.4	10.2	24.6 ± 3.051	9.6
A.m.carpatica[a]	21 ± 0.242	12 ± 0.57	10.1	10.1	33.7 ± 1.792	9.3
A.m.vostoca[b]	21.45					

[a]Selenge (1974–1995)
[b]Jaltsav (1976)

lower air temperature, central region plant bloom longevity is shorter than northern and eastern regions. Therefore, bee productivity is typically lower.

The main product from beekeeping is honey and its collecting season is July and August. The collected honey harvest goes to market for consumption, whereas some of it is exported. It typically comes from wild plants (polyfloral) or cultivated crops (monofloral). The honey from wild plants is highly sought after. Each year, Mongolian beekeepers produce approximately 60–150 tons of honey for national consumption (Jantsan 2013; Ulambayr 2016). The honey harvest varies with weather conditions and local blooming nectar plant intensity, which may differ from year to year. As a result, a colony can yield 5–60 kg per summer from nectar plants, but upwards of 100 kg from cultivated crops (Bulgan pers. comm.).

Early studies showed varying amounts of honey over the years (Table 9.3). Honey bees of all subspecies (e.g., *A. mellifera*, *A. m. caucasia*, and *A. m. carpatica*) were highly active from June 15th to July 15th in 1976 and from June 15th to July 25th in 1978 (Selenge 1978) (*A.m. vostoca* data for 1977 and 1978 not found). The winter of 1976 to 1977 was noted as being very cold, which influenced honey bee strength, development and nectar plants the following spring. Despite these explanations, studies of the honey and pollen plant bloom during these years were not found. However, honey production decreased sharply by 1977, as the author stated. Besides honey, beekeepers also produce pollen, as well as small amounts of propolis, royal jelly, and wax.

9.4.2.4 Preparation for Winter

By mid-August, beekeepers usually start to prepare for the winter. During this period honey bees usually require additional food to survive the winter, particularly if most honey was harvested.

Before beekeepers start to feed, they first control for the presence of young mated queens, honey bee population size, and comb quality. Usually by the end of August in the central region, the beginning of September for northern and eastern regions, and the middle of September for western region, beekeepers feed bee colonies sugar syrup at a 2:1 ratio. Ideally, it will represent no more than 30–50% of total food reserve for winter honey bees, depending on amount of honey left after harvest. Honey from cultivated crops is almost exclusively harvested in full because it can crystalize in the cells if left too long at winter house temperatures. (Yondonjamts pers. comm.). Beekeepers only feed pollen supplementally if stored pollen collected during summer is sufficient.

Along with feed management, beekeepers provide warm cushions inside the hive, and fix structures outside of hives to act as windbreakers. Successful overwintering of honey bee colonies is a key limitation to further developing beekeeping in Mongolia.

9.5 Pests and Diseases

Good practices and skills are necessary for keeping healthy livestock. Compared to developed countries, knowledge of bee disease and its management in Mongolia is in its infancy. During the socialist system period in the country, several bee farms were established in mountain forest and forest steppe zones of the country (Shagdar 1974). Based on these bee farms, the BBRU specialists have conducted some pathogen surveys on *Varroa destructor* (formerly *V. jacobsoni*), *Nosema* spp., and *Hafnia alvei*. According to the report of Selenge (1995), the first detected pathogen in Mongolia was *Nosema* spp. It was identified under light microscopy from 15 colonies of *A. m. mellifera, A. m. carpatica,* and *A. m. caucasia*. Colony symptoms observed included feces on the top of frames and inside of hive's front wooden plank, and mortality of honey bees during the winter. Laboratory analysis was conducted in samples (30 dead and 30 live bees in each sample) taken in April during 3 years (1976–1978). This revealed 40% *A. m. caucasia*, 32.5–35% *A. m. carpatica*, and 28–30% *A. m. mellifera* were infected. The *V. destructor* mite was first noted in 1986, and responsible for a high mortality of honey bees throughout the country (Selenge 1974–1995). The most recent survey of Mongolian honey bee pathogen infections was carried out during 2013–2014 (Tsevegmid et al. 2016). Samples were collected across nine distinct locations in Mongolia, from three different regions. Result showed that RNA viruses such as Sacbrood virus (SBV), Black queen cell virus (BQCV), Chronic bee paralysis virus (CBPV), Deformed wing virus (DWV), and *Apis mellifera* filamentous virus (FV) were present (Table 9.4). Bacterial infections such as *Melissococcus plutonius, Paenibacillus larvae*, other known viruses and *Acarapis woodi* were not found. The study also confirmed by molecular analyses that *V. destructor* mites represent the Korean haplotype, and that the *Nosema* species is *N. ceranae* rather than *N. apis*. The detected viruses (SBV, BQCV, DWV, and CBPV) were phylogenetically more similar to European clades, which suggests that the honey bees and their pathogens originated from Europe. Despite the close relationship between DWV and *V. destructor* in European countries (Dainat et al. 2012; Martin et al. 2012), this survey revealed no such close relationship in Mongolia.

For *V. destructor* management, new beekeepers use ready preparations that include amitraz and fluvalinate, such as Bipin, Varroapol, Fumisan, and Varatom. In

Table 9.4 The pests and pathogens of honey bees in Mongolia

Pathogens and pests		References
Viruses		
Black queen cell virus	+	Tsevegmid et al. (2016)
Chronic bee paralysis virus	+	
Deformed wing virus	+	
Sacbrood virus	+	
Microsporidian		
Nosema ceranae	+	Tsevegmid et al. (2016)
Mites		
Acarapis woodi	[a]/–	Tsedev (1982); Tsevegmid et al. (2016)
Varroa destructor	+	Tsevegmid et al. (2016)
Other		
Ascosphaera spp.	+	
Aunt	+	Khaliunaa pers. observation and Selenge pers. comm.
Hornets	+/–	
Field mouse	+	
Snake	+/–	
Forest bear	+	

+ detected; +/– rare

[a]*Acarapis woodi* recorded in the earlier work from colonies in the work Tsedev (1982), but described study not found in to our search. Meantime, the *Acarapis woodi* mites not detected (Tsevegmid et al. 2016)

contrast, experienced beekeepers prefer to use acetic or formic acids for the treatment and prevention for *V. destructor* and some other bacterial infections, respectively (Khaliunaa 1995, 2009, 2011).

Contrary to *V. destructor*, special medicaments are not widely used by Mongolian beekeepers to control nosemosis. At the beginning of the 1980s, fumagillin was suggested for *Nosema* treatment, but data on its use was not found. Conversely, beekeepers frequently use smoke from animal excrement to manage their colonies. In Mongolian this is called "argal," and was believed by historic nomad groups to contain medicinal properties (Delger 2013). As a result, it is used as a treatment against *V. destructor* and other ectoparasites. Argal mostly consists of plant material and includes methyl carbonates or mixtures of KNO_3. The methyl carbonate is thought to act as a sterilizer and have negative effects on bacteria and ectoparasites.

9.6 Other Problems of Beekeeping

Similar to other regions, Mongolian beekeeping faces many challenges. The most serious problems of beekeeping are climate change and pasture land degradation in Mongolia (Tsevegmid and Selenge 2015). Mountain forest areas suffer uncontrolled use, and sometimes fire (Gunin et al. 1999), whereas pasture quality can be reduced due to high density of grazing animals (CBD 2014). Therefore, beekeeping in Mongolia is always dependent on climate and local weather, particularly in relation to the reserve of foraging plants. To date, honey bees do not appear to have suffered from the effects of pesticides. This could be because agriculture in Mongolia does not heavily rely on pollination.

Additional problems can be described as secondary stresses for honey bees. This can include *Vespa* spp. (wasps) during the spring and autumn periods, in particular, during sugar syrup feedings. Though not common, large animals such as bears can be a threat to honey bee colonies as well. Forest bears usually attack honey bee colonies for honey and brood at night when apiaries are located near to large forests. This can result in the death of the queen, as well as destroyed hive equipment. In some cases, bears drag the hive to the forest. Several cases of bear attacks were historically recorded. Typically, bears destroyed only one hive at a time, but sometimes the destruction of more was observed (Selenge, pers. comm.).

Snakes do not attack honey bee colonies, but they will sometimes inhabit the beehives. It causes more stress for beekeepers rather than bees. It happens most frequently in the (mountainous) steppe zone, at the warmer locations, with rocks or the rugged topography of mountains. Snakes are calmer than bears, and they usually reside behind the hives or in the bottom of empty hives (personal observation). If snakes are found in hives, beekeepers very carefully remove them with a long stick from the bottom of the hives and move them to natural areas. Besides these, pests like *Aranei* spp. (spiders), *Formicidae* spp. (ants), and *Mus musculus domesticus* (house mouse) are very commonly observed in and nearby honey bee colonies and winter houses. Treatments against these pests are simple, as Mongolians have a respectful approach to nature and wildlife. For example, beekeepers fix up their bee hives to protect them from bears. They also use mouse-traps and manually remove ants nest and spiders that have settled in the hives.

9.7 Non-*Apis* Bees in Mongolia

Alongside managed bee colonies' crucial role for pollination of crops, the wild bees play a key role for the environment too (Williams 1991). In his book, Michener (2007) summarized in his book the 12 bee genera, including 16 subgenera, found in Mongolia. Also, Proshchalykin and Kupianskaya (2005) provided a revision of non-*Apis* bees found on the territory of Mongolia. This included bees from families of Colletidae, Halictidae, and Megachilidae.

9.8 Bumble Bees in Mongolia

As an important pollinator of wild and cultivated plants, nearly 30 species of bumblebees are recorded in Mongolia (Namkhaidorj et al. 2008). Their habitat is mostly in the northern, western, and central regions of the country, where good forage exists due to high levels of precipitation and good soil composition (CBD 2012; Kupianskaya et al. 2014). Among the bumblebee species distributed throughout Mongolia, *Bombus muscorum, Bombus modestus, Bombus sporadicus,* and *Bombus subbaicalensis* are noted very rare species. Therefore, the Government of Mongolia provides them special attention (Protocol 2012; Red Book of Mongolia 2013).

Despite these surveys, little is known about the interaction of bumble bees and the ecology of Mongolia and future work should focus on it.

9.9 Future Perspectives

9.9.1 Perspectives for Research

Mongolia is a landlocked country with a huge territory and rich ecological zones. In the country, honey bees primarily inhabit the mountain forest and forest steppe zones. In Mongolia, the relationship of climate and insects is not well described, but scientists are working to generate data.

Flora plays an important role for honey bees. Described studies on nectar and pollen producing plants demonstrate that their dynamic growth depends on local weather. The interaction between honey bee disease and local floral/fauna is an interesting area for investigation. Also, genetic research on bees that are well adapted to the harsh conditions of Mongolia could be of interest to other regions as well.

Furthermore, the field of honey bee pollination is untouched, although it is no doubt beneficial to the food industry (Gill 1991). An implementation of research toward honey bee or bumblebee and climate interaction in Mongolia would be interesting for the broad scientific community.

9.9.2 Perspectives for Rural Economy

Developing the rural economy of Mongolia is important. Transition to a market economy several decades ago brought plenty of new opportunities (Tsevegmid 2015). Poverty is slowly decreasing, but is not on par with other developed countries. Therefore, improving the economy in the countryside by way of developing beekeeping in these areas, and improving livelihoods, has merit. Regions particularly fit for beekeeping should be targeted, especially those where poverty is prevalent such as the northern and central regions.

9.9.3 Perspectives for Pollination Through Honey Bees

Pollination using managed bees is a worldwide practice (Gill 1991; Naumkin et al. 2004) to improve harvest (Breeze et al. 2014) and seed quality (Chuluunbaatar 1996). This is important to Mongolians as we are historically well-connected to land conservation. Despite this, overgrazing livestock has destroyed vegetation cover (GG 2015) and therefore limited forage availability for honey bees. To help restore vegetation cover, some people invite honey bee apiaries to their land where usually they make hay. This is because farmers observed that good hay production comes from areas where honey bees are present due to the effects of pollination (Bulgan per. Comm.).

Similarly, several attempts have been made to increase vegetable production. For example, a small-scale short-term experiment on turnips revealed that plots near to honey bee colonies produced 1221 seeds from 50 flowers compared to plots with no honey bees that produced 910 seeds from same number of flowers (Chuluunbaatar 1996). Furthermore, the study showed the seeds from the honey bee plots weighed 0.134 g/seed on average compared to 0.125 g/seed on average for seeds from the plots lacking honey bees.

Acknowledgement We would like to express our sincere gratitude for all the people who established in Mongolia beekeeping. Many thanks to the colleagues of the Institute of Bee Health University of Bern and Agroscope, Switzerland. Our special thanks to Mr. Bulgan Ts. and Mr. Tulgabat Ts. for encouragements, and Mr. Yondonjamts G. for kindly providing photos, figures for this chapter. Thank you all for your great enthusiasm and kind assistance.

References

AHS (2005–2006, 2016) Official website of Ministry of Food, Agriculture and Light Industry Mongolia. Animal husbandry section. http://mofa.gov.mn/exp/blog/7/4

AIACC (2006) AIACC final report. Climate change vulnerability and adaptation in the livestock sector of Mongolia, pp 84

Altantsetseg J, Badamkhand U, Khishigjargal N (2007) Some result of study on air temperature in Eastern region of Mongolia. In: Proceeding of joint symposium of Mongolia and Russia on climate change, pp 46–51

Avetisyan GA (1982) The beekeeping textbook, Moscow, pp 320 (in Russian)

Avetisyan GA, Cherevko YA (2001) The beekeeping, Moscow, pp 55–60 (in Russian)

Badarch G (1969) Geography of Mongolia. Ulaanbaatar Press, Ulaanbaatar, p 177. (in Mongolian)

Baljinnyam G (1974) Experience and investigation in honey bee industry in East region of Mongolia: research report of Shaamar Fruit Research Central Station, N3, pp 116–121 (in Mongolian)

Banzragch D (1964) Towards plant flora mountain forest steppe zone of Khangai mountain range. Bull Acad Sci Mongolia №2, pp 69–79 (in Mongolian)

Batima P, Natsagdorj L, Gomboluudev P, Erdenetsetseg B (2005) Observed climate change in Mongolia AIACC working paper no.12, pp 1–25

BMS (2010) Official website of Ministry of Food and Agriculture, Mongolia. Beekeeping management section Available at: http://mofa.gov.mn/livestock/index.php?option=com_content&view=category&id=80&layout=blog&Itemid=114&lang=en

Breeze TD, Vaissière BE, Bommarco R, Petanidou T, Seraphides N, Kozák L (2014) Agricultural policies exacerbate honey bee pollination service supply-demand mismatches across Europe. PLoS One 9(1):e82996

Buell PD (1999) Mongolian Empire and Turkicization: the evidence of food and food ways. In: Amitai-Preiss R (ed) The Mongol Empire and its legacy. Brill, Leiden, pp 200–223

CBD (2012) Convention on biological diversity. National Report of Mongolia Fifth Ed. 2012, pp 18–73

CBD (2014) Convention on biological diversity, National Report of Mongolia Fifth Ed. 2014, pp 64–81

Chadraa B (2000) Encyclopaedia of Mongolia, Ulaanbaatar, p 576 (in Mongolian)

Chuluunbaatar J (1996) Pollination by bees to turnip (vegetable). Scientific Report Bee Research unit. Mandal, p 41 (in Mongolian)

Chuluunjav C (2000) Insect tie. Ulaanbaatar, Urlakh erdem press, pp 16–25 (in Mongolian)

Dainat B, Evans JD, Chen YP, Gauthier L, Neumann P (2012) Dead or alive: deformed wing virus and Varroa destructor reduce the life span of winter Honey bees. J Appl Environ Microbiol 78(4):981–987

Danzanravjaa (mid of 1800) Lyric of the national song Mongolia "Ulemjiin chanar" (in Mongolian)

Dash D, Khaulanbek A, Mandakh N (2003) Ecosystem restoration and protection of the Gobi steppe zone's scientific basis. Institute of Geo ecology, Ulaanbaatar, Mongolia N2, pp 43–45 (in Mongolian)

Delger B (2013) Argal culture in Mongolian nomadic civilization. Mongolian Studies Section, Inner Mongolia University Library. http://www.shiqi.com.cn/s/ProductShow_other.asp?ClassID=85&ID=433

Dorjgotov D (2006) National Atlas of Mongolia, p 156 (in Mongolian)

Eskov EK (2002) Does anabiosis condition possible for overwintering honey bee? Russian J Beekeeping N2:11–12. (in Russian)

Ganaev AI (1965a) Far Eastern bees. Russian J Beekeeping 2:12–13. (in Russian)

Ganaev AI (1965b) Indian bees in the Far Eastern of Russia. Russian J Beekeeping 11:11–12. (in Russian)

Ganbold E (2000) Flora of North Mongolia: systematics, ecology, geography and history of development. Autoreferat of Doctor thesis, Sankt Peterburg, p 35 (in Russian)

GG (2015) National report on the rangeland health of Mongolia. Green gold project report, 2015, Ulaanbaatar, p 185

Gill RA (1991) The value of honey bee pollination to society. J Acta Horticulturae 288:62–68

Gunin PD, Vostokova EA, Dorofeyuk NI, Tarasov PE, Black CC (1999) Vegetation dynamics of Mongolia. Springer, Berlin, p 238

Jaltsav G (1964) Beekeeping development perspective in Mongolia. Scientific reports of the Animal Livestock Institute, vol 11–12, Ulaanbaatar, pp 124–135 (in Mongolian)

Jaltsav G (1974) Bee development in central and northern area of Mongolia. Scientific report of the Bee breeding research unit for 1974. Archive material (in Mongolian)

Jaltsav G (1975) Features of A.m. vostoca bees. Scientific report of the Bee breeding research unit for 1975. Archive material (in Mongolian)

Jaltsav G (1976) Features of wintering Russian Far east bee. Scientific report of the Bee breeding research unit for 1976. Archive material (in Mongolian)

Jambaajamts B (1986) Climate of Mongolia. Ulaanbaatar, p 268 (in Mongolian)

Jamts B (1974) The Gruzinan bee's biology and morphology. Bulletin Shaamar fruit Research central station Ed. 3, Ulaanbaatar, pp 65–75 (in Mongolian)

Jantsan T (2013) The taste of honey, vol 1315, Daily News, p 11

Khaliunaa T (1995) Notice of the Mongolian beekeeper's meeting, Shaamar, Selenge province (in Mongolian)

Khaliunaa T (2009) Notice of the Mongolian beekeeper's meeting, Darkhan city, Darkhan Province (in Mongolian)

Khaliunaa T (2011) Notice of the Mongolian beekeeper's meeting, Ulaanbaatar (in Mongolian)

Klochko RT, Luganski SV (2009) Kovitsan for honey bees. Russian J Beekeeping 3:11–13. (in Russian)

Kupianskaya AN, Proshchalykin MY, Lelej AS (2014) Contribution to the fauna of bumble bees (hymenoptera, apidae: *bombus* latreille, 1802) of the republic of Tyva, Eastern Siberia. Eurasian Entomol J 13(3):290–229. (In Russian)

Lee RE Jr (1989) Insect cold-hardiness: to freeze or not to freeze. Bioscience 39(5):308–313

Lukoyanov VD (1974) Beekeeping equipment. Moscow, COLOS, p 256. (in Russian)

Martin S, Highfield A, Brettell L, Villalobos E, Budge G, Powell M, Nikaido S, Schroeder DC (2012) Global honey bee viral landscape altered by a parasitic mite. Science 336: 1304–1306

MBM (1995–2013) Mongolian Beekeepers meeting reports Ulaanbaatar, Mongolia, p 16 (in Mongolian)

Michener D (2007) Bees of the world, 2nd edn. The Johns Hopkins University Press, Baltimore, MD, pp 192–798

Mongolian Statistical year book 2000–2014

Namkhaidorj G, Puntsagdulam J, Nansalmaa B, Myagmarsuren D (2008) Key to insect identification. Best Printing. Ulaanbaatar: Shinzhlėkh Ukhaany Akademiĭn Khėvlėl

Natsagdorj L, Jugder D, Chung YS (2003) Analysis of dust storms observed in Mongolia during 1937–1999. Atmos Environ 37:1401–1411

Naumkin VP, Velkova NI, Kulikov NI (2004) Insects on the Sinapis alba L. J Russian Beekeeping No 6:20–21

Ochir A (2003) History of Mongolia, vol 4. In Enkhtuvshin B (ed) Ulaanbaatar, p 424 (in Mongolian)

Ochirbat G (1975) Scientific reports of the honey bee research unit, Mandal, pp 9–16

Ochirbat G (1994) Biological resource for honey and pollen plants in Mongolia. Autoreferat for Doctor thesis, Ulaanbaatar, p 51 (in Russian)

Ochirbat G (2011) Research works (1970–2010), vol 1. General printing LLC, Ulaanbaatar, p 496 (in Mongolian)

Ostroverkhova NV, Konusova OL, Kucher AN, Sharakhov IV (2016) A comprehensive characterization of the honey bees in Siberia (Russia). In Chambo ED (ed) Beekeeping and bee conservation-advances in research. ISBN 978-953-51-2421-2

Otgonbileg K (1997) Potentiality use for beekeeping the cultured plantation plants in forest steppe zone of Mongolia. Autoreferat for PhD thesis. Ulaanbaatar, p 30 (in Russian)

Potts SG, Biesmeijer JC, Kremen C, Neumann P, Schweiger O, Kunin WE (2010) Global pollinator declines: trends, impacts and drivers. Trends Ecol Evol 25(6):345–353. https://doi.org/10.1016/j.tree.2010.01.007

Proshchalykin MY, Kupianskaya AN (2005) The bees (Hymenoptera, Apoidea) of the northern part of the Russian Far East. Far East Entomol 153:1–39

Protocol of Government of Mongolia from 1 July, 2012 Ulaanbaatar, Mongolia

Pullin AS (1996) Physiological relationships between insect diapause and cold tolerance: coevolution or coincidence? Eur J Entomol 93(2):121–129

Red Book of Mongolia (2013) Ulaanbaatar, Mongolia, p 250

Ruttner F (1988) Biogeography and taxonomy of honey bees, vol 192. Springer, Berlin

Sandag Sh (1971) Political foreign relationship of Mongolia during 1856–1919. In: History of Mongolia, vol 1. Ulaanbaatar, p 345 (in Mongolian)

Selenge, D (1974–1995) Research reports on honey bee adaptation in Mongolia. Archive material (in Mongolian)

Selenge D (1989) Perspectives beekeeping development in Mongolia, adaptation of A.m. carpathica in condition of Mongolia. Autoreferat for PhD thesis, Moscow, Russia (in Russian)

Selenge D (1994) Current research in beekeeping in Mongolia. Scientific report of the Beekeeping station for 1994. Archive material (in Mongolian)

Selenge D (2013) Bees in Mongolia. Oral presentation on the "Innovation and exhibition" conference of the Ministry of Science, Education and Culture. Ulaanbaatar, Mongolia (in Mongolian)

Shagdar E (1974) Opportunity to develop fruit and beekeeping in Mongolia. Bulletin Shaamar fruit Research central station Ed. 3, Ulaanbaatar, pp 3–12 (in Mongolian)

Sjursen H, Somme L (2000) Seasonal changes in tolerance to cold and desiccation in Phauloppia sp (Acari, Oribatida) from Finse, Norway. J Insect Physiol 46:1387–1396

Smith JM Jr (2000) "Dietary Decadence and Dynastic Decline in the Mongol Empire" Asian Topics in World History | Columbia University. J Asian History 34(1):12

Tatar Sarolta (2003) The festive beverages of the Khans. In: Proceedings of the 46 meeting of Permanent International Altaistic Conference, PIAC, 22–27 June, 2003 Ankara, Turkye, pp 457–461

Tsedev B (1982) Report on tracheal mite (*Acarapis woodi*) in honey bees. Scientific report of the Institute of Veterinary. Ulaanbaatar

Tsevegmid K (2015) Beekeeping in Mongolia. In: Proceedings of 44 international Apicultural congress, Daejeon, South Korea, 15–19 September 2015, p 121

Tsevegmid K Dooshin Selenge (2015) Prevalence and infection intensity of Nosema in honey bee (A. mellifera L.) colonies in Mongolia. In: Proceedings of 44th international Apicultural congress, Daejeon, South Korea, 15–19 September 2015, p 406

Tsevegmid K, Yañez O, Neumann P (2014) First report of honey bee viruses from Mongolia. In: Proceedings of 12th Asian Apicultural Association conference, Antalya, Turkey, 24–27 April 2014, p 14

Tsevegmid K, Neumann P, Yañez O (2016) The honey bee Pathosphere of Mongolia: European viruses in Central Asia. PLoS One 11(3):e0151164. https://doi.org/10.1371/journal.pone.0151164

Tudev L (2015) Mongolian Great Khaans. Urlakh erdem Press; Mongolia p 361 (in Mongolian)

Tuul Dooshin (2004) Research on soil composition in Central region of Mongolia Autoreferat for the PhD thesis. Darkhan, p 16 (in Russian)

Ulambayr G (2016) Beekeeping products, Newspaper "Today" vol 2314, p 17

Williams PH (1991) The bumble bees of the Kashmir Himalaya (Hymenoptera: Apidae, Bombini) – Bulletin of the British Museum. Nat Hist 60:1–204

WVIO (2011) An evaluation report on bee farming project. World Vision International Organization Mongolia, Ulaanbaatar, Mongolia, p 34

Chapter 10
Bee Diversity and Current Status of Beekeeping in Japan

Mikio Yoshiyama and Kiyoshi Kimura

Abstract Beekeeping in Japan is not a large industry; however, because of the role as pollinator, beekeeping is considered to be major agriculture sector. Beekeeping in Japan faces many problems as it does in European and North American counties. In this chapter, we will describe such problems focusing on bee diseases and parasites.

Keywords Pollination · *Apis cerana japonica* · *Apis mellifera* · Bee diseases Traditional beekeeping · Insecticide · History of beekeeping

10.1 Introduction

In comparison with other Asian countries, beekeeping is not a large industry in Japan (Fig. 10.1). Only about 5000 beekeepers register with local governments, and most of them keep less than 100 hives. In addition, most honey consumed in Japan is imported from China and other countries. Nonetheless, in Japan beekeeping is thought to be very important in agriculture due to the role of honey bee as pollinator in horticulture. Honey bees are used as pollinators for production of fruit such as apple and plum, and more than 200,000 bee hives have been introduced into

M. Yoshiyama (✉)
Honeybee Research Group, Animal Genetics Unit,
Division of Animal Breeding and Reproduction Research,
Institute of Livestock and Grassland Science, NARO,
Tsukuba, Ibaraki, Japan
e-mail: yoshiyam@affrc.go.jp

K. Kimura
Honeybee Research Group, Animal Genetics Unit,
Division of Animal Breeding and Reproduction Research,
Institute of Livestock and Grassland Science, NARO,
Tsukuba, Ibaraki, Japan

Biosphere Resource Science and Technology Program, Graduate School of Life and Environmental Sciences, University of Tsukuba, Tsukuba, Ibaraki, Japan
e-mail: kimura@affrc.go.jp

P. Chantawannakul et al. (eds.), *Asian Beekeeping in the 21st Century*,
https://doi.org/10.1007/978-981-10-8222-1_10

Fig. 10.1 Typical Japanese apiary

greenhouse production of crops such as strawberry. The peak of strawberry production is in January. Even though domestic honeys are traded in the market at higher prices than imported honey, and the demand for honey bees as pollinator remains steady, beekeeping still faces many problems in Japan, e.g., damage due to insecticides, bee diseases, parasitic mites, malnutrition, and lack of flowers, as it does in European and North American counties. In this chapter, we introduce Japanese beekeeping, focusing on the problem of bee diseases, which is considered to be the most important issue.

10.2 Honey Bee Diversity in Japan

Only one indigenous *Apis* species is distributed in Japan, the Japanese honey bee, *Apis cerana japonica* (Ruttner 1988; Engels 1999). Japanese honey bees are essentially a wild species (Fig. 10.2), but some are reared by beekeepers, and most managed hives are kept by hobbyists (Sasaki 1999; Yoshida 2000). After its introduction to Japan, the European honey bee (*Apis mellifera*) has dominated in beekeeping industry in Japan (Matsuura and Yamane 1984).

Japanese honey bees are distributed throughout the three main islands (e.g., Honshu, Kyushu, and Shikoku islands), but are not found on Hokkaido and Ryukyu islands. *A. cerana japonica* exhibit little or no genetic variation in their mitochondrial DNA (Takahashi et al. 2007). Genetically, the *A. cerana japonica* is closely related to *A. cerana* in Korea (Takahashi and Yoshida 2003).

Fig. 10.2 Japanese honey bee (*A. cerana japonica*) nest is often found in a cave. (Courtesy of Dr. Jun-ichi Takahashi (Kyoto Sangyo University))

The European honey bee was introduced in the late nineteenth century, as described below. Most European honey bees kept in Japan are *A. mellifera ligustica,* although some *A. mellifera carnica* have been imported from Balkan countries. Due to quarantine limitations in recent years, queens have been imported only from Australia and Slovenia. In 2015, 2127 queens and 748 queens were imported from Australia and Slovenia, respectively (Animal Quarantine Service, Ministry of Agriculture, Forestry and Fishery 2015). Whole colony import has not been performed within recorded history. Mitochondrial DNA surveys revealed only haplotype C1 (Takahashi et al. 2014), indicating that the genetic variation of *A. mellifera* in Japan is much smaller than in American countries and Australia (Oldroyd et al. 1995; Koulianos and Crozier 1996; Franck et al. 2001; Clarke et al. 2002).

10.3 Historical Records of Beekeeping

10.3.1 History of Japanese Beekeeping

The history of beekeeping using Japanese honey bee (*A. cerana japonica*) dates back to about 1400 years ago (Bureau of Animal Husbandry, Ministry of Agriculture and Forestry 1966; Sasaki 1999; Yoshida 2000). The first literature in which the word "honey bee" is used is the *Nihon-Shoki* (the second-oldest book of classical Japanese history). According to this book, in 643 AD, Yogi, a naturalized

Korean, tried to start beekeeping, but the attempt ended in failure. In later literature, the term "honey bee" was mentioned, revealing that honey was used as an important tribute. During this period, most honey was collected from natural wild hives; however, people started to capture and keep wild bees around 1000 years ago, representing the dawn of beekeeping in Japan. In the Edo period (ca. 1600–1860 AD), beekeeping became fully developed, and books on ecology and beekeeping technology were published. Although beekeeping using *A. cerana japonica* was not highly productive, beekeeping techniques were conducted widely (Fig. 10.3).

In the Meiji era (1868–1912 AD), Japan opened its doors to overseas and rushed towards westernization. The new government promoted aggressive measures in multiple industrial sectors, one of which was beekeeping. In 1877, an Italian strain of *A. mellifera* was first imported from the USA and maintained in a national agriculture experimental station, marking the beginning of a new era in beekeeping. Immediately thereafter, modern beekeeping began in the private sector. The beekeeping industry gained a great attention as a new industry during the period of industrialization following the Russo-Japanese War (1904–1905). Migratory beekeeping was also initiated around 100 years ago.

After World War II, in response to the high demand for sweet foods, beekeeping developed rapidly, with honey production peaking in the 1970s. In the 1960s, the demand of bees in greenhouse crops has been increasing. Starting in the 1980s, land development occurred and led to the change of natural environment. Honey bees

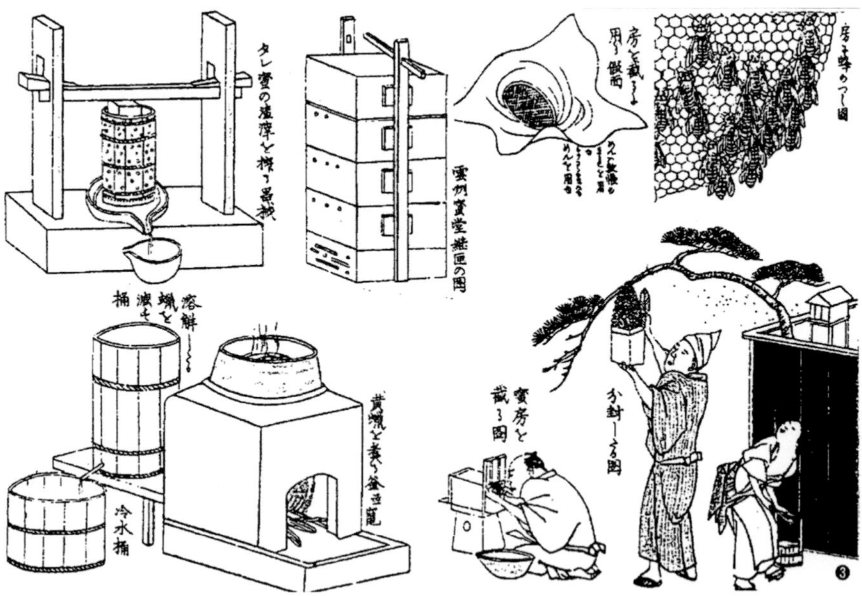

Fig. 10.3 Illustrations from *Text book of beekeeping* using Japanese honey bee before introduction of *Apis mellifera* (*Hachimitsu Ichiran* (Chart of Honey) 1872)

have higher risks to be affected by pesticides used in the farming practices. The cultivation conditions of agricultural land have also changed greatly, and the cultivation area of their former major nectar sources was reduced to about 11% for milk vetch and about 5% for rapeseeds, in comparison with that of in the year 1965. Imports of cheap honey sharply increased, prices of domestically produced honey were sluggish, and beekeepers aged. As a result of these factors, the number of beekeepers fell from 9499 in 1985 to 4790 in 2005.

10.3.2 Modern Beekeeping in Japan

Currently, some 200,000 colonies of bees are maintained in Japan, of which around 80,000 colonies are being used for pollination of greenhouse horticulture such as strawberry (MAFF 2016a). The beekeepers prepared hives (2–3 frames) for pollination before winter and introduced these small hives into the greenhouses for strawberry production (Fig. 10.4). These hives are kept in the houses until April–May.

Supply of pollination colonies is not sufficient to meet such demand, and beekeepers often need to supply bee colonies for both honey and pollination services (e.g., for melon) in the early spring. In 2009, the shortage of pollination hives has become a serious problem (Kimura 2011).

Fig. 10.4 A bee hive is installed in a greenhouse for pollination of strawberry production

On the other hand, honey produced within Japan accounts only for less than 10% of the honey consumed domestically. The price of domestically produced honey has remained high. In addition, numbers of both honey bee colonies and beekeepers started to decline in 1980. The rates of decline had leveled off since around 2000, and in the last few years, the numbers have slightly increased.

10.4 Bee Forage Resources

Due to Japan is located in a temperate zone, many flowers are available as honey and pollen sources in the spring. In Kanto (area around Tokyo), flowers start to bloom at the same time, starting at the end of March and peaking in May and June and the number of flowering species declines thereafter. Many species of flowers bloom in summer, but honey production is not high during this season due to the hot weather. Consequently, many beekeepers migrate to cooler regions in Northern Japan, e.g., Hokkaido and/or mountainous regions.

The most important flowers used as nectar sources by bees in Japan are listed in Table 10.1 (Japan Beekeeping Association 2005). Sixteen of the flower species are classified as "very important" plants, and 21 are classified as "important." Sasaki (2013) listed 647 species as pollen and nectar sources, of which 56.9% are native. Nevertheless, 37 major species of native honey plants drop to 46%, indicating that honey bees also depend on introduced species of plants as their food sources.

10.5 Local Knowledge of Honey Bees

As described above, traditional beekeeping using *A. cerana japonica* has a long history, but introduction of the *A. mellifera* caused a decline in *A. cerana japonica* beekeeping. However, in some parts of Japan, the traditional type of beekeeping still exists such as Tsushima Island, located in the Japan Sea between the Korean Peninsula and Kyushu island (Sasaki 1999; Yoshida 2000). *Hatto*, traditional beehives, can be seen throughout most of the island (Fig. 10.5).

There is no European beekeeping in Tsushima Island. In addition to traditional beekeeping, over the past 10 years *A. cerana japonica* beekeeping has become popular, with the formation of beekeeping hobby groups and the publication of many introductory books on beekeeping. Some groups have tried to introduce western-style methods to *A. cerana japonica* beekeeping (e.g., using Langstroth hives), whereas others have tried to improve traditional beekeeping. Many people believe that honey from *A. cerana japonica* has special medical effects than honey from *A. mellifera*, but no concrete scientific evidence supports this claim.

Under these circumstances, bee diseases have been transmitted from European honey bees into *A. cerana japonica*. For example, European foulbrood had not been reported to *A. cerana japonica*, but infected *A. cerana japonica* hives have been reported frequently over last few years (Uemura et al. 2013). In addition, infestation

Table 10.1 Major nectar source plants in Japan

	Scientific name	Introduced species	Flowering period
Very important[a]			
Satsuma Orange, Mikan	*Citrus unshiu*	○	Early Summer
Japanese Snowbell	*Styrax japonica*		Early Summer
Amur Corktree	*Phelodendron amurense*		Summer
Round Leaf Holly	*Ilex rotunda*		Spring
Japanese Chinquapin	*Eleutherococcus sciadophylloides*		Late Summer
Japanese Lime	*Tilia japonica*		Summer
Buckwheat	*Fagopyrum esculentum*		Summer-Fall
Longstalk Holly	*Ilex pedunculosa*		Early Summer
Romerillo	*Bidens pilosa*	○	Late Spring-Fall
Horsechestnut, Japanese Marronnier	*Aesculus turbinata*	○	Early Summer
Rapeseed	*Brasica rapa*	○	Spring
Falseacacia, Locustacacia	*Robiniapseudo-acacia*	○	Early Summer
Wax tree	*Rhus sucedanea*	○	Early Summer
White Clover	*Trifolium repens*	○	Summer
Apple	*Malus pumila*	○	Spring
Chinese Vetch	*Astragalus sinicus*	○	Spring
Important[a]			
Thistle	Genus *Cirsium*		Late Summer-Fall
Bastard Indigobush	*Amorpha fruticosa*	○	Fall
Japanese Knotweed	*Fallopia japonica*		Summer
Persimmon	*Diospyros kaki*		Early Summer
Japanese Prickly-ash	*Zanthoxylum ailanthoides*		Summer
Citrus fruits	Genus *Citreae*	○	Early Summer
Amur Corktree	*Phellodendron amurense*		Summer
Japanese Chestnut	*Castanea crenata*		Early Summer
Japanese Raisin Tree	*Hovenia dulcis*		Summer
Cherry Blossom	Sub-genus *Cerasus*		Spring
Cherry (fruit)	*Prunus avium*	○	Spring
Canada Goldenrod	*Solidago canadensis*	○	Fall
Caster Aralia	*Kalopanax septemlobus*		Early Summer
Dandelion	Genus *Taraxacum*	○	Spring
Glossy Privet	*Ligustrum lucidum*	○	Early Summer
Runner Bean	*Phaseolus coccineus*	○	Summer
Bee bee Tree	*Euodia daniellii*	○	Spring-Early Summer
Hairy Vetch	*Vicia villosa*	○	Late Spring
Safflower	*Carthamus tinctorius*	○	Summer
Japanese Clethra	*Clethra barbinervis*		Summer
American Tulip Tree	*Liriodendron tulipifera*	○	Spring

[a]Very important and Important Plants listed in "The Honeybee Plants of Japan" (Japan Beekeeping Association 2005)

Fig. 10.5 Traditional beekeeping in Tsushima Island, Japan. *Hatto* (hollow logs) are used for bee hives (Courtesy of Dr. Jun-ichi Takahashi (Kyoto Sangyo University))

by tracheal mites (*Acarapis woodi*) represents a serious problem for *A. cerana japonica* beekeeping (Maeda 2016). This mite was first reported in *A. cerana japonica* in 2011 (Kojima et al. 2011). In addition, Thai sacbrood causes serious problems in *A. cerana japonica* (Kojima et al. 2011) (see also Sect. 10.6).

10.6 Pests and Diseases of Honey Bees

Most of the diseases and pests affecting global apiculture are also present throughout Japan. In this section, we describe the presence of pathogens, parasites, and pests that influence the health of honey bee colonies in Japan, as well as methods for controlling these diseases and pests in Japanese apiculture. Main pests and diseases of honey bees in Japan are listed in Table 10.2.

10.6.1 Bacteria

10.6.1.1 American Foulbrood (*Paenibacillus larvae*)

American foulbrood (AFB), the most deleterious disease of honey bee broods (Bailey and Ball 1991; Genersch 2010), is a serious problem in apiculture all over the world. The causative agent is the Gram-positive, spore-forming bacterium

Table 10.2 Summary of pathogens, parasites, and pests associated with honeybees in Japan

Name of bee diseases	Disease incidence	Domestic animal infectious disease	Notifiable animal infectious disease	Listed in the OIE terrestrial Animal Health Code
Acarapisosis	O		O	O
American Foulbrood	O	O		O
European Foulbrood	O	O		O
Small hive beetle infection				O
Tropilaelaps				O
Varroasis	O		O	O
Nosemosis	O		O	
Chalkbrood disease	O		O	

Paenibacillus larvae (Genersch 2010). Young larvae become infected by ingestion of spores of *P. larvae* from contaminated food. In the larval midgut, the spores germinate, and the vegetative bacteria proliferate in the lumen, eventually breaching the midgut epithelium and invading the hemocoel (Yue et al. 2008; Ebeling et al. 2016). Spores of *P. larvae* have the ability to survive severely adverse environmental conditions and remain infective in hives for more than 35 years (Hasemann 1961). Due to this high degree of contagiousness in colonies and the high lethality in larvae, AFB is designated as a domestic animal infectious disease in Japan. AFB is subject to official control by the livestock hygiene service center. Affected colonies are subject to compulsory execution, mainly by incineration.

There are four genotypes (ERIC I–IV) of *P. larvae* strains (Genersch et al. 2006). In addition, Morrissey et al. (2015) developed a multilocus sequence typing (MLST) scheme to describe the genetic structure of *P. larvae*. Epidemiological studies showed that ERIC I and II are frequently isolated from AFB-infected hives. The two genotypes differ in virulence, with ERIC II strains more virulent against individual larvae and ERIC I strains less so (Genersch et al. 2005; Rauch et al. 2009; Genersch 2010). Recently, Hirai et al. (2016) genotyped Japanese isolates and detected the existence of ERIC I and II in Aichi prefecture in Japan. Apiten®, a formulation of the antibiotic mirosamicin, is commercially available as a prophylactic against for AFB (Kawashima 2000), and remains the only preventative drug approved by the Japanese government for control of the disease.

10.6.1.2 European Foulbrood (*Melissococcus plutonius*)

European foulbrood (EFB) is another bacterial disease of honey bee broods. In this case, the etiological agent is the Gram-positive lanceolate coccus *Melissococcus plutonius* (Forsgren 2010). *M. plutonius* is digested along with contaminated food during feeding by nurse bees. The bacteria multiply within the larval gut and kill larvae aged 4–5 days (Bailey and Ball 1991; Forsgren 2010). At present, EFB is widespread and present in most countries, with the exception of New Zealand (Ellis and Munn 2005). In recent years, the Ministry of Agriculture, Forestry and Fisheries has reported more than 100 confirmed cases of foulbrood disease annually (MAFF 2017a). However, the annual numbers of EFB cases are not known with precision because both AFB and EFB are referred to collectively as foulbrood diseases. Since the first suspected case was described in the 1950s (Azuma 1956), incidences of EFB have been sporadically reported. Recently, phenotypically and genetically atypical *M. plutonius* strains have been detected in both *A. mellifera* and *A. cerana japonica* manifesting clinical signs of EFB (Arai et al. 2012; Takamatsu et al. 2014).

10.6.2 Fungi

10.6.2.1 Nosemosis (*Nosema apis* and *Nosema ceranae*)

Nosemosis is one of the most prevalent worldwide diseases of honey bees caused by microsporidia. Two species of microsporidia, *Nosema apis* (Zander) and *Nosema ceranae* (Fries), are intestinal parasites in honey bee (Bailey and Ball 1991; Fries 2010). *N. apis* was first identified in the European honey bee, more than 100 years ago. This Nosema species became globally distributed, and is the primary cause of Nosemosis, whereas *N. ceranae* is a newly emerging microsporidian parasite first discovered in 1994 in the Asian honey bee *A. cerana* in a sample from China (Fries et al. 1996). Since then, *N. ceranae* has spread worldwide amongst European honey bee (Higes et al. 2006; Huang et al. 2007), and over the past two decades has become more prevalent (Klee et al. 2007; Paxton et al. 2007; Martín-Hernández et al. 2012). At the colony level, dysentery is a clinical sign of *N. apis* infection (Bailey 1967), which shortens the lifespans of both queens and adult bees (Wang and Moeller 1970), whereas no colony-level symptoms of *N. ceranae* infection have been observed (Higes et al. 2008). Sublethal doses of pesticides elevate the mortality of honey bees infected with *N. ceranae*; i.e., this pathogen may synergistically and negatively affect honey bee by increasing their susceptibility to stressors (Vidau et al. 2011; Aufauvre et al. 2012; Doublet et al. 2015). No commercial medication is made available and registered for this disease to date in Japan.

10.6.2.2 Chalkbrood (*Ascosphaera apis*)

Chalkbrood, a fungal disease of honey bee larvae caused by *Ascosphaera apis*, is currently found in honey bee colonies throughout the world (Aronstein and Murray 2010). Spores ingested by larvae germinate in the midgut lumen and penetrate the midgut epithelium. Subsequently, the fungal mycelium grows and penetrates the gut wall and breaks out of the posterior end of the larvae (Bailey and Ball 1991;

Aronstein and Murray 2010). Although Chalkbrood usually does not cause serious damages to colonies, it can cause a significant reduction in honey production (Heath 1982; Zaghloul et al. 2005). Furthermore, this disease is considered to be stress-related, and statistical evidence demonstrated that *N. ceranae* infection in spring and *Varroa* infestation in summer are associated with Chalkbrood outbreaks in the next season (Hedtke et al. 2011). The recommended approach is to have good management and sanitation in beekeeping.

10.6.3 Viruses

More than 20 different viruses have been isolated from honey bees (Chen and Siede 2007), and most of them can be transmitted by *Varroa*. The mites act as biological vectors for Kashmir bee virus (KBV), Sacbrood virus (SBV), acute bee paralysis virus (ABPV), Israeli acute paralysis virus (IAPV), and deformed wing virus (DWV) (Boecking and Genersch 2008).

10.6.3.1 Deformed Wing Virus (DWV)

DWV was initially discovered as a Japanese strain of Egyptian bee virus in adult honey bees from a *Varroa*-infested colony in Japan in 1982 (Ball 1983; Bailey and Ball 1991). DWV causes wing deformities and abdominal bloating in emerging adults, and is currently widely distributed in close relationship to *Varroa* (Allen and Ball 1996). In the absence of *Varroa destructor*, DWV is transmitted vertically via eggs and semen, as well as horizontally via larval food from nurse bees, and causes covert infection without any apparent symptoms (Yue and Genersch 2005; Chen et al. 2006; Yang and Cox-Foster 2007; Yue et al. 2007). Malformed appendages of emerging adults only arise when DWV is transmitted to pupal stages via *V. destructor* (de Miranda and Genersch 2010; Genersch and Aubert 2010). Overt DWV infection often results in death in the pupal stage, or dead and dying adult bees with shriveled wings harboring a high titer of virus particles, and is sometimes associated with bloated abdomens and discoloration (Chen et al. 2005). DWV plays an important role in colony collapse via a complex interaction between *Varroa* and honey bee population dynamics (Martin et al. 1998; Martin 2001).

10.6.3.2 Sacbrood Virus (SBV)

SBV has been first identified in the United States in 1913 (White 1913) and has been widely distributed in every continent (Ellis and Munn 2005). SBV affects both larvae and adults, but 2-day-old larvae are most susceptible to infections (Ball and Bailey 1997). Although SBV does not cause obvious signs of diseases in adult bees, affected bees may have shorter lifespans (Bailey 1969; Bailey and

Fernando 1972). Young larvae become infected through SBV-contaminated food, and the virus replicates within the larvae. SBV-infected larvae appear as water-filled sacs, giving the virus its name. SBV is more virulent in *A. cerana*, than in *A. mellifera*. The disease was first reported in *A. cerana* in Thailand in 1976. Due to differences in physiochemical and serological properties, this virus was designated as Thai Sacbrood Virus (TSBV) (Bailey et al. 1982). TSBV is widespread to Nepal and India, and more than 95% of affected colonies are killed (Rana et al. 1986; Verma et al. 1990). In Japan, infection by SBV has been reported in both *A. mellifera* and *A. cerana japonica* colonies. Several reports have described extensive brood removal from the hive, a behavior observed only in SBV-infected colonies of *A. cerana japonica* (Yamashita and Tanaka 2010; Kojima et al. 2011).

10.6.3.3 Chronic Bee Paralysis Virus (CBPV)

CBPV mainly affects adult honey bees (Bailey et al. 1963). The typical symptom is a trembling motion of bodies and wings, crawling on the ground due to lack of flight ability, and a hairless shiny body surface (Bailey et al. 1963, 1968). A field survey revealed that the frequency of CBPV infection was low, and no seasonal pattern appeared (Tentcheva et al. 2004). In Japan, CBPV infection is rare in *A. mellifera*, and CBPV could not be found in *A. cerana japonica* (Morimoto et al. 2012).

10.6.3.4 Kashmir Bee Virus (KBV)

KBV infection in honey bee populations is less prevalent, and was detected at low frequency in a field survey in France (Tentcheva et al. 2004). In Japan, the frequency of KBV is also considered to be low (Kojima et al. 2011). Currently, no chemotherapeutic agents are available for treating viral diseases of honey bees in Japan. Since *Varroa* can transmit viruses, prevention or control of mite infestation in the bee colonies is important for honey bee virus control strategies.

10.6.4 Mites

10.6.4.1 Varroasis (*Varroa destructor*)

The ectoparasitic mite *V. destructor* is one of the most devastating pests in worldwide apiculture (Anderson and Trueman 2000; Cornman et al. 2010). Of several haplotypes of *V. destructor*, only two (Japanese and Korean) are known to be capable of reproducing in *A. mellifera* (Anderson and Trueman 2000). In this species, the Japanese haplotype spread into Thailand and Brazil, and later into North

America (De Jong et al. 1982a; de Guzman et al. 1999). The Korean haplotype has a broader worldwide distribution than the Japanese haplotype, and is considered more pathogenic (Anderson and Trueman 2000). *V. destructor* directly causes serious damage to honey bees by feeding on the hemolymph of the larvae, pupae, and adult bees (Rosenkranz et al. 2010), decreasing body weight (De Jong et al. 1982b) and life span (Amdam et al. 2004). Furthermore, *V. destructor* can serve as a vector for several honey bee viruses, including KBV, SBV, ABPV, IAPV, and DWV (Boecking and Genersch 2008), thereby contributing to a decline in the honey bee population (Chen and Siede 2007; Wilfert et al. 2016). Without proper treatment, most bee colonies infected by *V. destructor* parasites collapse in less than 3 years (Rosenkranz et al. 2010). Only two acaricides are commercially available for controlling *Varroa* in Japan. Apistan®, made with the pyrethroid tau-fluvalinate, has been used for the past three decades. However, resistance of *V. destructor* to fluvalinate is widely documented (Thompson et al. 2002; Gracia-Salinas et al. 2006). In 2009, Apivar® Amitraz strips were approved for *Varroa* control and became available to beekeepers in Japan. However, some *Varro*a populations exhibited resistance to Amitraz (Elzen et al. 2000; Sammataro et al. 2005).

10.6.4.2 Acarapisosis (*Acarapis woodi*)

Acarapisosis is a disease of the adult honey bee *A. mellifera* and other Apis species caused by the tracheal mite *Acarapis woodi* (OIE 2008). *A. woodi* parasites reproduce in the tracheal tubes of adult honey bees, feeding on the hemolymph of hosts (Pettis and Wilson 1996). Heavy infestation by *A. woodi* causes colony death by reducing honey production (Eischen et al. 1989) and decreasing the hive's ability to thermoregulate (McMullan and Brown 2009). *A. woodi* was first discovered in the United Kingdom in 1919 (Rennie 1921). Since then, this mite has spread worldwide, except in Sweden, Norway, Denmark, New Zealand, and Australia, and the state of Hawaii (Sammataro et al. 2000). The first report of *A. woodi* in Japan was in the Japanese honey bee, *A. cerana japonica*, in 2010 (MAFF 2011). *A. woodi* is distributed widely in this species (Maeda 2015), and colonies infested by this mite exhibit elevated rates of mortality and colony demise during winter (Maeda and Sakamoto 2016). On the other hand, mite load in the European honey bee, *A. mellifera*, has been rarely observed in Japan (Kojima et al. 2011; Maeda 2015). No tracheal mite-specific chemotherapeutic agent is available for prevention of acarapisosis in Japan to date, and no commercial products have been registered for this purpose.

10.6.4.3 *Tropilaelaps* Mites

There is no record that *Tropilaelaps* species have been described in Japan thus far.

10.6.5 *Protozoans*

10.6.5.1 *Crithidia mellificae*

C*rithidia mellificae* is a trypanosomatid parasite of *A. mellifera*, first reported in Australia in 1967 (Langridge and McGhee 1967). *C. mellificae* are recognized as a contributory factor in honey bee decline, and has been found in *A. mellifera* throughout the world, including Japan (Morimoto et al. 2013; Ravoet et al. 2013; Schwarz et al. 2015). Infection by *C. mellificae* has been detected in the Asian honey bee, *Apis cerana cerana*, in China (Yang et al. 2013), but not in the Japanese honey bee *A. cerana japonica* (Morimoto et al. 2013).

10.6.5.2 *Lotmaria passim*

Lotmaria passim infection status of honey bees has not been extensively studied until recently. *L. passim*, and *L. passim* is currently the predominant trypanosomatid species in honey bee population around the world (Ravoet et al. 2015; Schwarz et al. 2015). *L. passim* was detected in an *A. mellifera* sample from Japan (Ravoet et al. 2015).

10.6.6 *Pests*

10.6.6.1 Small Hive Beetle

The small hive beetle (*Aethina tumida*) (SHB) is a parasite and scavenger of honey bee colonies native to sub-Saharan Africa, but is rare and therefore not especially problematic for honey bee colonies (Neumann and Ellis 2008). Since the discovery of SHB in 1996 in the southeastern United States, invasion by SHB has been reported in many other countries, including Canada, Mexico, Jamaica, Cuba, and across the east coast of Australia (Neumann et al. 2016). In 2014, an outbreak of SHB was recorded in the Philippines, where managed colonies of European honey bee spices were severely damaged, and most of them collapsed (Brion 2015). Thus, SHB poses considerable potential risks to honey bees in Japan, but to date there has been no record of SHB infestation in Japan.

10.6.6.2 Wax Moth

Two wax moth species, the greater wax moth *Galleria mellonella* and the lesser wax moth *Achroia grisella*, are ubiquitous pests of honey bee colonies worldwide (Ellis et al. 2013). The larvae of both moth species affect honey bee colonies by feeding on wax combs, pollen, stored honey and cast larval skins. In Japan, *Achroia innotata* tends to be a major pest in *A. cerana japonica*, and heavy infestation often causes the bees to abscond (Yoshida 2000). These moths can cause significant damage to stored beekeeping equipment, especially in weak or stressed colonies. Therefore, it is recommended that colonies should be kept strong and healthy by maintaining optimal sanitation, e.g., removal of wax and debris accumulated on the bottom of the hives.

10.6.6.3 Hornet

Hornets of genus *Vespa* are serious predators of honey bees in Asian countries (Matsuura 1988). In Japan, seven *Vespa* species are distributed, and among these, giant hornet (*Vespa mandarinia japonica*) causes most destructive damage on honey bee colonies. *Vespa mandarinia japonica* attack honey bee hives in groups of 10–20 workers, and often exterminate whole colonies (Matsuura and Yamane 1990). However, *A. cerana japonica* counter-attack against *V. mandarinia japonica* via a collective behavior called a "bee ball," enveloping the hornets and killing them with heat and carbon dioxide (Ono et al. 1995; Sugahara and Sakamoto 2009). On the other hand, catastrophic damage to hives is often observed in the hives of a non-native bee, *A. mellifera*. The defensive behavior of *A. mellifera* is not as well-organized, and is consequently less efficient (Ken et al. 2005; Abrol 2006). Yellow-legged hornet, *Vespa velutina* is indigenous to the Southeast Asia, but is currently expanding from its native area into southern Asia, and invaded South Korea in 2003 (Choi et al. 2012). In Japan, invasion by *V. velutina* was first recorded in 2012 on Tsushima Island (Sakai and Takahashi 2014). Commercial cardboard coated with sticky substances can be used to capture hornets. Attachment of a hornet trap device in front of the hive entrance can also decrease the damage from *Vespa* attacks.

10.6.7 Vertebrate Predators

10.6.7.1 Bears

In Japan, there are two bear species: Japanese black bear, *Ursus thibetanus japonicus*, which resides on the main Honshu island and Shikoku island; and Hokkaido brown bear, *U. arctos yesoensis*, a bigger bear, which resides in Hokkaido island (Tsuruga et al. 1994; Kurihara 2010). Statistical information about damage to bee colonies associated with bears in Japan has been reported annually. In 2011, a total of 1551 colonies were attacked by bears, resulting in estimated losses of 53,580,000 Japanese yen (Japan Beekeeping Association 2011). The use of electric fences for bear abatement is preventive to some degree, but damage to hives can still occur.

10.6.8 *Other Problems of Beekeeping*

10.6.8.1 Distribution of Beehives

As mentioned earlier, supply of honey bees in Japan is chronically insufficient. The reason that this has become such a significant issue has to do primarily with the issues related to distribution (Kimura 2011). Populations of pollinator honey bees have temporarily decreased in some areas. However, as of spring in 2009, the number of honey bee colonies throughout the nation did not markedly decrease, but the imbalances in distribution made it difficult to provide honey bees in areas that did not have enough of them. Moreover, in spring, beekeepers already start preparing for honey extraction, and cannot afford to prepare honey bees as pollinators. To address this issue, the Ministry of Agriculture, Forestry, and Fisheries has launched a system for adjusting the supply and demand of honey bees (MAFF 2016b). This system is intended "to determine the number of bees that are available for supply, as well as investigate shortages of bees, in each and every prefecture; to adjust supply and demand in each prefecture; and in the event of persistent shortage, to provide the necessary information to the parties involved, so that adjustment of supply and demand can be made beyond the borders of prefectural borders." This may have been proven effective, and we have not experienced large-scale honey bee shortages since 2010.

10.6.9 *Pesticides*

Agricultural damage to honey bees caused by pesticides has been an issue for many years. Recently, the effects of neonicotinoid-based pesticides on honey bees have been of concern, and some stakeholders have pointed out that these agents are the primary causes of honey bee decline. However, the effects of pesticides on honey bees have not been fully characterized (Kimura 2011). Some of the insecticide damage to bees in Japan occurs in rice paddies. The development of spotted rice grains is caused by stink bugs of sub-order Heteroptera. In summer, insecticides are widely used to control this problem. Honey bees visit paddies to collect nectar and pollen from nearby flowers, and are exposed to these chemicals while also collecting pollen from rice. Contaminated bees return to the hive, causing piles of dead bees to accumulate in front of the hive entrance (Kimura et al. 2014). In addition, many worker bees are dead before returning to the hives. In order to avoid insecticide exposure to the worker bees, the government is promoting the following measures (MAFF 2017b):

1. Active communication between farmers and beekeepers
2. Relocation of hives from spraying areas (by beekeepers)
3. Use of agricultural chemicals which are difficult for bees to be exposed (e.g., granules) (by farmers)

The government continues to promote these measures and also verifies the effectiveness of countermeasures.

10.6.10 Aging

Not only in beekeeping, but throughout agriculture, the aging of farmers is a serious problem in Japan. The average age of beekeepers was 72 years in 2009 (Kimura unpublished survey), and it is safe to assume that this population continues to age.

10.7 Other Non-*Apis* Species Found in the Region

There are 15 bumblebee species in Japan (Kinoda et al. 2013). Also, the buff-tailed bumblebee *Bombus terrestris* was introduced into Japan from Europe in 1991 for pollination of tomato plants.

B. terrestris is widely used for pollinating tomatoes grown in greenhouses in Japan. This practice risked biological invasion by this species, and indeed, *B. terrestris* has become naturalized, especially in Hokkaido (Matsumura et al. 2004; Inoue et al. 2008). Accordingly, the Ministry of Environment classified *B. terrestris* as invasive and requiring regulation by law; and, its use was permitted on the condition it be used only for agriculture and strict measures to avoid escape.

The Ministry has tried to replace *B. terrestris* with *Bombus ignitus* and other native Japanese bumblebees. However, 60,000 hives of buff-tailed bumblebees are still distributed annually to meet the farmer need for pollination service (ME and MAFF 2017).

The horned-face bee (horn-faced mason bee), *Osmia cornifrons*, is a solitary bee indigenous to Northern Asia. In Japan, this species provides up to 80% of the apple pollination (Maeta 1990, 1993).

This *Osmia* species is used for pollination in 80% of the apples in Aomori prefecture (the northernmost prefecture of Honshu island, where about half of the apples in Japan are produced). The use of *O. cornifrons* is increasing in Japan since *O. cornifrons* achieves higher pollination rates than that of honey bees. In addition, this species is stingless, making it safe to handle. Nevertheless, infestation by acrid mites has become a problem as *O. cornifrons* has expanded (Yamada 1986).

10.8 Future Perspectives

Considering the limitation of nectar sources and the aging of beekeepers, it is unlikely that honey production will expand in the future. On the other hand, the demand for honey bees for use as pollinators remains high. Thus, the most important consideration for Japanese beekeeping is maintaining a sufficient supply of bee colonies for both honey production and pollination service.

References

Abrol DP (2006) Defensive behaviour of *Apis cerana* F. against predatory wasps. J Apic Sci 50(2):39–46
Allen MF, Ball BV (1996) The incidence and world distribution of honey bee viruses. Bee World 77:141–162
Amdam GV, Hartfelder K, Norberg K et al (2004) Altered physiology in worker honey bees (Hymenoptera: Apidae) infested with the mite *Varroa destructor* (Acari: Varroidae): a factor in colony loss during overwintering? J Econ Entomol 97:741–747
Anderson DL, Trueman JW (2000) *Varroa jacobsoni* (Acari: Varroidae) is more than one species. Exp Appl Acarol 24(3):165–189
Animal Quarantine Service, Ministry of Agriculture, Forestry and Fishery (2015) Annual report of animal quarantine 2015. Animal Quarantine Service, Ministry of Agriculture, Forestry and Fishery, Tokyo. 145pp (In Japanese)
Arai R, Tominaga K, Wu M et al (2012) Diversity of *Melissococcus plutonius* from honeybee larvae in Japan and experimental reproduction of European foulbrood with cultured atypical isolates. PLoS One 7:e33708
Aronstein KA, Murray KD (2010) Chalkbrood disease in honey bees. J Invertebr Pathol 103:S20–S29. https://doi.org/10.1016/j.jip.2009.06.018
Aufauvre J, Biron DG, Vidau C et al (2012) Parasite–insecticide interactions: a case study of *Nosema ceranae* and fipronil synergy on honeybee. Sci Rep 2(326):1–7. https://doi.org/10.1038/srep00326
Azuma R (1956) About foulbrood of honeybee. J Jpn Vet Med Assoc 9:255–259
Bailey L (1967) *Nosema apis* and dysentery of the honeybee. J Apic Res 6:121–125
Bailey L (1969) The multiplication and spread of sacbrood virus of bees. Ann Appl Biol 63:483–491
Bailey L, Ball BV (1991) Honey bee pathology, 2nd edn. Academic Press, London. 193pp
Bailey L, Fernando EFW (1972) Effects of sacbrood virus on adult honey-bees. Ann Appl Biol 72:27–35
Bailey L, Gibbs AJ, Woods RD (1963) Two viruses from adult honeybees (*Apis mellifera* Linnaeus). Virology 21:390–395
Bailey L, Gibbs AJ, Woods RD (1968) The purification and properties of chronic bee-paralysis virus. J Gen Virol 2:251–260
Bailey L, Carpenter JM, Woods RD (1982) A strain of sacbrood virus from *Apis cerana*. J Invertebr Pathol 39:264–265
Ball BV (1983) The association of *Varroa jacobsoni* with virus diseases of honey bees. Exp Appl Acarol 19:607–613
Ball BV, Bailey L (1997) Viruses. In: Morse RA, Flottum K (eds) Honeybee pest, predators, & diseases. The A. I. Root Co., Medina, pp 11–31
Boecking O, Genersch E (2008) Varroasis—the ongoing crisis in bee keeping. J Consum Protect Food Safety 3(2):221–228
Brion ACB (2015) Small hive beetle poses threat to bee industry. In: The Philippine Star. http://www.philstar.com/agriculture/2015/02/22/1426217/small-hive-beetle-poses-threat-bee-industry. Accessed 10 Sept 2017
Bureau of Animal Husbandry, Ministry of Agriculture and Forestry (1966) History of livestock development. Chuou-Kouron, Tokyo, 1843 (in Japanese)
Chen YP, Siede R (2007) Honey bee viruses. Adv Virus Res 70:33–80
Chen YP, Higgins JA, Feldlaufer MF (2005) Quantitative real-time reverse transcription–PCR analysis of deformed wing virus infection in the honeybee (*Apis mellifera* L.) Appl Environ Microbiol 71:436–441
Chen YP, Pettis JS, Collins A et al (2006) Prevalence and transmission of honeybee viruses. Appl Environ Microbiol 72:606–611
Choi MB, Martin SJ, Lee JW (2012) Distribution, spread, and impact of the invasive hornet *Vespa velutina* in South Korea. J Asia Pac Entomol 15:473–477

Clarke KE, Rinderer TE, Franck P et al (2002) The Africanization of honeybees (*Apis mellifera* L.) of the Yucatan: a study of a massive hybridization event across time. Evolution 56(7): 1462–1474

Cornman RS, Schatz MC, Johnston JS et al (2010) Genomic survey of the ectoparasitic mite *Varroa destructor*, a major pest of the honey bee *Apis mellifera*. BMC Genomics 11:602. https://doi.org/10.1186/1471-2164-11-602

de Guzman LI, Rinderer TE, Stelzer JA (1999) Occurrence of two genotypes of *Varroa jacobsoni* Oud. in North America. Apidologie 30:31–36

De Jong D, Morse RA, Eickwort GC (1982a) Mite pests of honey bees. Annu Rev Entomol 27:229–252

De Jong D, de Jong PH, Gonçalves LS (1982b) Weight loss and other damage to developing worker honeybees from infestation with *V. jacobsoni*. J Apic Res 21:165–216

de Miranda J, Genersch E (2010) Deformed wing virus. J Invertebr Pathol 103:S48–S61

Doublet V, Labarussias M, de Miranda JR et al (2015) Bees under stress: sublethal doses of a neonicotinoid pesticide and pathogens interact to elevate honey bee mortality across the life cycle. Environ Microbiol 17:969–983. https://doi.org/10.1111/1462-2920.12426

Ebeling J, Knispel H, Hertlein G et al (2016) Biology of *Paenibacillus larvae*, a deadly pathogen of honey bee larvae. Appl Microbiol Biotechnol 100(17):7387–7395

Eischen FA, Cardoso-Tamez D, Wilson WT et al (1989) Honey production of honey bee colonies infested with *Acarapis woodi* (Rennie). Apidologie 20(1):1–8

Ellis JD, Munn PA (2005) The worldwide health status of honey bees. Bee World 86:88–101

Ellis JD, Graham JR, Mortensen A (2013) Standard methods for wax moth research. J Apic Res 52(1):1–17

Elzen P, Baxter J, Spivak M et al (2000) Control of *Varroa jacobsoni* Oud. resistant to fluvalinate and amitraz using coumaphos. Apidologie 31:437–441

Engel MS (1999) The taxonomy of recent and fossil honey bees (Hymenoptera: Apidae: Apis). J Hymenopt Res 8:165–196

Forsgren E (2010) European foulbrood in honey bees. J Invertebr Pathol 103:S5–S9. https://doi.org/10.1016/j.jip.2009.06.016

Franck P, Garnery L, Loiseau A et al (2001) Genetic diversity of the honeybee in Africa: microsatellite and mitochondrial data. Heredity 86(4):420–430

Fries I (2010) *Nosema ceranae* in European honey bees (*Apis mellifera*). J Invertebr Pathol 103:S73–S79. https://doi.org/10.1016/j.jip.2009.06.017

Fries I, Feng F, Da Silva A et al (1996) *Nosema ceranae* n. sp. (Microspora, Nosematidae), morphological and molecular characterization of a microsporidian parasite of the Asian honey bee *Apis cerana* (Hymenoptera, Apidae). Eur J Protistol 32:356–365

Genersch E (2010) American foulbrood in honeybees and its causative agent, *Paenibacillus larvae*. J Invertebr Pathol 103:S10–S19

Genersch E, Aubert M (2010) Emerging and re-emerging viruses of the honey bee (*Apis mellifera* L.) Vet Res 41(6):54. https://doi.org/10.1051/vetres/2010027

Genersch E, Ashiralieva A, Fries I (2005) Strain-and genotype-specific differences in virulence of *Paenibacillus larvae* subsp. *larvae*, a bacterial pathogen causing American foulbrood disease in honeybees. Appl Environ Microbiol 71:7551–7555

Genersch E, Forsgren E, Pentikäinen J et al (2006) Reclassification of *Paenibacillus larvae* subsp. *pulvifaciens* and *Paenibacillus larvae* subsp. *larvae* as *Paenibacillus larvae* without subspecies differentiation. Int J Syst Evol Microbiol 56:501–511

Gracia-Salinas MJ, Ferrer-Dufol M, Latorre-Castro E et al (2006) Detection of fluvalinate resistance in *Varroa destructor* in Spanish apiaries. J Apic Res 45(3):101–105

Hasemann L (1961) How long can spores of American foulbrood live? Am Bee J 101:298–299

Heath LAF (1982) Development of Chalkbrood in a honey bee colony; Chalkbrood pathogens: a review. Bee World 63(3):119–135

Hedtke K, Jensen PM, Jensen AB et al (2011) Evidence for emerging parasites and pathogens influencing outbreaks of stress-related diseases like chalkbrood. J Invertebr Pathol 108(3):167–173. https://doi.org/10.1016/j.jip.2011.08.006

Higes M, Martín R, Meana A et al (2006) *Nosema ceranae*, a new microsporidian parasite in honeybees in Europe. J Invertebr Pathol 92:81–83
Higes M, Martín-Hernandez R, Botias C et al (2008) How natural infection by *Nosema ceranae* causes honeybee colony collapse. Environ Microbiol 10:2659–2669
Hirai Y, Suzuki T, Inaba N et al (2016) Existence of *Paenibacillus larvae* genotypes ERIC I-ST2, ERIC I-ST15 and ERIC II-ST10 in the western region of Aichi prefecture, Japan. J Vet Med Sci 78(7):1195–1199. https://doi.org/10.1292/jvms.16-0041
Huang W, Jiang J, Chen Y et al (2007) A *Nosema ceranae* isolate from the honeybee *Apis mellifera*. Apidologie 38:30–37
Inoue MN, Yokoyama J, Washitani I (2008) Displacement of Japanese native bumblebees by the recently introduced *Bombus terrestris*(L.) (Hymenoptera: Apidae). J Insect Conserv 12:135–146
Japan Beekeeping Association (2005) The honeybee plants of Japan. The Japan Beekeeping Association, Tokyo. 333pp. (in Japanese)
Japan Beekeeping Association (2011) Damage situation in apiculture by bears. The Bee Journal in Japan Volume 577. The Japan Beekeeping Association, Tokyo. (in Japanese)
Kawashima M (2000) Apiten, the preventive drug against American foulbrood. Honeybee Sci 21(2):55–60. (In Japanese; English abstract)
Ken T, Hepburn HR, Radloff SE et al (2005) Heat-balling wasps by honeybees. Naturwissenschaften 92:492–495
Kimura K (2011) Investigation to determine the cause of honey bee colony loss and prompt measures. Nougyou 1544:25–35. (In Japanese)
Kimura K, Yoshiyama M, Saito K et al (2014) Examination of mass honey bee death at the entrance to hives in a paddy rice production district in Japan: the influence of insecticides sprayed on nearby rice fields. J Apic Res 53(5):599–606. https://doi.org/10.3896/IBRA.1.53.5.12
Kinoda K, Tamakizawa K, Ito M (2013) The bumblebees of Japan. Hokkaido University Press, Sapporo. 194pp. (In Japanese)
Klee J, Besana AM, Genersch E et al (2007) Widespread dispersal of the microsporidian *Nosema ceranae*, an emergent pathogen of the western honey bee, *Apis mellifera*. J Invertebr Pathol 96:1–10
Kojima Y, Toki T, Morimoto T et al (2011) Infestation of Japanese native honey bees by tracheal mite and virus from non-native European honey bees in Japan. Microb Ecol 62(4):895–906. https://doi.org/10.1007/s00248-011-9947-z
Koulianos S, Crozier RH (1996) Mitochondrial DNA sequence data provides further evidence that the honeybees of Kangaroo Island, Australia are of hybrid origin. Apidologie 27: 165–174
Kurihara T (2010) Records of recent bear witnesses in Kyushu Island, Japan. Mamm Sci 50:187–193. (In Japanese; English abstract)
Langridge D, McGhee R (1967) *Crithidia mellificae*: an acidophilic trypanosomatide of honey bee *Apis mellifera*. J Protozool 14:485–487
Maeda T (2015) Infestation of honey bees by tracheal mites, *Acarapis woodi*, in Japan. J Acarol Soc Jpn 24(1):9–17. (In Japanese; English abstract)
Maeda T (2016) Effects of tracheal mite infestation on Japanese honey bee, Apis cerana japonica. J Acarol Soc Jpn 25(S1):109–117.
Maeda T, Sakamoto Y (2016) Tracheal mites, *Acarapis woodi*, greatly increase overwinter mortality in colonies of the Japanese honeybee, *Apis cerana japonica*. Apidologie 47:762. https://doi.org/10.1007/s13592-016-0434-x
Maeta Y (1990) Utilization of wild bees. Farm Japan 24:13–19
Maeta Y (1993) Utilization of *Osmia cornifrons* for apple pollination. In: Inoue T, Kato M (eds) Animals attracted by flowers. Heibonsha, Tokyo, pp 195–232. (in Japanese)
Martin SJ (2001) The role of *Varroa* and viral pathogens in the collapse of honeybee colonies: a modelling approach. J Appl Ecol 38:1082–1093
Martin SJ, Hogarth A, van Breda J et al (1998) A scientific note on *Varroa jacobsoni* Oudemans and the collapse of *Apis mellifera* colonies in the United Kingdom. Apidologie 29:369–370

Martín-Hernández R, Botías C, Bailón EG et al (2012) Microsporidia infecting *Apis mellifera*: coexistence or competition. Is *Nosema ceranae* replacing *Nosema apis*? Environ Microbiol 14:2127–2138

Matsumura C, Yokoyama J, Washitani I (2004) Invasion status and potential ecological impacts of an invasive alien bumblebee, *Bombus terrestris* L. (Hymenoptera: Apidae) naturalized in southern Hokkaido, Japan. Glob Environ Res 8:51–66

Matsuura M (1988) Ecological studies on vespine wasps (Hymenoptera: Vespidae) attacking honeybee colonies. Appl Entomol Zool 23:428–440

Matsuura M, Yamane S (1984) Ethology of Vespinae. Hokkaido University Press, Sapporo. 428pp. (in Japanese)

Matsuura M, Yamane S (1990) Biology of the Vespine wasps. Springer, Berlin. 323pp

McMullan JB, Brown MJF (2009) A qualitative model of mortality in honeybee (*Apis mellifera*) colonies infested with tracheal mites (*Acarapis woodi*). Exp Appl Acarol 47(3):225–234

Ministry of Agriculture, Forestry and Fisheries MAFF (2016a) Situation over beekeeping (in Japanese). http://www.maff.go.jp/j/chikusan/kikaku/lin/sonota/attach/pdf/bee-3.pdf

Ministry of Agriculture, Forestry and Fisheries MAFF (2016b) System for ensuring the stability of honey bees for pollination (in Japanese). http://www.maff.go.jp/j/chikusan/gijutu/mitubati/attach/pdf/index-3.pdf

Ministry of Agriculture, Forestry and Fisheries MAFF (2017a) The outbreak situation of the monitoring epidemic (in Japanese). http://www.maff.go.jp/j/syouan/douei/kansi_densen/kansi_densen.html

Ministry of Agriculture, Forestry and Fisheries MAFF (2017b) Influence of chemicals on honey bees (in Japanese). http://www.maff.go.jp/j/nouyaku/n_mitubati/honeybee.html

Ministry of Environment ME and Ministry of Agriculture, Forestry and Fisheries MAFF (2017) Usage policy of alternative species of alien bumblebee (in Japanese). http://www.maff.go.jp/j/press/seisan/kaki/attach/pdf/170421_11-1.pdf

Morimoto T, Kojima Y, Yoshiyama M et al (2012) Molecular identification of chronic bee paralysis virus infection in *Apis mellifera* colonies in Japan. Virus 4(7):1093–1103

Morimoto T, Kojima Y, Yoshiyama M et al (2013) Molecular detection of protozoan parasites infecting *Apis mellifera* colonies in Japan. Environ Microbiol Rep 5:74–77. https://doi.org/10.1111/j.1758-2229.2012.00385.x

Morrissey BJ, Helgason T, Poppinga L et al (2015) Biogeography of *Paenibacillus larvae*, the causative agent of American foulbrood, using a new multilocus sequence typing scheme. Environ Microbiol 17:1414–1424

Neumann P, Ellis JD (2008) The small hive beetle (*Aethina tumida* Murray, Coleoptera: Nitidulidae): distribution, biology and control of an invasive species. J Apic Res 47(3): 181–183

Neumann P, Pettis JS, Schäfer MO (2016) Quo vadis *Aethina tumida*? Biology and control of small hive beetles. Apidologie 47:427–466

Office International des Épizooties, OIE (2008) Manual of standards for diagnostic test and vaccines for terrestrial animals. Chapter 2.2.1. Acarapisosis of honeybees. OIE—World Organisation for Animal Health. http://www.oie.int/fileadmin/Home/eng/Health_standards/tahm/2.02.01_ACARAPISOSIS.pdf

Oldroyd BP, Cornuet JM, Rowe D et al (1995) Racial admixture of *Apis mellifera* in Tasmania, Australia: similarities and differences with natural hybrid zones in Europe. Heredity 74:315–325. https://doi.org/10.1038/hdy.1995.46

Ono M, Igarashi T, Ohno E et al (1995) Unusual thermal defence by a honeybee against mass attack by hornets. Nature 377:334–336

Paxton RJ, Klee J, Korpela S et al (2007) *Nosema ceranae* has infected *Apis mellifera* in Europe since at least 1998 and may be more virulent than *Nosema apis*. Apidologie 38:558–565

Pettis JS, Wilson WT (1996) Life history of the honey bee tracheal mite (Acari: Tarsonemidae). Ann Entomol Soc Am 89(3):368–374

Rana BS, Garg ID, Khurana SM et al (1986) Thai sacbrood virus of honeybees (*Apis cerana indica* F) in northwest Himalayas. Indian J Virol 2:127–131

Rauch S, Ashiralieva A, Hedtke K et al (2009) Negative correlation between individual-insect-level virulence and colony-level virulence of *Paenibacillus larvae*, the etiological agent of American foulbrood of honeybees. Appl Environ Microbiol 75:3344–3347

Ravoet J, Maharramov J, Meeus I et al (2013) Comprehensive bee pathogen screening in Belgium reveals *Crithidia mellificae* as a new contributory factor to winter mortality. PLoS ONE 8(8):e72443.

Ravoet J, Schwarz RS, Descamps T et al (2015) Differential diagnosis of the honey bee trypanosomatids *Crithidia mellificae* and *Lotmaria passim*. J Invertebr Pathol 130:21–27. https://doi.org/10.1016/j.jip.2015.06.007

Rennie J (1921) Isle of Wight disease in hive bees—acarine disease: the organism associated with the disease—*Tarsonemus woodi*, nsp. Earth and Environ Science Trans Royal Soc Edinburgh 52(4):768–779

Rosenkranz P, Aumeier P, Ziegelmann B (2010) Biology and control of *Varroa destructor*. J Invertebr Pathol 103:S96–S119

Ruttner F (1988) Biogeography and taxonomy of honeybees. Springer, Heidelberg. 284pp

Sakai Y, Takahashi J (2014) Discovery of a worker of *Vespa velutina* (Hymenoptera: Vespidae) from Tsushima Island, Japan. Kontyu (new series) 17:32–36. (in Japanese)

Sammataro D, Gerson U, Needham G (2000) Parasitic mites of honey bees: life history, implications, and impact. Annu Rev Entomol 45:519–548

Sammataro D, Untalan P, Guerrero F et al (2005) The resistance of *Varroa* mites (Acari: Varroidae) to acaricides and the presence of esterase. Int J Acarol 31:67–74

Sasaki M (1999) Wonders of the Japanese honey bee. Kaiyusha, Tokyo. 191pp (in Japanese; English abstract)

Sasaki M (2013) Bee's eye view of flowering plants—nectar and pollen source plants and related honeybee products. Kaiyusha, Tokyo. 413pp. (in Japanese; English abstract)

Schwarz RS, Bauchan G, Murphy C et al (2015) Characterization of two species of Trypanosomatidae from the honey bee *Apis mellifera*: *Crithidia mellificae* Langridge and McGhee, 1967 and *Lotmaria passim* n. gen., n. sp. J Eukaryot Microbiol 62:567–583

Sugahara M, Sakamoto F (2009) Heat and carbon dioxide generated by honeybees jointly act to kill hornets. Naturwissenschaften 96:1133–1136

Takahashi J, Yoshida T (2003) The origin of Japanese honey bee *Apis cerana japonica* inferred from mitochondrial DNA. Honeybee Sci 24(2):71–76. (in Japanese; English abstract)

Takahashi J, Yoshida T, Takagi T et al (2007) Geographic variation in the Japanese islands of *Apis cerana japonica* and in *A. cerana* populations bordering its geographic range. Apidologie 38:335–340

Takahashi J, Takeuchi M, Matsumoto M et al (2014) Genetic structure of apicultural honeybee Apis Mellifera in Japan. Bull Res Inst Adv Technol Kyoto Sangyo Univ 13:25–37. (in Japanese; English abstract)

Takamatsu D, Morinishi K, Arai R et al (2014) Typing of *Melissococcus plutonius* isolated from European and Japanese honeybees suggests spread of sequence types across borders and between different *Apis* species. Vet Microbiol 171(1–2):221–226. https://doi.org/10.1016/j.vetmic.2014.03.036

Tentcheva D, Gauthier L, Zappulla N et al (2004) Prevalence and seasonal variations of six bee viruses in *Apis mellifera* L. and *Varroa destructor* mite populations in France. Appl Environ Microbiol 70:7185–7191

Thompson HM, Brown MA, Ball RF et al (2002) First report of *Varroa destructor* resistance to pyrethroids in the UK. Apidologie 33(4):357–366

Tsuruga H, Mano T, Yamanaka M et al (1994) Estimate of genetic variations in Hokkaido brown bears (Ursus arctos yesoensis) by DNA fingerprinting. Jpn J Vet Res 42(3–4):127–136

Uemura T, Sakashita N, Morinishi, K et al (2013) Cases of occurrence of European foulbrood (in Japanese). http://www.pref.kagawa.lg.jp/chikusan/eisei/H25/25-6.pdf

Verma LR, Rana BS, Verma S (1990) Observations on *Apis cerana* colonies surviving from Thai sacbrood virus infestation. Apidologie 21:169–174

Vidau C, Diogon M, Aufauvre J et al (2011) Exposure to sublethal doses of fipronil and thiacloprid highly increases mortality of honeybees previously infected by *Nosema ceranae*. PLoS One 6:e21550. https://doi.org/10.1371/journal.pone.00215

Wang DL, Moeller FE (1970) Comparison of the free amino acid composition in the haemolymph of healthy and *Nosema*-infected female honey bees. J Invertebr Pathol 15:202–206

White G (1913) Sacbrood, a disease of bees. US Dept Agr Bur Entomol Circ 169:1–5

Wilfert L, Long G, Leggett HC et al (2016) Deformed wing virus is a recent global epidemic in honeybees driven by *Varroa* mites. Science 351(6273):594–597

Yamada M (1986) Control techniques of acarid mites on *Osmia cornifrons* [Hymenoptera: Megachilidae] pollinating apple. Agric Hortic 61(3):425–429. (in Japanese)

Yamashita T, Tanaka S (2010) Colony collapse of Japanese honeybee in Yamaguchi prefecture. Honeybee Sci 28(2):73–80. (in Japanese; English abstract)

Yang X, Cox-Foster D (2007) Effects of parasitization by *Varroa destructor* on survivorship and physiological traits of *Apis mellifera* in correlation with viral incidence and microbial challenge. Parasitology 134:405–412

Yang B, Peng G, Li T et al (2013) Molecular and phylogenetic characterization of honey bee viruses, Nosema microsporidia, protozoan parasites, and parasitic mites in China. Ecol Evol 3:298–311. https://doi.org/10.1002/ece3.464

Yoshida T (2000) Methods of rearing and ecology of Japanese honey bee. Tamagawa University Press, Machida, Tokyo. 135pp (in Japanese)

Yue C, Genersch E (2005) RT-PCR analysis of deformed wing virus in honeybees (*Apis mellifera*) and mites (*Varroa destructor*). J Gen Virol 86:3419–3424

Yue C, Schroder M, Gisder S et al (2007) Vertical-transmission routes of deformed wing virus of honeybees (*Apis mellifera*). J Gen Virol 88:2329–2336

Yue D, Nordhoff M, Wieler LH et al (2008) Fluorescence in situ hybridization (FISH) analysis of the interactions between honeybee larvae and *Paenibacillus larvae*, the causative agent of American foulbrood of honeybees (*Apis mellifera*). Environ Microbiol 10:1612–1620

Zaghloul OA, Mourad AK, El Kady MB et al (2005) Assessment of losses in honey yield due to the chalkbrood disease, with reference to the determination of its economic injury levels in Egypt. Commun Agric Appl Biol Sci 70(4):703–714

Ministry of Agriculture, Forestry and Fisheries MAFF (2011) The outbreak situation of the monitoring epidemic (in Japanese). http://www.maff.go.jp/j/syouan/douei/kansi_densen/kansi_densen.html

Chapter 11
Beekeeping in Vietnam

Pham Hong Thai and Tran Van Toan

Abstract There are six honeybee (*Apis laboriosa, Apis dorsata, Apis mellifera, Apis cerana, Apis andreniformis, Apis florea*), eight stingless bee (*Trigona laeviceps, Trigona ventralis, Trigona pagdeni, Trigona gressitti, Trigona fuscobalteata, Trigona carpenteri, Trigona scintillans, Trigona iridipennis*), and two bumble bee (*Bombus haemorrhoidalis, Bombus breviceps*) species known in Vietnam. All of them are native to the country, with the exception of *A. mellifera*. Today nearly 1,500,000 *A. mellifera* and *A. cerana* honeybee hives are managed in Vietnam. These bees, along with *A. dorsata*, are responsible for the majority of honeybee products produced in the country. Vietnam is the second largest honey exporter in Asia, with a total of ~48,000 tons distributed internationally in 2014. Domestically, honey is also an important commodity. Furthermore, *A. cerana* plays an important role in poverty alleviation in mountainous and remote areas of Vietnam due to beekeeping development projects sponsored by both government and nongovernmental organizations. Major bee diseases that infect Vietnamese honeybees include Sacbrood disease (SBV), European foulbrood disease (EFB), Nosema, and parasitic mites (*Tropilaelaps mercedesae* and *Varroa destructor*). Most of these diseases can be resolved with biocontrol methods.

Keywords Honeybee · Stingless bee · Bumble bee · Honeybee product · Honey export · Beekeeping development · Bee disease · Conservation

P. H. Thai (✉)
Research Center for Tropical Bees and Beekeeping, Vietnam National University of Agriculture, Hanoi, Vietnam

Faculty of Agronomy, Department of Entomology, Vietnam National University of Agriculture, Hanoi, Vietnam
e-mail: phthai@vnua.edu.vn

T. Van Toan
Research Center for Tropical Bees and Beekeeping, Vietnam National University of Agriculture, Hanoi, Vietnam

P. Chantawannakul et al. (eds.), *Asian Beekeeping in the 21st Century*,
https://doi.org/10.1007/978-981-10-8222-1_11

11.1 Diversity of Honeybees in Vietnam

In Vietnam, there are six honeybee species present, five of which are indigenous such as *Apis laboriosa*, *Apis dorsata*, *Apis cerana*, *Apis andreniformis*, and *Apis florea*.

*11.1.1 Rock Honeybees (*Apis laboriosa*)*

Rock honeybees (*Apis laboriosa*) were discovered nesting in Son La and Hoa Binh provinces (about 20°47′N and 104°48′E, at an altitude of 1270 m) in northwestern Vietnam in 1996 (Trung et al. 1996). They were also found in Dienbien and Laichau provinces of northwestern Vietnam at an elevation of 1000 m above sea level. They migrate annually to establish their combs in the mountainous provinces of Son La and Hoa Binh in April, and leave their comb empty for overwintering or for another flowering location in August. This honeybee species has the largest worker bee body size. The forewing length of *A. laboriosa* is about 8.5% longer than that of *A. dorsata*, the length of sternite 3 and the proboscis of *A. laboriosa* is 7.4% and 5.7%, respectively, longer than *A. dorsata*. The tergites of *A. laboriosa* worker bees are brown-black as compared to a bronze color in *A. dorsata* (Trung et al. 1996).

*11.1.2 Giant Honeybee (*Apis dorsata*)*

Worker body size of the giant honeybee *A. dorsata* is smaller than *A. laboriosa* (Trung et al. 1996). Their nesting behavior consists of occupying one open-air comb on a tree branch or rock cliff. The *A. dorsata* distributes across all Vietnam, with the exception of the Red River Delta (Chinh et al. 1995). The *A. dorsata* population in the Melaleuca Forest of Vietnam has been well studied (Thai 1994; Chinh et al. 1995; Tan et al. 1999; Chinh et al. 2004; Tan 2007). According to Tan (2007), the width of *A. dorsata* combs are 43–162 cm, and the height 23–90 cm. About 3–4 weeks after nesting, a colony can store 4.1 ± 2.6 kg of honey. Thickness of worker brood comb is ~3.3 cm, whereas drone brood comb is ~3.7 cm. Areas containing honey comb is ~19.0 cm wide.

*11.1.3 Cavity Nesting Asian Honeybee (*Apis cerana*)*

Apis cerana is distributed across all provinces of Vietnam, except in the U Minh Forest of Ca Mau province (Thai 2008). It has been reported that there are two subspecies of *A. cerana* in Vietnam: *A. cerana cerana* in the North and *A. cerana indica* in the South (Ruttner 1988; Hepburn et al. 2001; Radloff et al. 2005, 2010; Abrol 2013). However, DNA and morphological analyses reported that *A. cerana cerana* is distributed only in the Dong Van Karst plateau (Global geological park) in the northern Ha Giang province (Thai 2008). Some biological characteristics such as proboscis length, forewing length and width, Basitarsus length and width, queen laying capacity, and colony strength suggest that *A. c. cerana* workers in Dong Van district of Ha Giang province are significantly larger than that of *A. c. indica* in Ha Tay province (Now Hanoi city) (Toan 2012).

*11.1.4 Red Dwarf (*Apis florea*) and Black Dwarf (*Apis andreniformis*) Honeybees*

Apis florea is distributed throughout Vietnam, but it is more frequently found in the South. Normally, they nest with a single comb in the open air like *A. dorsata*. Both species build their nests on branches, typically in areas with more dense, darker foliage. *Apis andreniformis* has a stronger defensive behavior than *A. florea*, particularly when intruders come within 2 m of the nest. Comb size of *A. andreniformis* is smaller than that of *A. florea* (15.20 × 18.90 cm and 16.51 × 19.71 cm, respectively); however, worker cells of *A. florea* are wider than *A. andreniformis* (3.13 mm and 2.91 mm, respectively) (Thai et al. 1997). More than 10 years ago, it was difficult to find any red dwarf bee nests on the campus of Vietnam National University of Agriculture. However, in the last 5 years, colony density has increased. This change is likely due to horticultural plants being planted around campus instead of vegetables receiving pesticide applications. Even then, population of these species, particularly *A. andreniformis*, appears to be reduced in Vietnam. According to Quyen et al. (2001), the numbers of colonies in a farm of U Minh Forest reduced between 1997 and 1999.

*11.1.5 European Honeybees (*Apis mellifera*)*

Apis mellifera were first introduced into Vietnam in 1887 by the French (Rialan 1887), but did not become established until it was reintroduced in the 1960s in the south of the country; it expanded across all Vietnam after 1975. Commercial beekeeping with European honeybees accounts for over 70% of honeybee colonies in Vietnam. Farmers keep *A. mellifera* in all provinces, with some even migrating annually from North to South (Dung 1996; Thai 2008) (Fig. 11.1). Vietnam has legally imported some subspecies (races) of *A. mellifera* such as *A. m. ligustica*, *A. m. carpatica*, *A. m. caucasia*, *A. m. mellifera*, and *A. m. carnica* (Thai 2008).

Fig. 11.1 The exotic honeybee (*A. mellifera*). (**a**) A strong colony and (**b**) beehives placed in parallel lines

11.2 Historical Records and Recent Development of Beekeeping in Vietnam

11.2.1 Traditional Beekeeping

In Vietnam, traditional beekeeping with the Asian honeybee *A. cerana* has been practiced for a long time by farmers. In the eighth century, Mr. Pham Le, who was Mandarin of the Agricultural Ministry of Vietnam, wrote documents about traditional beekeeping techniques with *A. cerana* (Chinh 1996). In the eighteenth century, Le Qui Don, a poet and great scholar in feudal Vietnam described some biological characteristics of *A. cerana* in the encyclopedia "Van dai loai ngu" (Crane 1999). At the third Apimondia Conference in 1902, a report on rafter techniques in beekeeping with a report was presented on *A. dorsata* in the Melaleuca Forest in Southern Vietnam (Crane 1999). Later, Toumanoff (1933) and Toumanoff and Nanta (1933) described beekeeping in Tonkin of North Vietnam, where honeybees were kept in both horizontal and upright log hives. Even today, techniques of sustainable honey harvesting and rafter beekeeping for *A. dorsata* are still applied in the Melaleuca Forest of Southern Vietnam. Honey harvested from traditional logs and *A. dorsata colonies* are preferred by domestic consumers.

11.2.2 Modern Beekeeping

According to Rialan (1887), *Apis mellifera* were introduced into Vietnam by the French in 1887. However, they did not survive due to the then-called parasitic mites *Varroa jacobsoni* and *Tropilaelaps clareae* (Woyke 1996). Beekeeping techniques with moveable frame hives were introduced to the North of Vietnam in 1960; this also greatly assisted in developing beekeeping with the native Asian honeybee *A. cerana* (Chinh 2012). During the same period, exotic *A. mellifera* colonies were successfully reintroduced into the south of Vietnam from Hong Kong. After more than 50 years, this honeybee species has adapted well to the climate and flower sources in country (Tam et al. 2010).

11.2.3 Recent Beekeeping Developments in Vietnam

The two honey bee species kept commercially in Vietnam are *A. mellifera* and *A. cerana*. For the native honeybee *A. cerana*, two subspecies that exist are *A. c. cerana* and *A. c. indica. Apis c. cerana* is distributed on the Dong Van plateau, Ha Giang province in the North of Vietnam, whereas *A. c. indica* is distributed throughout the rest of the country (Thai 2008). For the European honeybee, *A. mellifera ligustica* (Italian bees) was the first imported subspecies to Vietnam, but afterwards others such as *A. m. carpatica* and *A. m. caucasia* were also imported from Russia. They were most likely killed by honey bee mites. In the last decade, *A. m. carnica* from Germany and *A. m. ligustica* from New Zealand and Austria were also successfully imported into Vietnam.

Linh (2015) reported that Vietnam has a total of 1,500,000 honey bee colonies kept by 33,000 beekeepers. Among these are 1,100,000 *A. mellifera* colonies and 400,000 *A. cerana* colonies, which produced ~42,000 tons of honey; ~90% was exported. A small proportion of honey produced is consumed domestically, with honey prices ranging massively from 4 to 50 USD/L. The price of exported honey is only 2 USD/L.

11.3 Bee Forage Resources

11.3.1 The Importance of Nectar and Pollen Plants for Honeybees

The natural carbohydrate source of honeybees is nectar, collected by foragers from plants, transported to the hive, and finally stored in cells as honey (Brodschneider and Crailsheim 2010). The only natural protein source for honey bees is pollen (Crailsheim 1990; Chinh 2012). Habitat with ample supply of pollen and nectar forage for bees is essential for the growth of colonies and their health. Therefore, it is necessary to place apiaries in areas that have sufficient flower resources (Chinh 2012).

11.3.2 Main Nectar Plants in Vietnam

The honeybee visits blooming plants for collecting nectar and pollen, but not every plant resource provides both these resources. Some plants produce only nectar while others just produce pollen. The plants in Vietnam that give nectar or both nectar and pollen include longan, lychee, coffee, acacia, and rubber. Plants that give only pollen are referred to as pollen plant resources (Chinh 2012). Vietnam is located in the eastern edge of the Indochinese peninsula, encompassing an area in the center of Southeast Asia that stretches from 8°34′ to 23° 23′ North. Due to dramatic differences in climate because of latitude and topography, there is rich and diverse vegetation. Flowers bloom throughout the year, allowing Vietnamese apiculture to develop. Table 11.1 contains the main floral sources in Vietnam.

Table 11.1 Primary pollen and nectar plant sources for honeybees in Vietnam (Chinh 2012; Chinh et al. 2012)

Vietnamese name	Common name	Scientific name	Harvest season	Amount		Distribution
				Nectar	Pollen	
Bạc hà dại	Wild mint (*Elsholtzia*)	*Elsholtzia cypriani* Pavol.	10–12	+++	++	Hà Giang only (North)
Bạch đàn trắng	Red gum	*Eucalyptus camaldulensis* D.	4	++	++	North, Central
Bí đỏ	Pumpkin	*Cucurbita pepo* L.	2–5	+	++	North, Central, South
Bí xanh	Green gourd	*Benincasa cerifera* Savi./*Benincasa hispida*	2–4	+	++	North, Central, South
Bưởi	Pomelo	*Citrus grandis* Oshek.	2–3	+	++	North, Central, South
Cà phê chè	Arabica coffee	*Coffea arabica* L.	11–3	++	++	North, Central South
Cà phê mít	Coffea liberica	*Coffea excelsa* Achev.	9–10	++	++	Central, South
Cà phê vối	Robusta coffee	*Coffea robusta* Lindens.	11–2	++	++	North, Central South
Cam	Orange	*Citrus sinensis* (L.) Osbeck.	2–3	+	++	North, Central, South
Cao su	Rubber tree	*Hevea brasiliensis* Muell.	2–4	+++*	+	North, Central, South
Cây chân chim	Umbrella plant	*Schefflera octophylla* (Lour.) Harms	4–5	+++	+	North, Central
Chanh	Rangpur/ lemandarin	*Citrus limonia* Osbeck.	1–2	+	++	North, Central, South
Chè	Tea	*Thea sinensis* Seem.	9–12	+	++	North, Central, South
Chôm chôm	Rambutan	*Nephelium lappaceum* L.	3–5	+++	+	South
Cây đơn kim (Đơn buốt)	Black-jack, beggar-ticks, cobbler's pegs, and Spanish needle	*Bidens pilosa* L.	Year-round	+++	++	North, Central
Cỏ lào	Siam weed, Christmas bush, devil weed, camfhur grass	*Eupatorium odoratum* L.	12–1	++	+	North, Central, South

Table 11.1 (continued)

Vietnamese name	Common name	Scientific name	Harvest season	Amount		Distribution
				Nectar	Pollen	
Dẻ Yên Thế	Chinkapin	*Castanopsis boisii* Hickel.	11–12	++	+	North, Central
Dưa chuột	Cucumber	*Cucumis sativus* L.	3, 10	+	++	North, Central, South
Dưa hấu	Watermelon	*Citrullus lanatus* Mats.	4–5	++	++	North, Central, South
Dừa	Coconut	*Cocos nucifera* L.	Cả năm	+++	+	North, Central, South
Đay cách	Kenaf	*Hibiscus cannabinus* Var.	4–7	++	+	North, South
Điều	Cashew	*Anacardium occidentale* L.	11–12	+++*	–	South
Hướng dương dại (cúc quỳ)	Mexican sunflower	*Tithonia diversifolia* Gay.	10–11	++	+	Central, South
Keo tai tượng	Black wattle, Hickory wattle, Mangium, and Forest mangrove	*Acacia mangium* Willd	4–7	+++*	+	North, Central, South
Lúa	Rice	*Oryza sativa* L.	4, 9	–	++	North, Central, South
Ngô	Corn	*Zea mays* L.	4–12	–	+++	North, Central, South
Nhãn	Longan	*Euphoria longan* Steud	3–4	+++	+	North, Central
Sắn	Cassava	*Manihot esculenta* C.	10–11, 5–8	+++*	+	North, Central, South
Sắn dây	Kudzu	*Pueraria tonkinensis* Gagnep.	11–1	++	+	North, Central, South
Sen	Lotus	*Nelumbo nucifera* Gaertn.	5–8	–	++	North, Central, South
Sòi đất	Mountain Tallowtree (Tallow tree)	*Sapium discolor* Muell.	4–5	+++	+	North, Central
Sú	Black mangrove, River mangrove or Khalsi	*Aegiceras corniculatum* (L.) Blanco	4–5	++	+	North
Táo ta	Jujube	*Ziziphus mauritiana* Lam.	9–10	+++	+	North

(continued)

Table 11.1 (continued)

Vietnamese name	Common name	Scientific name	Harvest season	Amount		Distribution
				Nectar	Pollen	
Tràm	Cajuput tree	*Melaleuca leucadendron*	1–4, 6–8	+++	++	South
Trinh nữ cao	Shameplant/ sensitive plant, sleepy plant	*Mimosa pudica* L.	10–11	–	++	North, Central, South
Vải chua	Litchi	*Litchi* sp.	2	++	+	North
Vải nhỡ	Litchi	*Litchi* sp.	2–3	+++	+	North, Central
Vải thiều	Lychee	*Litchi chinensis* Sonn.	3–4	+++	+	North, Central
Vẹt	Oriental mangrove	*Bruguiera gymnorrhiza* (L.) Lam.	6–7	+++	++	North
Vối rừng	Jamun tree	*Eugenia jambolana* Roxb.	4–5	++	++	North
Vừng	Sesame	*Sesamum indicum* L.	5–8	++	+	North, Central, South

Note: +++: a lot; ++: average; +: little; -: no; * secreted honeydew from leave

11.4 Local Knowledge on Honeybees

Vietnam has a long history of traditional beekeeping, so it is a natural and comfortable transition for Vietnamese farmers to move into modern beekeeping. Since the 1960s, when modern beekeeping techniques were first introduced into Vietnam, more than 50 years of knowledge and experience has advanced the Vietnamese beekeeping industry. Thanks to the support of government and nongovernmental organizations, beekeeping extension programs are conducted regularly. Every year training courses about beekeeping are organized for beginners who live in different areas of the country in an attempt to expand and develop beekeeping. Professional beekeepers also attend a variety of short courses on food safety and sanitation, as well as best management practices for prevention and treatment of bee diseases; Viet GAP (Vietnamese Good Apiculture Practices) has also been applied to practical beekeeping. As a result of these efforts, Vietnamese beekeepers are extremely knowledgeable about how to produce bee products since only those of high quality are purchased by honeybee companies for export. A lot of projects for beekeeping development are funded by NGOs to train economically strained farmers in mountainous and remote areas. The results of such projects help many farmers maintain stable incomes. Traditional beekeeping techniques with log hives are still used by farmers, even though the resulting honey yield is lower. However, employing traditional beekeeping means that farmers do not need to invest in expensive equipment, and that they can sell their honey at a higher price because of the demand for traditional products.

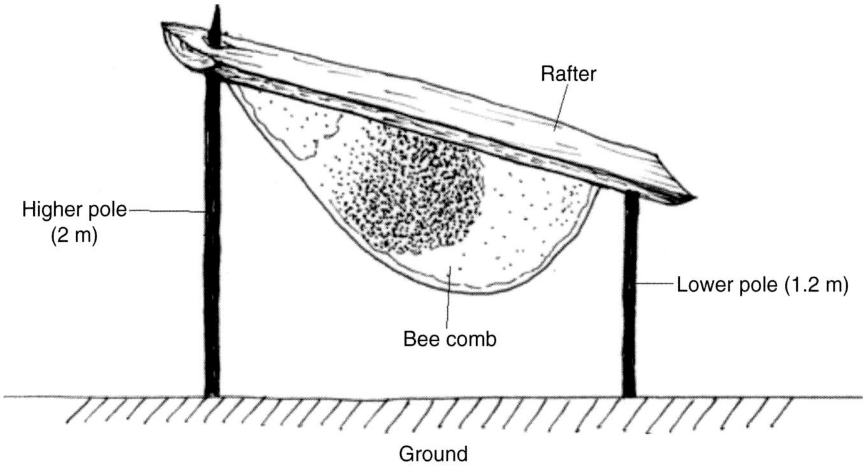

Fig. 11.2 Rafter beekeeping with *Apis dorsata* in Vietnam

Vietnam is not only known for traditional beekeeping, but also for rafter beekeeping with giant honeybees *A. dorsata* in U Minh Melaleuca Forest, Ca Mau province in southern Vietnam. Rafter beekeeping is an old age concept where honey-hunters lure bees to build nests on tree-poles—rafters—that mimic large tree branches. This works particularly well in degraded forest areas where the only suitable nesting sites for these bees are the rafters (Fig. 11.2), which when placed conveniently at human eye-level, provide easy access to the comb (Chinh et al. 1995; Tan et al. 1997; Gupta et al. 2014). Previously during honey harvesting season in U Minh Melaleuca Forest, people often used torches to burn the nests, which resulted in forest fires that caused a reduction in the number of colonies that nested. Today, smoke is used by residents of U Minh to dispel bees, which is safer for both people and forests. Every year there are two rafter seasons in the U Minh Melaleuca Forest—in rainy and dry seasons. Water content of honey harvested during the rainy season is higher than in the dry season. Even though the price of honey from the Melaleuca Forest is expensive, local people and tourists prefer this honey. However, due to the higher water content, especially in honey harvested during rainy seasons, honey stored for a long time experiences a reduction in quality. That is why it is recommended that the honey collected from Melaleuca should be used within 1 year (Chinh et al. 1995).

11.5 Diseases and Pests of Honeybees

11.5.1 European Foulbrood

European foulbrood (EFB) is a bacterial brood disease caused by the Gram-positive bacterium *Melissococcus plutonius*. In addition, it has been found that the presence of the bacteria *Bacillus alvei, Bacillus laterosporus,* and *Enterococcus faecalis* do not cause disease, but influence the odor and consistency of dead brood (Shimanuki and Knox 1991). According to Chinh (2012), EFB was discovered in Vietnam in 1969 as a result of importation of honeybee colonies from foreign countries. In Vietnam, *M. plutonius* and *B. alvei* were found on a specimen of EFB sampled from *A. cerana* colonies, but these bacteria belong to another strain compared to that found on the honeybee *A. mellifera* Chinh (2012). The general symptoms of an EFB affected colony are spotty sealed brood, with an intermixture of capped and uncapped cells. A heavily infested colony may have few capped cells. EFB kills young larvae, with dead larvae positioned in an unusual twisted or coiled position in the cell. Larvae change color, from white to yellowish and then to dark brown. Dead larvae drop down and subsequently dry up, making them easy to remove by worker bees (Chinh 2012).

In the laboratory, EFB can be diagnosed by identifying the infection of *M. plutonius*. Symptoms of EFB are rather similar to other larval diseases, so it can be difficult to diagnose (Bailey 1981).

To prevent EFB, the beekeeper should keep a populated colony with a young, healthy queen. Colonies should be relatively equal in strength, sufficient in honey and pollen reserves, and placed in a dry location (Chinh 1990). A popular way to control EFB in *A. mellifera* colonies is to cage the queen for 4–5 days to prevent diseased colony from absconding or to transfer and isolate diseased colonies to new floral sources, combined with removing infected combs and feeding sugar syrup to bees (Chinh 2012).

11.5.2 Sacbrood Disease

Sacbrood is a brood disease caused by a virus. Sacbrood virus (SBV) was first described by White on *A. mellifera* in the USA in 1913 (Bailey et al. 1964). SBV is one of many insect viruses generally referred to as picornavirus-like (Grabensteiner et al. 2001). At present, SBV strains have been identified, including sacbrood virus on *A. mellifera* (SBV) (Bailey et al. 1964), Thai sacbrood virus (TSBV), Chinese sacbrood virus (CSBV) (Bailey et al. 1982; Zhang et al. 2001), and Korean sacbrood virus (SBV-Kor) (Lee et al. 2010). In Vietnam, two strains of the virus have been found in *A. cerana*—CSBV in the north and TSBV in the south. According to Chinh (1996), an epidemic of sacbrood disease in *A. cerana* colonies in Vietnam originated from a high honey producing line of *A. cerana* colonies imported from China. From 1974 to 1978, SBV spread from the northern provinces to southern areas of Vietnam, damaging 90% of *A. cerana* colonies (Chinh 2012).

Diseased brood comb is sunken, often displaying uncapped pupae and many sac-like larvae seemingly twisted in their cells. The infected larvae change in color from pearly white to pale yellow (Chinh 1990). Shortly after death they dry out into gondola-shaped scales (Bailey 1975). The segmented lines of brood are not clear, and dead larvae are typically odorless (Chinh 1990). In the laboratory, SBV can be diagnosed by using electronic microscopy, serology, and immunology. In recent years, multiplex RT-PCR/RFLP techniques have been effectively applied to diagnose early SBV infection (Trung et al. 2010; Thai et al. 2010; Duong et al. 2015; Thu et al. 2016). Sacbrood can be reliably diagnosed in the field by the characteristic symptoms produced in developing brood of both *A. mellifera* and *A. cerana* (Ball 1999; Toan et al. 2014a, b).

To prevent disease, colonies must be headed with a good queen, an adequate number of workers, and enough stored honey (Chinh 1990). Combs of diseased colonies should not be combined with a healthy one. Infected colonies must be handled immediately to avoid spreading the disease. Before reusing empty hives, old equipment should be cleaned by washing with water and then exposed to the sun (Chinh 1990).

At present, and in common with nearly all viruses found in animals, there are no known direct treatments for virus infections in bees. According to Lan et al. (1998), herbal extracts are used to treat sacbrood disease on *A. cerana* colonies in Vietnam. They showed that S-95, a product extracted from plants, might have potential to treat diseased honeybees because colonies appeared asymptomatic 3 months post treatment. However, this experiment must be replicated with a higher sample size in order to determine the effectiveness of extracted products. Currently, biological controls are widely applied (Chinh 2012). The principle of these methods is to make the colony broodless for 8–10 days, which will starve the virus of resources (Bailey 1981). There are two methods to create a broodless colony. One way is to replace the queen of a diseased colony with a queen cell or virgin queen; they should be reared from a disease resistant colony. Another option is to cage the laying queen of the diseased colony for 8–10 days. Both methods should be combined with eleminating the brood diseased combs. Then, sugar syrup should be fed to bees 3–4 times continuously or these colonies should be moved to new place with better floral sources. Using these methods, more than 90% of diseased colonies typically do not display symptoms again. However, the effectiveness of these treatments depends on the weather and flower sources. Furthermore, sometimes simply caging the queen can result in a recurrence of the disease; in this instance, it is recommended to replace the queen with another bred from resistant stock (Toan et al. 2014a, b).

There is some evidence that strain of honeybee may differ in susceptibility to sacbrood virus (Bailey 1967). Colonies headed by imported queens showed a significant increase in larval mortality due to SBV than local colonies maintained under the same conditions (Ball 1999). Some case studies in Vietnam demonstrated that *A. cerana* breeding efforts significantly reduced SBV infection rate. Queens were reared from colonies that had survived an outbreak, and were used to replace queens in the infected colonies (Chinh 2012). Disease resistance in *A. cerana* was applied to a closed selection population from 1989 to 1996, and after four generations the SBV infection rate was reduced from 23.1 to 2.3%. After six generations, the infection rate of the experimental population and the controls were 3.2 and 26.7%, respectively (Chinh et al. 1996). Furthermore, hybridizing *A. c. cerana* with *A. c. indica* resulted in lower infection rates compared to each pure subspecies alone (Toan 2012).

11.5.3 Nosema Disease

Nosema disease is an adult honeybee disease causing significant economic losses to beekeepers worldwide (KLee et al. 2007; Toan et al. 2014a, b). In Vietnam, this disease is found in both the native Asian honeybee *A. cerana* and the exotic European honey bee *A. mellifera* (Toan et al. 2014a, b). The Microsporidia species *Nosema apis*, which causes Nosema disease, was first described in 1907 by Zander. *Nosema apis* was considered the sole perpetrator of Nosema disease in *A. mellifera* for a long time. However, in 1994, a new species, *Nosema ceranae*, was discovered in *A. cerana* in China (Fries et al. 1996). This species has since spread to *A. mellifera* in many regions of the world (Chen and Huang 2010). In Vietnam, according to Koeniger et al. (1997), adult bees that died of diarrhea in Son La province were infected by *N. apis.* In 2006, Nosema spores were detected in *A. cerana* sampled from Hung Yen and Ha Nam provinces, and *A. mellifera* sampled from Gia Lai province. The results showed that Nosema species found in both *A. mellifera* and *A. cerana* populations was *N. ceranae* (Klee et al. 2007). *A. mellifera* samples collected from 2007 to 2009 in five provinces—Son La, Bac Giang, Nghe An, Gia Lai and Dong Nai—showed similar results (Tam et al. 2010). In *A. mellifera, N. ceranae* harms intestinal epithelial cells of worker bees, resulting in malnutrition, significantly shortened worker lifespan, and reduced number of combs (from 6 combs down to 5.4 combs/colony after 20 days and from 7.6 combs down to 4.2 combs after 50 days of infection); this leads to a significantly reduced honey yield. The disease usually begins in April, rising in late spring/early summer every year. The number of spores per bee is highest around September, and then decreases in the months leading up to the end of the year (Tam 2010).

The life cycle of *N. apis* occurs within adult intestinal epithelial cells. Spores pass through the midgut and esophagus down to the intestines, where they germinate and multiply rapidly; within 6–10 days the midgut is infected with high levels of spores (Fries 1988). Like *N. apis, N. ceranae* parasitizes primarily adult bees but does not show typical symptoms. However, it is the prime suspect when bee numbers are reduced, bees consume more food, have distended bellies and crawl slowly, lose sting reaction, indiscriminately defecate, and die in front of the entrance (Fries et al. 2013).

Nosema disease can be diagnosed under the microscope at 250–500× magnification. *N. ceranae* spores are oval shaped, with a dark interior and fluorescence; they are 4.439–4.637 μm long and 2.501–2.576 μm wide, and are smaller than *N. apis* spores (Tam 2010). Molecular techniques are also developed for detection of *Nosema* spp. in bees, and are PCR-based (Fries et al. 2013). In Vietnam, Tam et al. (2010) designed multi-specific primers for detection of both *N. apis* and *N. ceranae* in honeybees.

There are no typical symptoms for Nosema disease in honeybee colonies. Therefore, it is normally confused with other diseases. Beekeepers should use best management practices when caring for colonies. This includes cleaning bee boxes, removing old combs and drying them in direct sunlight, feeding sugar syrup to bees during dearth seasons, uniting weak colonies, replacing old queens (1–2 times a year) reared from a strong colony free of Nosema spores or from a disease resistant strain, and preventing bees from robbing honey (Tam 2010).

11.5.4 **Varroa destructor** *and* **Tropilaelaps mercedesae** *Mites*

11.5.4.1 *Varroa destructor*

Currently, four *Varroa* mite species—*Varroa destructor, Varroa jacobsoni, Varroa underwoodi*, and *Varroa rindereri*—have been found on the honeybee. Among these, *V. destructor* (formerly *Varroa jacobsoni*) initially parasitized its native host *Apis cerana* but started to parasitize *Apis mellifera* wherever it had access to it (Ritter 1981). In Vietnam, *V. destructor* was first found with the honeybee *A. mellifera* in 1968 (Stephan 1968). According to Boot et al. (1997), *A. mellifera* colonies in Vietnam were heavily infested with hundreds to thousands of mites per colony, whereas *A. cerana* colonies contained only tens of mites at most. About 80% of mites from *A. mellifera* colonies reproduced when introduced into worker cells of either *A. mellifera* or *A. cerana*, and this percentage was similar to reproduction in naturally infested worker cells of *A. mellifera* (Boot et al. 1999). Individual bees infested with *V. destructor* mites during their development usually survive to emergence, but may show signs of physical or physiological damage (Hanh and Tan 1991). These signs include a shorter lifespan, reduced weight, or shrunken and deformed wings (MAAREC 2005). If one to two adult mites infest a brood cell, young bees will usually emerge without visible damage and are normal in appearance. However, as the number of mites in a cell rises, so to do the harmful effects. In some cases, 12 *V. destructor* mites can be found in worker cells and up to 20 *V. destructor* mites in drone cells (Hoang 2002). A worker bee parasitized by one mite had longevity of 18 days, but only 9 days when parasitized by two mites (Woyke 1990). According to MAAREC (2005), individual honeybees heavily infested with more than a few adult mites usually become visibly crippled or die in their cells without emerging. Beside, in heavily Varroa infested colonies, nearly 100% of adult workers may be infected with deformed wing virus (DWV) and have high virus titers even without showing symptoms (de Miranda et al. 2010).

11.5.4.2 *Tropilaelaps mercedesae*

Up to now, four Tropilaelaps species have been discovered—*Tropilaelaps mercedesae, Tropilaelaps clareae, Tropilaelaps konigerum, Tropilaelaps thaii.* The development of Tropilaelaps is quite similar to that of Varroa. The tropilaelaps mite life cycle also consists of four developmental stages: the egg, two eight-legged nymph stages (protonymph and deutonymph), and the adult (Ha and Dung 1991). Female tropilaelaps mites enter the brood cell just before the cell is sealed (Dung 1990).

Total developmental time is about 9 days (Woyke 1985b); however, in Vietnam, Ha and Dung (1991) reported that this time was only 5.5–6.1 days.

Both Varroa and Tropilaelaps mites infect the honeybee *A. mellifera*, and cause economic losses to beekeepers in Vietnam (Hanh and Tan 1991; Ha and Dung 1991). According to Woyke (1987), in honeybee *A. mellifera* colonies examined in southern Vietnam an average of 4.9% of brood cells were infested by Varroa and 46.2% by Tropilaelaps. In Vietnam, however, the actual infection rate of Varroa and Tropilaelaps was from 1.82 to 3.1% and from 4.06 to 4.98%, respectively (Huan 1991). Boot et al. (1997) reported that successful reproduction rates of Varroa mites taken from *A. mellifera* colonies or *A. cerana* colonies were 80% and 10%, respectively, in either bee species after they had been artificially transferred into worker brood cells. They proposed this to be due to a mite-specific difference in reproductive traits of Varroa from both species of bee.

To prevent Varroa and Tropilaelaps mite infestations in honeybee colonies, a general rule of thumb is to retain the balance between host and parasite (Thanh and Tu 1991). In Vietnam, queens of diseased colonies were caged for 21 days resulting in 100% mite reduction (Woyke 1985b). However, this method should only be applied during the honey season, and in an area where there is an abundant supply of nectar and pollen.

To limit miticide resistance and to increase the effect of treatments, formic acid and Marjoram oil are combined to control *V. destructor* with 99% efficacy (Long 1998). Hoang (2002) reported that 65% formic acid will kill 91.19% *V. destructor* and 96.54% of *T. mercedesae.* At present, most Vietnamese beekeepers use only formic acid to control both *V. destructor* and *T. mercedesae.* Per beehive, a piece of foam, measuring 8 × 10 cm, is soaked with 10–15 mL of 85% formic acid solution and placed immediately on top or below the combs.

11.5.5 Pests and Predators

11.5.5.1 Insects

Wax Moth

The greater wax moth, *Galleria mellonella* L., is the most serious insect pest of unprotected honeybee comb throughout the warm areas of the world (Shimanuki and Knox 1991). This moth is thought to have evolved with honeybees in Asia (Morse 1975), and it now commonly inhabits the nest of *A. cerana, A. dorsata, A. florae, and A. mellifera* (Ruttner 1988; Singh 1962). According to Chinh (2012), there are two kinds of wax moths that harm honeybees—the greater *G. mellonella* and the lesser *Achroia grisella.* The former adult females and males are 20 mm and 15 mm, respectively, whereas the lesser wax moths are 10 mm and 13 mm, respectively (Chinh 2012). A couple of days after emerging, adult females and males mate with each other at night. Then females enter the beehive to lay eggs (Shimanuki et al. 1992). The life cycle of the wax moth is about 4–5 weeks (Chinh 2012).

Wax moths cause heavy damage to beekeeping in Vietnam. When larvae tunnel through comb, they destroy pollen, honey, and brood cells. Therefore, *A. cerana* colonies that are heavily damaged by the larvae of wax moth abscond (Chinh 2012). To prevent and control wax moths, beekeepers keep their beehives full of pollen and honey stores, and free of brood disease. In the dearth season, old combs are eliminated. The beehives should be regularly cleaned, and contain a reduced beehive entrance.

Wasps

There are different wasp species that are both social and solitary. In Vietnam, wasps harm both *A. mellifera and A. cerana* colonies, mainly from July to November. They catch worker honeybees on flowers or at hive entrances. Wasps collect both brood and adult honeybees to feed to their own brood (Chinh 2012), and attack honeybee colonies intensively before or after periods of bad weather. When honeybee colonies are attacked by wasps, weak *A. cerana* colonies tend to abscond and some weak *A. mellifera* colonies are killed (Chinh 2012). Generally, the Asian honeybee *A. cerana* is able to protect its colonies better than those of the European honeybee *A. mellifera.* Ono et al. (1987) and Ichino and Okada (1994) reported that *A. c. japonica* showed a distinct balling reaction against workers of the predatory hornet *Vespa simillima xanthoptera.* This resulted in the killing of hornets by heat and asphyxiation induced by the ball. The temperature of the bee ball increased up to 46 °C, which is hot enough to kill hornets but has little effect on *A. cerana* worker bees. Gaps of the beehive should be covered to prevent wasps from going into colonies. During the period when wasps attack intensively, the entrance of beehives should be narrowed. A bamboo broom can be used to kill wasps in front of the entrance of beehives. It has been found that scout wasps come early in the morning to look for bees. If these scout wasps are killed, other nestmates would not be aware of the beehive location (Chinh 2012).

11.6 Other Problems of Beekeeping

11.6.1 Antibiotic Residue

When honeybee colonies are attacked by pests and diseases, beekeepers often use medication for prevention and treatment. Incorrect use of the kind of drugs or drug dosage can result in residues in bee products, and can cause drug-resistant honeybees. Some antibiotic residues exceed the permitted limit in bee products, adversely affecting the health of consumers and creating difficulties to export honey. According to Hoa and Tuat (2012), 72 samples of honey collected from apiaries in some southern provinces (Dong Nai, Gia Lai, Binh Phuoc, Lamdong, Binh Duong, Lam Dong, and Dak Lak) from 2007 to 2010 contained antibiotics that belonged to the group B1, such as enrofloxacin, tylosin, streptomycin, sulfadiazine, and tetracyclines. Residues of chloramphenicol (a group of banned substances, A6) were also found in 2007 and 2008, but in subsequent years were no longer detected.

11.6.2 Pesticide Residues

According to Dong et al. (2012a, b), honeybees collect pollen and nectar from a variety of plants, including agricultural crops where farmers have been using pesticides to control pests and diseases. These are potential sources of pesticide exposure. Hoa and Tuat (2012) reported that chemicals originating from the organic phosphorus group were found in honey samples collected from the southern and central provinces of Vietnam between 2007 and 2010. Residues of pesticides in honey samples tended to increase during that period; however, these chemicals were reduced and could not be found in honey samples in later years. Another issue relevant to exported honey from Vietnam to the United States is Carbendazim residues in honey (Dong et al. 2012a, b). In 2017, the Minister of Agriculture and Rural Development signed Decision No. 03/2017/QD-BNN-BVTV dated 03/01/2017 entitled "Putting plant protection drugs containing Carbendazim ingredients out of the list of the plant protection drugs that are allowed to be used in Vietnam."

11.7 Other Non-*Apis* Species Found in the Region

11.7.1 Stingless Bees

Besides the six true honeybee species (*Apis*) in Vietnam, there are over eight species of stingless bees distributed across the country (except the Red River Delta): *Trigona laeviceps*, *Trigona ventralis*, *Trigona pagdeni*, *Trigona gressitti*, *Trigona fuscobalteata*, *Trigona carpenteri*, *Trigona scintillans*, and *Trigona iridipennis*. Stingless bee nesting sites are found in holes of trees, pillars, or house walls (Chinh 2004). Meliponiculture, or beekeeping with stingless bees, is found in some provinces, including Lai Chau, Son La, Can Tho, Tien Giang, and Ho Chi Minh City. One particular farmer in Khanh Hoa province maintains a 1000 stingless bee boxes (Fig. 11.3). Stingless bees serve as important pollinators in green houses for sweet pepper, cucumber, and melons.

Fig. 11.3 The stingless bee colonies kept in the wooden boxes by a farmer in Cam Hai Tay commune, Cam Ranh Town, Khanh Hoa province

11.7.2 Bumble Bees

In Vietnam, bumblebee research is very limited, and only some initial publications of the list of species and their distribution existing. According to Long et al. (2012), four species from the genus Bombus (*Bombus campestris Panze, Bombus Smith funerarius, Bombus magretti Griboro, Bombus trifasciatus Smith*) have been found in Vietnam, in the provinces of Hoa Binh, Phu Tho, Bac Giang, and Vinh Phuc. Williams and Repsen (2016) reported that two species of bumble bee (*Bombus haemorrhoidalis* (= *montivolans*) and *B. breviceps*) were also found in all of the mountains of North Vietnam, 900 m above sea level these species are potential to employ commercially for pollination of tomatoes in greenhouses.

11.8 Future Perspectives

Although modern beekeeping techniques have been applied in Vietnam since the early 1960s, it was still considered a minor industry in the 1990s. Since 2002, the number of beekeepers, honeybee colonies, and honey production has all continuously increased. According to Linh (2015), Vietnam is considered a potentially large honey exporting country and is one of the largest honey exporters on the world market. Vietnam ranks sixth in the world and second in Asia (after China) in the export of honey. In 2014, the United States imported 165,584 tons of honey from different countries around the world. Of which, 46,530 tons were imported from Vietnam. This made Vietnam the largest honey exporter to the United States (Linh 2015). In 2016, Vietnam continues to be the leading country exporting honey to the USA with 38,115 tons (Flottum 2017).

The main honey export market of Vietnam is the USA and European countries. The demand for imported honey to these markets is expected to increase in future years. However, regulations on the quality of the honey exported to these markets are also very strict. To meet these rigorous demands, the Ministry of Agriculture and Rural Development of Vietnam (MARD) issued circulars and decrees to improve the quality of honeybee products in general, and honey in particular. In addition, the Vietnam Beekeepers Association (VBA) also conducts training to improve knowledge in bee research and beekeeping development. In recent years, domestic consumer interest in honeybee products such as pollen, royal jelly, propolis, and single flower honeys (coffee, longan, lychee) has increased. These products are now sold widely in retail stores and supermarkets. Moreover, companies producing Vietnamese honey are developing very rapidly, such as Apidona, Vinapi, and Dakhoney. In the future, these companies will improve packaging and design while promoting diversification of products and advertising to increase knowledge of the nutritional value of honey for human health. In the domestic market, consumers are more concerned with the safety of food products, particularly those derived from natural products such as honey. Therefore, the demand for these products will continue to increase in the future.

Acknowledgements We would like to thank Mrs. Katrina Klett (Fulbright grantee - 2015) and Dr. David Dodge (The Honey Bee Breeding, Genetics and Physiology – ARS – USDA) who have spent much time correcting English in the manuscript of this chapter. We are also grateful to Dr. Lilia De Guzman (The Honey Bee Breeding, Genetics and Physiology – ARS – USDA) for encouraging us to finish this chapter.

References

Abrol DP (2013) Asiatic honeybee *Apis cerana*: biodiversity conservation and agricultural production. Springer, Heidelberg

Bailey L (1967) The incidence of virus diseases in the honey bee. J Appl Bacteriol 60:43–48

Bailey L (1975) Recent research on honey bee viruses. Bee World 56:55–64

Bailey L (1981) Honeybee pathology. Academic, London

Bailey L, Gibbs AJ, Woods RD (1964) Sacbrood virus of the larvae of honeybee (*Apis mellifera* L.) J Virol 23:425–429

Bailey L, Carpenter JM, Woods RD (1982) A strain of sacbrood virus from *Apis cerma*. J Invertebr Pathol 39:264–265

Ball BV (1999) Sacbrood. In: Colin ME, Ball BV, Kilani M (eds) Bee disease diagnosis. CIHEAM, Zaragoza, pp 91–97. http://om.ciheam.org/om/pdf/b25/99600239.pdf. Accessed 15 Sept 2017

Boot WJ, Tan NQ, Dien PC, Van Huan L, Van Dung N, Long LT, Beetsma J (1997) Reproductive success of Varroa jacobsoni in brood of its original host, Apis cerana, in comparison to that of its new host, A. mellifera (Hymenoptera: Apidae). Bull Entomol Res 87(02):119–126

Boot WJ, Calis JNM, Beetsma J et al (1999) Natural selection of *Varroa jacobsoni* explains the different reproductive strategies in colonies of *Apis cerana* and *Apis mellifera*. Exp Appl Acarol 23:133–144

Brodschneider R, Crailsheim K (2010) Nutrition and health in honey bees. Apidologie 41:278–294

Chen Y, Huang Z (2010) *Nosema ceranae*, a newly identified pathogen of *Apis mellifera* in the USA and Asia. Apidologie 41:364–374

Chinh PH (1990) A disease of bee larvae *A. cerana*. Agriculture Hanoi, Vietnam

Chinh PH (1996) A number of research and technical solutions to improve productivity and qualities native honeybee *Apis cerana* in northern Vietnam. Dissertation, Hanoi Agricultural University

Chinh TX (2004) Reproduction in eusocial bees (Apidae: Apini, Meliponini). Dissertation, Utrecht University

Chinh PH (2012) Keeping domestic honey bee *Apis cerana* in Vietnam. Agriculture Hanoi

Chinh PH, Thai PH, Minh NH, Tan NQ (1995) Raftering: a traditional technique for honey and wax production from *Apis dorsata* in Vietnam. Beekeep Dev 36:8–9

Chinh PH, Lap PV, Thang DV et al (1996) Evaluation of a closed population breeding program with *Apis cerana cerana* after six years of selection. In: 3rd Asian apiculture association conference, Hanoi, p 47

Chinh PH, Tan NQ, Thai PH (2004) Some studies on biological characteristics of Apis dorsata in Southern Vietnam. In: Proceeding of the 7th AAA conference on Bees for New Asia Laguna, Philippines 23–27 Feb 2004, pp 75–85

Chinh PH, Otis GW, Tam DQ et al (2012) Keeping domestic bee *Apis cerana* in Vietnam. Agriculture

Crailsheim K (1990) The protein balance of the honey bee worker. Apidologie 21:417–429

Crane E (1999) The world history of beekeeping and honey hunting. New York Press

de Miranda JR, Dainat B, Locke B et al (2010) Genetic characterization of slow bee paralysis virus of the honeybee (*Apis mellifera* L.) J Gen Virol 91:2524–2530

Decision No. 03/2017/QD-BNN-BVTV dated 03/01/2017. http://www.ppd.gov.vn/uploads/news/2017_01/QD03%20-%20ngay%2003.01.2016%20loai%20bo%20thuoc%20BVTV%20chua%20Carbendazim%20Benomyl.pdf. Accessed 17 Sept 2017

Dong PV, Hoa BTP, Hang NT et al (2012a) The situation of implementation of a monitoring program safety hygiene honey 2012. J Sci Technol Beekeep 1:43–48

Dong PV, Hoa BTP, Hang NT et al (2012b) The situation of implementation of a monitoring program safety hygiene honey 2012. J Sci Technol Beekeep 2:45–52

Dung NV (1990) Report on the study result Tropilaelaps on *Apis mellifera*. In: Proceedings of the conference on the Vietnam national honeybee disease. pp 17–21

Dung PX (1996) Researching on some morphological and biological characteristics of Italian honeybees in Vietnam to improve breeding. Dissertation, Hanoi University of Agriculture

Duong BTT, Thu HT, Linh MT et al (2015) Study on the distribution of Black Queen Cell virus infecting honeybees in Vietnam. J Biotechnol 13:435–440

Flottum K (2017) Imports, exports, production and consumption from 2016. Bee Culture. Http://www.beeculture.com/u-s-honey-industry-report-2016. Accessed 28 Oct 2017

Fries I (1988) Infectivity and multiplication of *Nosema apis* Z. in the ventriculus of the honey bee. Apidologie 19:319–328

Fries I, Feng F, Silva AD et al (1996) *Nosema ceranae* n. sp. (Microspora, Nosematidae), morphological and molecular characterization of a microsporidian parasite of the Asian honey bee *Apis cerana* (Hymenoptera, Apidae). Eur J Protistol 32:356–365

Fries I, Chauzat MP, Chen YP et al (2013) Standard methods for Nosema research. J Apic Res 52:1–28

Grabensteiner E, Ritter W, Carter MJ, Davison S et al (2001) Sacbrood virus of the honeybee (*Apis mellifera*): rapid identification and phylogenetic analysis using reverse transcription. Clin Diagn Lab Immunol 8:93–104. https://doi.org/10.1128/CDLI.8.1.93-104.2001

Gupta RK, Reybroeck W, van Veen JW, Gupta A (2014) Beekeeping for poverty alleviation and livelihood security: Vol. 1. Technological aspects of beekeeping. Springer, Dordrecht

Ha TD, Dung NV (1991) Some biological characteristics of *Tropilaelaps clareae* and treatment methods. In: Proceeding of the 1st Vietnam national conference on entomology, Ha Noi, pp 132–142

Hanh NV, Tan NQ (1991) The situation survey of parasite mite and treatment methods using in some apiaries of Northern Vietnam provinces. In: Proceeding of the 1st Vietnam national conference on entomology, Agriculture, Ha Noi, pp 110–127

Hepburn HR, Smith DR, Radloff SE, Otis GW (2001) Infraspecific categories of *Apis cerana*: morphometric, allozymal and mtDNA diversity. Apidologies 32:3–23

Hoa BPT, Tuat CV (2012) Study and evaluation of toxic chemicals and antibiotics in honey in some southern provinces. J Sci Technol Vet 19:67–75

Hoang LM (2002) Research on some biological characteristic, epidemiology and prevention and treatment measures *Varroa jacobsoni* and *tropilaelaps clareae* on Italian bees (*Apis mellifera ligustica*) in Daklak imported from New Zealand. Dissertation, Hanoi University of Agriculture

Huan LV (1991) The survey of infection rate and control methods for *Varroa jacobsoni* and *Tropilaelaps clareae* and some characteristic of the two parasite mites in the southern Vietnam. In: Proceeding of the 1st Vietnam national conference on entomology, Ha Noi, pp 128–132

Ichino H, Okada I (1994) Japanese honey bees living in hornet nest. Honeybee Sci 15:123–124

Klee J, Besana AM, Genersch E et al (2007) Widespread dispersal of the microsporidian *Nosema ceranae* an emergent pathogen of the western honeybee, *Apis mellifera*. J Invertebr Pathol 96:1–10

Koeniger N, Koeniger G, Long LT (1997) Some notes on bee diseases: Nosema disease in imported honey bees *Apis mellifera* in Viet Nam. J Sci Technol Beekeep 3:21–23

Lan NK, Long LT, Ban DD, Toan TV (1998) Experimental control of Sacbrood Disease on *Apis cerana* using By-products S-95 of Medical Plants. J Sci Technol Beekeep 4:31–34

Lee CH, Choe SE, Kang SW (2010) GenBank accession number HQ322114

Linh VT (2015) Report on EU honey market. http://mutrap.org.vn/index.php/vi/explore/bao-cao-nghien-cuu/finish/45/7435. Accessed 28 Oct 2017

Long LT (1998) Killing parasite mite *Varroa jacobsoni* on *Apis mellifera*. J Sci Technol Beekeep 1:29–33

Long KD, Hoe LX, Hoa DT, Phong PH (2012) A preliminary study on bees (Hemenoptera: Apoidea: Apiformes) from northern central Vietnam. J Biol 34:419–426

MAAREC (Mid-Atlantic Apicultural Research Extension Consortium) (2005) Varroa mite. http://agdev.anr.udel.edu/maarec/wp-content/uploads/2010/03/Varroa_Mites_PMP2.pdf. Accessed 17 Sept 2017

Morse RA (1975) Honey bee pests, predators and diseases. Cornell University Press, Ithaca

Ono M, Okada I, Sasaki M (1987) Heat production by balling in the Japanese honeybee, *Apis cerana japonica* as a defensive behavior against the hornet, *Vespa simillima xanthoptera* (Hymenoptera: Vespidae). Experientia 43:1031–1032

Quyen PB, Niem NV, Thai PH, Lam NV (2001) Honey bee and the decrease in number of honeybee colonies in the U Minh melaleuca forests. In: Proceeding of international workshop on biology, vol 1, pp 170–175

Radloff SE, Hepburn HR, Hepburn C et al (2005) Multivariate morphometric analysis of the *Apis cerana* population of mainland southern Asia. Apidologies 36:127–139

Radloff SE, Hepburn C, Hepburn HR et al (2010) Population structure and classification of *Apis cerana*. Apidologies 41:589–610

Rialan C (1887) Note on *Apis mellifica* in the province of Thai-Nguyen. Bulletin Comitooj d'aaxtudes agricole Annam et Tonkin 2:63–73

Ritter W (1981) Varroa disease of honeybee *Apis mellifera*. Bee World 62:141–153

Ruttner F (1988) Biogeography and taxonomy of honeybees. Springer, Berlin

Shimanuki H, Knox DA (1991) Diagnosis of honey bee disease. USDA Agricultural Handbook Ah-690

Shimanuki H, Knox DA et al (1992) Diseases and pests of honey bees. In: Graham JM (ed) The hive and the honey bee. Dadant and Sons, Hamilton, pp 1083–1151

Singh S (1962) Beekeeping in India. Indian Council of Agricultural Research, New Delhi

Stephan WA (1968) Mites: a beekeeping problem in Vietnam and India. Bee World 49:119–120

Tam DQ (2010) Research on the situation of arising, harmfulness, prevention and control microsporidia species Nosema ceranae parasite on Honeybees *Apis mellifera* Linnaeus in Viet Nam. Dissertation, Hanoi University of Agriculture

Tam DQ, Anh HK, Toan TV et al (2010) Determination of two microsporidia species spores *Nosema apis* and *Nosema ceranae* harm on honey bees in Viet Nam. J Agric Rural Dev 15:72–78

Tan NQ (2007) Biology of *Apis dorsata* in Vietnam. Apidologie 38:221–229

Tan NQ, Chinh PH, Thai PH, Mulder V (1997) Rafter beekeeping with *Apis dorsata*: some factors affecting the occupation of rafters by bees. J Apic Res 36(1):49–54

Tan NQ, Mardan MB, Thai PH, Chinh PH (1999) Observations on multiple mating flights of Apis dorsata queens. Apidologie 30:339–346

Thai PH (1994) The honey production of *A. dorsata* in U Minh Forest. J Sci Technol Beekeep 3:26–28

Thai PH (2008) A study morphological molecular biology (mitochondrial DNA) of the domestic bee populations *Apis cerana* Fabricius and direction on the use of genetic resources in domestic honeybees breeding selection in Viet Nam. Dissertation, Hanoi University of Agriculture

Thai PH, Trung LQ, Dung PX (1997) Some comparative characteristics between *Apis florea* and *Apis andreniformis* in Minh Hai province. J Sci Technol Beekeep 1:13–17

Thai PH, Cuong HV, Giang NV et al (2010) Molecular detection of Sacbrood virus and deformed wing virus infected on *Apis mellifera* in northern Viet Nam. In: Proceeding of the 7th Vietnam national conference on entomology, Ha Noi, pp 892–897

Thanh DH, Tu TM (1991) Some research results on two parasite mite species *Varroa jacobsoni* and *Tropilaelaps clareae* in honey bees. In: Proceeding of the 1st Vietnam national conference on entomology, Ha Noi, pp 104–109

Thu HT, Lien NTK, Linh MT et al (2016) Prevalence of bee viruses among *Apis cerana* populations in Vietnam. J Apic Res 55:379–385

Toan TV (2012) Study on some biological and ecological characteristics of the hybrid between Dongvan indigenous honeybee breed (*Apis cerana cerana* Fabricius) with the local indigenous honeybee breeds (*Apis cerana indica* Fabricius) in some provinces in the north of Viet Nam. Dissertation, Hanoi University of Agriculture

Toan TV, Tam DQ, Lee ML et al (2014a) Nosema disease on honey bee *Apis cerana* in Viet Nam. J Apic 29:1–7

Toan TV, Lee ML, Sim HS et al (2014b) Biological mitigation of Sacbrood disease on *Apis cerana* colonies. J Apic 29:181–186

Toumanoff C (1933) Documentation on beekeeping in Annam. Bull Econ Indochine 36:169–175

Toumanoff C, Nanta J (1933) Enquiry on beekeeping in Tonkin. Bull Econ Indochine 36:1015–1048

Trung LQ, Dung PX, Ngan TX (1996) A scientific note on first report of *Apis laboriosa* F Smith, 1871 in Vietnam. Apidologie 27:487–488

Trung QL, Tam DQ, Toan TV et al (2010) Application Multiplex RT-PCR/RFLP for early detection and genetics relationship research of sacbrood viruses on honey bees in Viet Nam. J Agric Rural Dev 1:34–40

Williams P, Jepsen S (2016) IUCN BBSG - Bumblebee Specialist Group Report 2015. https://www.researchgate.net/publication/299633368_IUCN_BBSG_-_Bumblebee_Specialist_Group_Report_2015. Accessed 7 Mar 2018

Woyke J (1985) Further investigations into control of parasite bee mite *Tropilaelaps clareae* without medication. J Apic Res 24:250–254

Woyke J (1987) Infestation of honeybee (*Apis mellifera*) colonies by the parasite mites *Varroa jacobsoni* and *Tropilaelaps clareae* in south Vietnam and results of chemical treatment. J Apic Res 26:64–67

Woyke J (1990) *Varroa jacobsoni* and *Tropilaelaps clareae* in the report summarizing the work of veterinary period 1974-1990 and orientation for veterinary activities in Vietnam Apiculture 1990-1995, pp 79–86

Woyke J (1996) The background and history of apicultural scientific activities in Vietnam. In: Proceeding of the 3rd AAA conference on bee research and beekeeping development Hanoi, Vietnam, 6–10 Oct 1996, pp xxvi–xxxii

Zhang J, Feng J, Liang Y et al (2001) Three dimensional structure of the Chinese Sacbrood virus. Sci China C Life Sci 44:443–448. https://doi.org/10.1007/BF02879612

Chapter 12
Bee Diversity and Current Status of Beekeeping in Thailand

Panuwan Chantawannakul

Abstract Traditional honey bee hunting and beekeeping are crucial to the economic and spiritual lives of Thais. Bee products such as honey, brood, and royal jelly are regarded as healthy foods and frequently used as traditional medicine. In this chapter, honey bee diversity in Thailand, traditional hunting, and beekeeping are described. The giant and dwarf honey bees are harvested by hunting, only the Asian cavity nesting honey bee (*Apis cerana*) is domesticated and maintained in the traditional hives for harvesting honey and other bee products. The introduced species, the European honey bee (*Apis mellifera*) are kept in the modern box hives. By sharing food sources and habitat, the honey bees have also shared parasites and diseases. The ectoparasitic mites (both *Varroa destructor* and *Tropilaelaps mercedesae*) were jumped from *A. cerana* and *A. dorsata* respectively to the *A. mellifera*. The parasitic mites have become widespread and serious cause of colony loss in Thailand. In addition, microbial diseases (e.g., bee viruses, and *N. ceranae*) also can be detected in both native and introduced honey bee species. Other factors contributing to honey bee declines are also described.

Keywords Thai · Beekeeping · Asian honey bees · *Apis cerana* · Bee diseases · Bee pests

12.1 Diversity and History of Beekeeping in Thailand

It is well-known that Southeast Asia has the largest diversity of honey bees covering all ten species. Half of them are found in Thailand: *Apis cerana*, *Apis andreniformis*, *Apis florea*, *Apis dorsata*, and the introduced *Apis mellifera*. They are regarded

P. Chantawannakul
Bee Protection Laboratory, Department of Biology, Faculty of Science, Chiang Mai University, Chiang Mai, Thailand

International College of Digital Innovation, Chiang Mai University, Chiang Mai, Thailand

Environmental Science Research Center, Faculty of Science, Chiang Mai University, 50200 Chiang Mai, Thailand

P. Chantawannakul et al. (eds.), *Asian Beekeeping in the 21st Century*,
https://doi.org/10.1007/978-981-10-8222-1_12

as one of the key pollinators maintaining the forest ecosystem and agricultural crops. The giant honey bee, *A. dorsata*, frequently hangs their open air nests on the branches of tall trees, cliffs, and nowadays has adapted themselves to nest on high concrete buildings (Wongsiri et al. 1996). The giant honey bee resides from the southern coastal flats of Narathiwat province in the south of Thailand to forests in mountain areas in the northern part of Thailand (Wongsiri et al. 2014). They frequently nest their colonies in dense aggregation for a few months, migrate, and frequently come back to their nest sites in the next season. In Northern Thailand, they migrate to the lowland from November to April (dry season) and migrate to forests in highland from May to June (wet season). The migratory distance is up to 100 km. These colonies in the aggregation sites were not related as mother–daughter suggesting that rapid increase of the bee colonies during flowering events most likely occur by swarms from other areas, not the reproductions of the bee colonies (Rattanawannee et al. 2013). Since the single large comb nest (1.5 × 1 m) can accommodate adult bee populations of 70,000 individuals (Morse and Laigo 1969). The nests therefore hold up to 17 kg (Buawangpong et al. 2014). The giant honey bee, therefore, requires a strong support for their heavy combs. It was reported that the wax from *A. dorsata* was the strongest and stiffest wax when compared to the other species of honey bees (Buchwald et al. 2006).

A bee tree, one of the trees that giant honey bee prefers to nest in Thailand, is *Kompassia alaccenesi*. It is a tropical rainforest tree and large aggregations of giant honey bees can be found, sometimes more than 100 colonies occupy one tree (Wongsiri et al. 2000). The diurnal foraging patterns of *A. dorsata* workers were observed in Chiang Mai province and it showed that the average temperature at flight initiation was 23.0°C. However, the flight activity was not dependent on temperature but ambient daylight. Their first fights started at dawn reflecting the sensitivity of eyes that only required dim light to fly and forage. This leads to better competition on food sources as many flowers produce nectar at night and bloom in early morning. In addition, the colony composition (number of brood) also affects the foraging behavior (Saraithong et al. 2012).

The two dwarf Asian honey bee species (*A. florea* and *A. andreniformis*) have open air nests and can be found in branches of small trees such as longan trees (*Dimocarpus longan*). *A. florea* is present in most parts of Thailand but *A. andreniformis* is rarely present and has never been found in central Thailand (Wongsiri et al. 2014). It also has been reported that this species of honey bee has become rare due to deforestation. *A. andreniformis* is present from coastal flats, foothill areas of Chanthaburi province and mountains up to 1600 m altitude in the northern part of Thailand while *A. florea* is found below 1000 m (Wongsiri et al. 1996).

A. andreniformis and *A. florea* often have similar preferences on nest sites. *A. andreniformis* nested at heights averaging about 6 m while *A. florea* nested at heights averaging about 4 m above the ground (Rinderer et al. 2002). Similar to giant honey bees, the tendency of colonies of these species to establish their nest sites near existing nest sites of colonies of the same species might be able to increase the chances that the newly selected nest sites are near suitable floral resources and assure that they can find potential mates within their mating range for a colony's future reproductive success. Therefore, the nest aggregation of these bees is frequently observed. Based on the microsatellite analysis of *A. florea* in woodlands in Phitsanulok province, the results showed that the colonies of *A. florea* at aggregation sites were not related as mother and daughter, suggesting that the colonies were not related to each other genetically (Wattanachaiyingcharoen et al. 2002). Nevertheless, the colonies may avoid areas containing nests of the other species. This may also help diminish interspecific interference with mating that may arise from the species having similar sex pheromones (Rinderer et al. 2002). Multivariate morphometric studies of *A. florea* and *A. andreniformis* in both mainland and two of the islands of Thailand were homogenous and fall into one group (Chaiyawong et al. 2004; Rattanawannee et al. 2007). However, in the case of *A. andreniformis* it was found that the body size increased from east to west and south to north. In addition, based on genetic analysis using mitochondrial cytochrome oxidase subunit b encoding gene, Thai *A. andreniformis* can be grouped in two distinct groups: one is the mainland which has less genetic variation and the other group is in Chiang Mai province and Phuket Island. Nevertheless, geographical signal or correlation was not shown from the study (Rattanawannee et al. 2007).

The cavity nesting Asian honey bee seems to have more variation. Based on morphometric analysis, *A. cerana*, which is native to Thailand, can be separated into four groups (Northern to Central Thailand, Southern Thailand, Samui Island, and Phuket Island (Sylvester et al. 1998)).The distinctness of the *A. cerana* in Samui Island was also evidenced by genetic analysis (Pramual 1994).

The European honey bee (*A. mellifera*) was firstly imported from Australia in early 1950s for research purposes (Akratanakul 2000; Wongsiri 1989). Later in 1970s, a large number of *A. mellifera* colonies were introduced from Taiwan to Chiang Mai and Lamphun in the North of Thailand by Taiwanese beekeepers for beekeeping industry (Wongsiri et al. 1995). Later *A. mellifera* have been imported from Russia and Europe (Kavinseksan et al. 2015). Since it provides more honey yield and was also easier to manage when compared to the other cavity nesting Asian honey bee *A. cerana* which is native to Thailand. The beekeeping industry has expanded and it has now more than 300,000 colonies distributed throughout Thailand, of which most apiaries are located in the North of Thailand. However, the up-to-date information on genetic variance of *A. mellifera* is still understudied.

Honey bees are social insects that not only produce food but also play an important part in religious beliefs and cultures. The history of Asian honey bees in Thailand is ancient and there are numerous depictions of honey being offered to Buddha as can be seen in temples in Thailand today. The picture displays a monkey offering a harvested colony of *A. florea* to the Buddha. This carving depicts an occasion when the Buddha went to the jungle for meditation. The elephant and monkey brought him food which were bananas and dwarf honey bee comb. Honey has also been recorded as an important food in sacred texts of Buddhism. Many rituals during Buddhism ceremonies have involved mixing honey as one of ingredients and sharing communally. In Buddhism, monks are permitted to consume honey as a tonic in the evenings, while only consuming one or two meals a day before noon. Bee wax is also used to make candles for monk meditation and prayer at night time, during the rainy season when monks are confined to their temples.

12.2 Bee Hunting and Beekeeping Practices in Thailand

Thai villagers still maintain gathering practices to exploit wild honey bees. They are opportunistically harvested or actively sought out by honey hunters. Harvesting wild honey from the giant honey bee is a destructive operation as it requires cutting the comb. The harvesting period falls during March–April which is a dry season and where the moisture in the honey is low and less likely to be fermented. Hunters observe the bees from the time of their arrival at the tree and often choose hives which are full of honey. In particular, hunters harvest honey immediately prior to the colony's next season for migration which can be observed through bee behavior such as when the queen's brood laying rate is reduced or paused. Hunters often prefer to take down the bee hives at night, especially moonless ones, or early mornings. In local cultures, it is also believed that bee sites or bee trees have angels living in and protecting the trees. Before the operation, some hunters will pray or chant for permission and for a safe and successful operation. They use bamboo ladders to climb trees or rope wrapped around the trunk and climb the trees and light their torch and brush the comb. In Northern Thailand, honey and the brood are commonly consumed as traditional dishes. Honey is filtered with white cotton and the brood is cut into small squares. For the bee brood, it can be mixed with chicken eggs and steamed in banana leaves or fried as food.

It is relatively easy to harvest the hive products of dwarf honey bees since they nest in lower sites such as tree branches in orchards and are not aggressive. Hunters shake nests vigorously and once the queen is airborne the adult worker bees will follow and leave the comb. Hunter often cut the supporting branch and bind it into a bamboo tripod. The upper part of the comb which contains honey may be cut out for the honey and the colony left for 6–8 weeks to rebuild the comb and replenish the honey and then the honey is harvested again if the colony has not absconded due to the human disturbance. About 40,000–50,000 nests are sold annually in local Thai markets (Chen et al. 1998) (Fig. 12.1a). At present, wild honey is still popular as it is perceived to be a pure and healthy product.

The cavity nesting Asian honey bee in Thailand is still kept in the traditional way by using bait hives. Bait hives are hollow log hives or simple man-made containers built of local materials such as concrete hive resembling the hollow log hive. Honey can be harvested by opening hives and cutting out the honey comb (Fig. 12.1b). The cavity nesting Asian honey bee, *A. cerana*, has been kept through a traditional approach for a long time to produce honey and wax for trade items. The Tai, ancestors of the modern Thai-Lao-Shan ethnolinguistic group, have practiced beekeeping with *A. cerana* in Xishuang Banna of the southern Yunnan Province in China for more than 1000 years (Wongsiri et al. 1987). Thai farmers often keep them in their agricultural areas such as the coffee plantations in Doi Saket, Chiang Mai, and they can harvest about 2.4 kg/colony/year which is much lower than that of the European honey bee 35–40 kg/colony/year. Even though *A. cerana* bees produce less honey than the European honey bee, it is important to note that they are more resistant to diseases and pests (Chantawannakul et al. 2016), adapt well in the mountainous areas, and require lower levels of investment. In addition, cavity nesting Asian

Fig. 12.1 (**a**) Combs of *Apis florea* (15 cm wide) (by Wannapha Mookhploy) (**b**) Combs of *Apis cerana* (20–25 cm wide) (by Sasiprapa Krongdang)

honey bees have been reported to be more efficient pollinators of various fruit and vegetable crops than the European honey bee (Hepburn and Radloff 2011). The European honey bee is being managed by keeping bee in a standard bee box and moving the bee boxes to the locations where flowers bloom such as longan, lychee, sunflower, sesame, coffee, bitter bush, and para rubber. Monofloral honey in Thailand mainly depends on the availability of the flower resources. Our study showed that floral origin, honey bee species, and post-collection processing are contributing factors of biomedical properties (antimicrobial and antioxidant activities) in addition to their fragrance, the specific aroma of Thai honey. Coffee honey from *A. cerana* had the highest antioxidant properties amongst Thai honey. Longan honey was the second best in terms of biomedical properties (Pattamayutanon et al. 2015). Most honey samples tested contained less than 21% moisture, and HMF content and total acidity were mostly low indicating good honey quality in the accepted range for international standards (Wanjai et al. 2012).

Apart from honey, bee products are brood, propolis, bee wax, royal jelly, bee pollen, and bee venom. Only honey, royal jelly, and bee pollen are widely used in commercial products. Nevertheless, honey is still regarded more as a medicine than a food. Bee products that are consumed by local people vary greatly across the region depending on local cultures and the sphere of religious influence. Royal jelly and bee pollen are known to be consumed as part of food supplements and medicine alternatives.

Bee propolis, which is natural resin that honey bees collect from plants in order to block holes and cracks, and make entrance of the hive, is not yet used as a commercial product in Thailand. Chemical profiles of Thai propolis collected from *A. mellifera* in Northern Thailand demonstrated similar composition but different level of sugars and sugar derivatives, triterpenes, and phenolic lipids. The most active compounds that exhibited antimicrobial activities in the propolis collected from Prayao province were from mango (*Mangifera indica*) (Sanpa et al. 2017). Propolis extract from Nan province was found to be able to suppress cancer cell growth through induction of apoptosis pathway (Khacha-ananda et al. 2016). By oral administration of the propolis granule produced from the propolis extracts, the tumor-bearing mice could live longer, thereby suggesting that the propolis can be further developed as a natural supplement with the synthetic chemotherapy drugs for cancer treatment (Khacha-ananda et al. 2016). Bee pollen is also used as a food supplement. Pollen from tea trees showed the highest antioxidant properties when compared to Mimosa, coffee, and lotus pollen.

12.3 Bee Forage Resource

Honey bees depend on plant nectar and pollen for the carbohydrates, proteins, and fat in their diet. Nectar and pollen vary in protein content amongst plant species and this affects the pollen foraging behavior of honey bees. In Northern Thailand, *Mimosa pudica* also called sensitive plant is regarded as one of the major pollen sources for honey bees (Sinpoo et al. 2017). Many other plant species are pollen sources for honey bees such as tea, coffee, and lotus. Bee pollen from *Biden pilosa* L. was the most abundant during the period of investigation at Chiang Mai University (Chantaphanwattana 2017). Although the plant species defined the protein composition in bee pollen, new findings from our research showed that plant species did not have any consequence on the protein in royal jelly indicated by the same protein patterns in all royal jelly samples.

A location suitable for good beekeeping is one in which nectar and pollen producing plants grow abundantly and with a relatively long blooming season. To manage such productive foraging environments in Thailand, considering that we have large diversity of plants and blooming seasons, migratory beekeeping is commonly conducted to ensure the bee colonies gain adequate nectar and pollen. Research to compile floral sources in Thailand and the identification of areas suitable for beekeeping are still understudied. However, the floral calendar is one of the most useful tools of the apicultural extension worker and beekeepers to estimate time and location throughout the year for their beekeeping business. This actually mimics a migratory strategy of feral colonies of native bees which move during the seasons for floral sources. During beginning of dry season (November through January), *Eupatorium odoratum* is a honey plant in the North, Kapok (*Ceiba pentandra*) in the northeast part, and Sunflower (*Helianthus annuus*) in the central part of Thailand. Late January through March, Lychee (*Litchi chinensis*) and Rambutan (*Nephelium lappaceum*) are honey plants. Later during the year, Longan (*Dimocarpus longan*) starts to bloom in the north of Thailand and Para rubber in the south until April. During the wet season (June– October), *Mimosa* plants are the main source of pollen. These plants are also foraged by the native Asian honey bees apart from European honey bee. For instance, *E. odoratum* is foraged by *A. cerana*, *A. florea*, and *A. dorsata*. During the flowering season of *E. odoratum*, *A. dorsata* migrates to Chiang Mai and Mae Hong Son city in the northern part of Thailand

(Thapa and Wongsiri 1997). During the study in 2009 in the Nan province where *A. cerana*, *A. dorsata*, and *A. florea* were present, the results revealed that the major pollen source for the three species of honey bees were from *Mimosa pudica* L., *Mimosa pigra*, *Celosia argentea* L., *Zea mays* L., *Wedelia trilobata* L., and *Syzygium malaccense* L. The most abundant pollen source was *M. pudica* L., perhaps due to its flower morphology, long blooming season, and wide distribution (Suwannapong et al. 2013).

12.4 Diseases and Parasites of Honey Bees in Thailand

Bee health issues have been intensively investigated since 2002 both in European honey bee and Asian honey bees (Chantawannakul et al. 2016). Six honey bee viruses (Deformed wing virus (DWV), Acute bee paralysis virus (ABPV), Sacbrood virus (SBV), Kashmir bee virus (KBV), Chronic bee paralysis virus (CBPV), and Black queen cell virus (BQCV) were surveyed using Reverse Transcriptase-Polymerase Chain Reaction (RT-PCR) in *A. mellifera*. Bee samples were positive for DWV, ABPV, SBV, and KBV but not CBPV, and BQCV. DWV was the most widespread and ABPV was the second most prevalent. Kashmir bee virus was found only in the Lampang province where high infestation of *Varroa destructor* mite occurred (Sanpa and Chantawannakul 2009). Later, BQCV and IAPV were also detected in *A. mellifera* colonies in Northern Thailand. In native Asian honey bees, BQCV was the most prevalent amongst *A. dorsata*, *A. cerana*, and *A. florea* (Mookhploy et al. 2015). It is interesting to note that the BQCV isolates from different species of *Apis* in Thailand were more phylogenetically associated with geographic origin than host bee species from which the isolates were obtained (Mookhploy et al. 2015).

Nosema ceranae was found to be prevalent in *A. mellifera*, *A. cerana*, *A. florea*, and *A. dorsata*. However, *Nosema apis* was not found in any samples from honey bees and bumble bees in Thailand. Interestingly, based on polar tube protein (PTP1) gene, phylogenetic analysis showed *N. ceranae* isolates from the cavity nesting species (*A. mellifera* and *A. cerana*) formed a single clade which separated from the open air nesting honey bees (*A. dorsata* and *A. florea)* (Chaimanee et al. 2011). This reflects the genetic lineage of the hosts and suggests ongoing coevolution of the pathogens and its hosts.

American foulbrood has not been reported in Thailand. European foulbrood and Chalkbrood have been found sporadically especially in rainy season in *A. mellifera* colonies (Budge et al. 2010; Theantana and Chantawannakul 2008). Based on molecular methods, the 16S rRNA gene sequence of *Melissococcus plutonius*, the EFB pathogen, was present in *A. florea* larvae collected in Chiang Mai province (Saraithong et al. 2014).

We also found that 74.4% of *Ascosphaera apis* (Chalkbrood pathogen) produced the β-*N*-acetylglucosaminidase that can breakdown the cuticle component inside the larval body (Theantana and Chantawannakul 2008). This might be involved in the

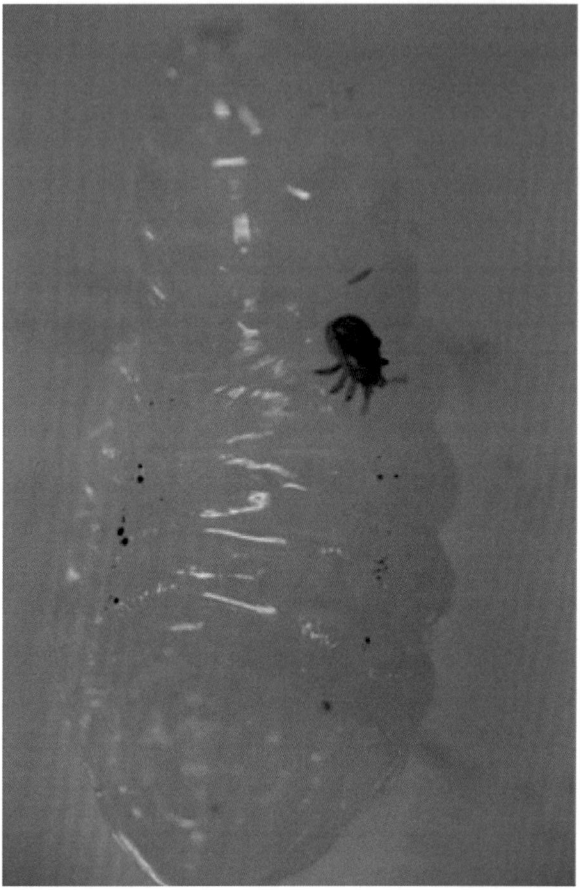

Fig. 12.2 *Tropilaelaps mercedesae* on *Apis mellifera* larvae (brown dots represent wound punctured by *T. mercedesae*) (by Patcharin Phokasem)

mechanism of penetration in Chalkbrood disease. It is known that endosymbiotic bacteria are prevalent in arthropods, *Arsenophonus* spp. were present in *A. mellifera* (24.2%), *A. dorsata* (8.3%), and *A. florea* (8.3%), but not in *A. cerana* while *Wolbachia*, *Spiroplasma*, and *Rickettsia* could not be found in bee samples collected from both wild and managed colonies of the four honey bee species in Chiang Mai and Phatthalung, Thailand (Yañez et al. 2016).

Amongst the most serious problem in Thai beekeeping is parasitic mite infestation. Two species of mites that are commonly found in *A. mellifera* colonies are *Varroa destructor* and *Tropilaelaps mercedesae* (Chantawannakul et al. 2016). These two mites can concurrently infest *A. mellifera* colonies, even though this is rare (<0.1%). In Thailand, *T. mercedesae* was the more dominant brood parasite of *A. mellifera* (Buawangpong et al. 2015) (Fig. 12.2). The dominance of *Tropilaelaps* over *Varroa* may be due to their higher reproductive success. The study demonstrates that although both mites in *A. mellifera* colonies examined in Northern Thailand produced similar numbers of progeny, the proportion of nonreproductive *T. mercedesae* (29.8 ± 3.9%) was lower than that of *V. destructor* (49.5 ± 5.9%).

The effects of *T. mercedesae* on *A. mellifera* were similar to *Varroa* mites by shortening the life span of the host, reducing emergence weight, and heightening infection of DWV, leading to more clinical symptoms of wing deformity (Khongphinitbunjong et al. 2015; Khongphinitbunjong et al. 2016). *T. mercedesae* infestation also alters immune response in *A. mellifera* (Khongphinitbunjong et al. 2015). The mites also were found to harbor genetic materials of two viruses, DWV and BQCV (Khongphinitbunjong et al. 2015) while *V. destructor* had more viruses. It was reported that a single mite (*V. destructor*) collected from *A. mellifera* hives in Northern Thailand was simultaneously carrying up to five viruses (KBV, ABPV, DWV, SBV, and BQCV) (Chantawannakul et al. 2006). Infestation by *Varroa* has been associated with KBV, DWV, and IAPV (Chen et al. 2004; Chen and Siede 2007; Di Prisco et al. 2011; Martin et al. 2012; Tentcheva et al. 2004). *Tropilaelaps* mites can be spread by natural causes (e.g., drifting, robbing, and swarming) and by migratory beekeeping. Thai beekeepers often observe the "bald brood" condition in which the pupal caps are removed and the developing pupae are exposed when the colonies are highly infested by *Tropilaelaps* mites (Pettis et al. 2013).

Mite infestation is also present in Asian honey bees however, with lesser levels of infestation. The indigenous hosts of *Tropilaelaps* spp. are the giant Asian honey bees (*A. dorsata*, *A. laboriosa*, and *A. breviligula*). *T. koenigerum* was reported on *A. dorsata* (Tangjingjai et al. 2003). *T. mercedesae* (one adult) was also found in *A. cerana* brood in Thailand (Anderson and Morgan 2007). Because of the migratory nature of Asian honey bees that disrupts the brood cycle and reduces mite populations in the colonies, *Tropilaelaps* populations are generally low in their natural giant honey bee hosts. This is the case even though *T. mercedesae* does not exhibit a sex preference but experiences reproductive success in both drone and worker broods of *A. dorsata* (2% vs. 4.8% infestation prevalence) (Buawangpong et al. 2013). Apart from this, our study displays that the three species of honey bees (*A. mellifera*, *A. cerana*, and *A. dorsata*) had varied degrees of grooming behavior to *Tropilaelaps* in the phoretic stage. *A. cerana* and *A. dorsata* showed faster behavioral responses to the presence of *T. mercedesae* than did *A. mellifera* in cage experiments. Also, a higher proportion of injured mites was observed in *A. cerana* and *A. dorsata* than those which fell from *A. mellifera*. *A. cerana* exhibited the highest behavioral resistance owing to its body shaking in response to the presence of mites (Khongphinitbunjong et al. 2012).

The first described *Varroa* mite of *A. cerana* was *V. jacobsoni* in Java in 1904. Anderson and Trueman (2000) reclassified *V. jacobsoni* into two species, *V. jacobsoni* and *V. destructor*. Nine mitochondrial COI haplotypes (mitochondrial genotypes) from *V. jacobsoni* and seven haplotypes from *V. destructor* were reported. All *V. destructor* known to have colonized *A. mellifera* have one of two mtDNA haplotypes (Korean haplotype and Japan–Thailand haplotype) (Anderson and Trueman 2000). During June–August 2001, 77 samples of *Varroa* mites collected from the feral and semidomesticated *A. cerana* colonies from north (Chiang San, Chiang Rai province) to south (Krasay Sin, Songkhla province) of Thailand. Four haplotypes of *V. jacobsoni* (Malaysia, NorthThai1, NorthThai2, and Samui1 haplotypes) were found. *V. jacobsoni* (Malaysia haplotype) was found in the south of the peninsula Thailand, and the Samui1 haplotype was present on Samui Island. *V. jacobsoni* (NorthThai1 haplotype and NorthThai2 haplotype) were distributed in the north and central regions, and in part of the peninsular region of Thailand. *V. destructor* (Vietnam haplotype) were restricted to the mountain region of Northern Thailand (Warrit et al. 2006). Although *V. destructor* is regarded as a major parasitic mite that causes *A. mellifera* colony loss worldwide (Anderson and Trueman 2000), *A. cerana* is found to be resistant to *V. destructor*. It has been proposed that the resistance mechanism is social immunity; for example, workers detect and remove infested brood, thereby interrupting its reproduction. In addition, the development of infested *A. cerana* was significantly delayed compared to *A. mellifera* and altruistic suicide of immature bees constitutes an analogue of apoptosis as it prevents the spread of infestation by sacrificing parts of the whole organism (Page et al. 2016).

In highly infested honey bee colonies, especially without any chemical treatment, the colonies usually collapse in a few months. Chemical treatment for both mite species are similar using acaricides (e.g., fluvalinate). Some lower cost chemicals have been used such as sulfur, tobacco, and camphor (Burgett and Kitprasert 1990; Camphor et al. 2005; Kongpitak et al. 2008). Safer products such as formic acid, thymol, and lemon grass oil are also being used (Booppha et al. 2010). Nevertheless, Integrated Pest Management (IPM) is recommended for mite control.

Mites in the genus *Euvarroa* are known to parasitize dwarf Asian honey bees. *Euvarroa wongsirii* was first reported and described from Thai *A. andreniformis* (Lekprayoon and Tangkanasing 1991). *Euvarroa sinhai* was also reported on *A. mellifera* and *A. cerana* workers in Thailand (Koeniger et al. 1993). However, *Euvarroa* infestation in dwarf Asian honey bees is naturally low.

12.4.1 Other Pests

Another serious pest of *A. mellifera* beekeeping in Thailand is an Asian hornet (*Vespa velutina*). Beekeepers usually kill hornets by hitting them in front of the hives, nest destruction using insecticides or biocide gas if they can locate their nests. Sometimes, beekeepers use food bait mixed with insecticides (e.g., meat, sugar syrup) (Chantawannakul et al. 2016). Ants also attack honey bees for food in Thailand, both *A. mellifera* and native Asian honey bees (e.g., Waver ants are found to disturb *A. florea*) (Seeley et al. 1982).

12.5 Other Beekeeping Problems

Natural habitat for wild bees has been greatly reduced due to deforestation and permanent cropping. The expansion of cash crops (rubber, cabbage, and corn) is also greatly increasing in Thailand. The downside of this operation is that it greatly reduces plant diversity because villagers invade forested areas and practice monoculture-dominated agriculture leading to a decline of food sources and habitats for wild bees. In addition, the negative impacts of modern agricultural interventions, e.g., use of chemical fertilizers, pesticides, and environmental pollution have been reported in many countries including those in Southeast Asia and have caused a decline of honey bee populations (Aizen and Feinsinger 1994; Partap and Partap 1997, 2002; Ricketts et al. 2008). However, pesticides (e.g., herbicides, fungicides, formamidines, organophosphates, pyrethroids, and neonicotinoids) were found in different levels in pollen samples collected from *A. mellifera* colonies, in both agricultural and non-agricultural locations in Chiang Mai, Lampang, Phayao, and Phrae in Northern Thailand from 2015 to 2016. However, overall, the detection of agricultural chemicals was low. Many areas of Northern Thailand remain a relatively clean environment in which to keep honey bees and produce bee products. Climate change is also another factor that affects honey bee diversity in the region (Brown and Paxton 2009). The risks of the region are warming and drying trends and extreme temperatures and precipitation. According to surveys, the most viable beekeeping operations in Chiang Mai, Thailand are medium sized ones (101–1000 hives), which account for about 66% of the total number of beekeepers. Colony operation costs include local pest control, sugar that is used to feed colony during dearth periods, land rental costs, and transport expenses for trucks and other heavy equipment due to the migratory nature of the business. Small-scale beekeeping operations remain sideline enterprises, whereas medium- and large-scale operations can develop into full-time occupations. The pollination service business does not exist as it does in other countries such as the USA. Nevertheless, other factors such as the input of capital must be taken into consideration. Beekeeping management techniques, marketing practices, and seasonal factors such as climate, nectar flow, etc. are important variables as they influence the investment and decision-making of the beekeepers. Some local beekeepers cannot obtain high yields of honey. Factors that contribute to low honey yields are a lack of beekeeping knowledge, diseases and pests, as well as agricultural pollutants (Chantawannakul et al. 2016).

12.6 Stingless Bees and Bumble Bees

Stingless bees are widespread over tropical and some subtropical regions of the world including Thailand. Honey that is made commercially is not only from honey bees but also from stingless bees (*Tetragonula laeviceps*) (Fig. 12.3a) especially Chanthaburi and Trat provinces. Total moisture contents of stingless bee honeys were higher than that of *Apis* species. However, they can inhibit the growth of 13 species of bacteria (*Klebsiella pneumoniae*, *Listeria monocytogenes*, *Micrococcus luteus*, *Proteus mirabilis*, *Pseudomonas aeruginosa*, *Staphylococcus epidermidis*, *Streptococcus pyogenes*, *Staphylococcus aureus*, methicillin-resistant *S. aureus*, *Serratia marcescens*, *Salmonella typhimurium*, *Bacillus cereus*, and *Escherichia coli*) (Suntiparapop et al. 2012). Propolis collected from *Tetragonula laeviceps* and *Tetrigona melanoleuca* showed high antibacterial activities. One of the main components is mangostin, a natural xanthone which can be isolated from various parts of the mangosteen tree (*Garcinia mangostana*). The mangosteen trees are widespread in Southeast Asia. The pericarp is known to be used in Thai indigenous medicine for skin infection (Sanpa et al. 2015).

Fig. 12.3 (**a**) Hive entrance of *Tetragonula laeviceps* (by Sasiprapha Krongdang) (**b**) bumble bee (*Bombus breviceps*) forages flowers in Doi Inthanon, the highest mountain in Thailand (by Chainarong Sinpoo)

In pollination services, bumble bees play an important role to maintain the natural and agricultural ecosystem especially in mountainous regions (e.g., tomato, pepper, melon, cucumber, and strawberry). Parasites that affect the bumble bee population were investigated. During surveys in Chiang Mai, Mae Hongson, Chiang Rai, and Nan, *Bombus breviceps*, *Bombus montivagus*, *Bombus eximius*, and *Bombus haemorrhoidalis* were found (Fig. 12.3b). *Nosema bombi* and *N. ceranae* was present in the three species of *Bombus* (*B. montivagus*, *B. haemorrhoidalis*, and *B. breviceps*).

12.7 Future Perspectives

Honey and other bee products are important to both modern beekeeping and the traditions that Thais practice. Diseases and pests of honey bees have been intensively investigated and documented (Chantawannakul et al. 2016). However, Thai beekeepers are facing many problems both economically and technically. Parasitic mites remain the most serious problem to Thai beekeeping especially *Tropilaelap*s mites. The wild bees have been affected by many environmental factors with expansion of cities and the labor shift from agricultural to industrial sectors. Therefore, strategic plans are needed to maintain habitats for honey bees and non-*Apis* bees (bumble bees and stingless bees) to secure human food sources. More research will be needed to understand wild bees and their resistance mechanism against bee diseases and parasites. This body of knowledge can then be applied to *A. mellifera* beekeeping.

Most of the world including Southeast Asia is experiencing the risks and opportunities of economic globalization. Human agro-economic activities not only affect the global bee population, but in some cases can disrupt the evolutionary processes of honey bee species. Habitat change and the loss and fragmentation associated with agricultural intensification are fundamental risks for native honey bees (Kremen et al. 2002; Murray et al. 2009). Therefore, the conservation of bee ecosystems and subsequently bee diversity is the greatest challenge that we have ever encountered. The "pollinator crisis" is not far-fetched. Honey bee losses, in turn, will affect both wild flowers and cultivated crops which are dependent on pollination service by honey bees (Potts et al. 2010). The increase of European honey bee populations in the region for commercial purposes might compensate the pollination service provided by feral honey bees; however, they will not cover everything. Introduced honey bee species, in this case, the European honey bees, have never been thoroughly assessed in the Southeast Asian region for example, in terms of food and nest site competition. This area urgently requires the attention of researchers. Environmental costs warrant recognition and considerations of agriculture and conservation policies are developed in addition to effective measures for putting these policies into practice. As mentioned above, the conservation of honey bee species in Southeast Asia not only depends on biological factors, but also social, political, and economic ones. New technologies have introduced new regimes of extraction and commodification that directly impact the region's key pollinators. The beekeeping industry in Southeast Asian nations requires a holistic approach including input from scientists, policy makers, beekeepers, and business partners. As such, a constructive and innovative modus operandi is essential to promote the economic vitality of the beekeeping industry. Further, it's important to maintain the balance of wild bees to ensure sufficient pollinators in the forest ecosystem and guarantee human food sources as human societies and their activities continue to impact the region's ecosystems and services.

Acknowledgement P.C. acknowledge Thailand research fund (RSA 6080028) and Chiang Mai University fund.

References

Aizen MA, Feinsinger P (1994) Forest fragmentation, pollination, and plant reproduction in a Chaco dry forest, Argentina. Ecology 75:330–351. https://doi.org/10.2307/1939538

Akratanakul P (2000) Apiculture development in Thailand. In: Proceedings of the 7th international conference on tropical bees: management and diversity, vol 38. Apidologie, pp 395–398

Anderson DL, Morgan MJ (2007) Genetic and morphological variation of bee-parasitic Tropilaelaps mites (Acari: Laelapidae): new and re-defined species. Exp Appl Acarol 43:1–24. https://doi.org/10.1007/s10493-007-9103-0

Anderson DL, Trueman JW (2000) *Varroa jacobsoni* (Acari: Varroidae) is more than one species. Exp Appl Acarol 24:165–189

Booppha B, Eittsayeam S, Pengpat K, Chantawannakul P (2010) Development of bioactive ceramics to control mite and microbial diseases in bee farms. Adv Mater Res 93:553–557. https://doi.org/10.4028/www.scientific.net/AMR.93-94.553

Brown MJF, Paxton RJ (2009) The conservation of bees: a global perspective. Apidologie 40:410–416

Buawangpong N, Khongphinitbunjong K, Chantawannakul P, Burgett M (2013) *Tropilaelaps mercedesae*: does this honey bee brood mite parasite exhibit a sex preference when infesting brood of the adapted host *Apis dorsata*? J Apic Res 52:158–159. https://doi.org/10.3896/IBRA.1.52.3.07

Buawangpong N, Saraithong P, Khongphinitbunjong K, Chantawannakul P, Burgett M (2014) The comb structure of *Apis dorsata* F. (hymenoptera: Apidae): 3-dimensional architecture and resource partitioning. Chiang Mai J Sci 41:1077–1083

Buawangpong N, de Guzman LI, Khongphinitbunjong K, Frake AM, Burgett M, Chantawannakul P (2015) Prevalence and reproduction of *Tropilaelaps mercedesae* and *Varroa destructor* in concurrently infested *Apis mellifera* colonies. Apidologie 46:779–786. https://doi.org/10.1007/s13592-015-0368-8

Buchwald R, Breed MD, Greenberg AR, Otis G (2006) Interspecific variation in beeswax as a biological construction material. J Exp Biol 209:3984–3989. https://doi.org/10.1242/jeb.02472

Budge GE et al (2010) The occurrence of *Melissococcus plutonius* in healthy colonies of *Apis mellifera* and the efficacy of European foulbrood control measures. J Invertebr Pathol 105:164–170. https://doi.org/10.1016/j.jip.2010.06.004

Burgett DM, Kitprasert C (1990) Evaluation of Apistan as a control for *Tropilaelaps clareae* (Acari: Laelapidae), an Asian honey bee brood mite parasite. Am Bee J 130(1):51–53

Camphor ESW, Hashmi AA, Ritter W, Bowen ID (2005) Seasonal changes in mite (*Tropilaelaps clareae*) and honeybee (*Apis mellifera*) populations in Apistan treated and untreated colonies. Apiacta 40:34–44

Chaimanee V, Chen Y, Pettis JS, Scott Cornman R, Chantawannakul P (2011) Phylogenetic analysis of *Nosema ceranae* isolated from European and Asian honeybees in Northern Thailand. J Invertebr Pathol 107:229–233. https://doi.org/10.1016/j.jip.2011.05.012

Chaiyawong T, Deowanish S, Wongsiri S, Sylvester HA, Rinderer TE, de Guzman L (2004) Multivariate morphometric study of *Apis florea* in Thailand. J Apic Res 43:123–127. https://doi.org/10.1080/00218839.2004.11101122

Chantaphanwattana T (2017) Identification of bee pollens collected from honey bee (*Apis mellifera*) hives in Chiang Mai University. Dissertation, Chiang Mai University

Chantawannakul P, Ward L, Boonham N, Brown M (2006) A scientific note on the detection of honeybee viruses using real-time PCR (TaqMan) in Varroa mites collected from a Thai honeybee (*Apis mellifera*) apiary. J Invertebr Pathol 91:69–73. https://doi.org/10.1016/j.jip.2005.11.001

Chantawannakul P, de Guzman LI, Li J, Williams GR (2016) Parasites, pathogens, and pests of honeybees in Asia. Apidologie 47:301–324. https://doi.org/10.1007/s13592-015-0407-5

Chen YP, Siede R (2007) Honey bee viruses. In: Maramorosch K, Shatkin A (eds) Advances in virus research, vol 70. Elsevier, Amsterdam, pp 33–80. https://doi.org/10.1016/S0065-3527(07)70002-7

Chen PP et al (1998) Honey bees and other edible insects used as human food in Thailand. Am Entomol 44:24–29

Chen Y, Zhao Y, Hammond J, Hsu HT, Evans J, Feldlaufer M (2004) Multiple virus infections in the honey bee and genome divergence of honey bee viruses. J Invertebr Pathol 87:84–93. https://doi.org/10.1016/j.jip.2004.07.005

Di Prisco G, Pennacchio F, Caprio E, Boncristiani HF Jr, Evans JD, Chen Y (2011) *Varroa destructor* is an effective vector of Israeli acute paralysis virus in the honeybee, *Apis mellifera*. J Gen Virol 92:151–155. https://doi.org/10.1099/vir.0.023853-0

Hepburn HR, Radloff SE (2011) Biogeography of the dwarf honeybees, *Apis andreniformis* and *Apis florea*. Apidologie 42:293–300. https://doi.org/10.1007/s13592-011-0024-x

Kavinseksan B, Wongsiri S, de Guzman LI et al (2015) Absence of infestation from recent swarms of in Thailand. J Apic Res 42(3):49–50. https://doi.org/10.1080/00218839.2003.11101091

Khacha-ananda S, Tragoolpua K, Chantawannakul P, Tragoolpua Y (2016) Propolis extracts from the northern region of Thailand suppress cancer cell growth through induction of apoptosis pathways. Investig New Drugs 34:707–722. https://doi.org/10.1007/s10637-016-0392-1

Khongphinitbunjong K, de Guzman LI, Burgett MD, Rinderer TE, Chantawannakul P (2012) Behavioral responses underpinning resistance and susceptibility of honeybees to *Tropilaelaps mercedesae*. Apidologie 43:590–599. https://doi.org/10.1007/s13592-012-0129-x

Khongphinitbunjong K, de Guzman LI, Tarver MR, Rinderer TE, Chantawannakul P (2015) Interactions of *Tropilaelaps mercedesae*, honey bee viruses and immune response in *Apis mellifera*. J Apic Res 54:40–47. https://doi.org/10.1080/00218839.2015.1041311

Khongphinitbunjong K, Neumann P, Chantawannakul P, Williams GR (2016) The ectoparasitic mite *Tropilaelaps mercedesae* reduces western honey bee, *Apis mellifera*, longevity and emergence weight, and promotes deformed wing virus infections. J Invertebr Pathol 137:38–42. https://doi.org/10.1016/j.jip.2016.04.006

Koeniger N, Koeniger G, De Guzman LI, Lekprayoon C (1993) Survival of *Euvarroa sinhai* Delfinado and Baker (Acari, Varroidae) on workers of *Apis cerana* Fabr, *Apis florea* Fabr and *Apis mellifera* L in cages. Apidologie 24:403–410

Kongpitak P, Polgár G, Heine J, Bayer Healthcare AG (2008) The efficacy of Bayvarol® and Check Mite® in the control of *Tropilaelaps mercedesae* in the European honey bee (*Apis mellifera*) in Thailand. Apıacta 43:12–16

Kremen C, Williams NM, Thorp RW (2002) Crop pollination from native bees at risk from agricultural intensification. Proc Natl Acad Sci 99:16812–16816. https://doi.org/10.1073/pnas.262413599

Lekprayoon C, Tangkanasing P (1991) *Euvarroa wongsirii*, a new species of bee mite from Thailand. Int J Acarol 17:255–258. https://doi.org/10.1080/01647959108683915

Martin SJ et al (2012) Global honey bee viral landscape altered by a parasitic mite. Science 336:1304–1306. https://doi.org/10.1126/science.1220941

Mookhploy W, Kimura K, Disayathanoowat T, Yoshiyama M, Hondo K, Chantawannakul P (2015) Capsid gene divergence of black queen cell virus isolates in Thailand and Japan honey bee species. J Econ Entomol 108:1460–1464. https://doi.org/10.1093/jee/tov102

Morse RA, Laigo FM (1969) The potential and problems of beekeeping in the Philippines. Bee World 50:9–14. https://doi.org/10.1080/0005772X.1969.11097235

Murray TE, Kuhlmann M, Potts SG (2009) Conservation ecology of bees: populations, species and communities. Apidologie 40:211–236

Page P, Lin Z, Buawangpong N, Zheng H, Hu F, Neumann P, Chantawannakul P, Dietemann V (2016) Social apoptosis in honey bee superorganisms. Sci Rep 6:27210. https://doi.org/10.1038/srep27210

Partap U, Partap T (1997) Managed crop pollination: the missing dimension of mountain agricultural productivity. Mountain Farming Systems' Discussion Paper, MFS 97. ICIMOD, Kathmandu

Partap U, Partap T (2002) Warning signals from the apple valleys of the Hindu Kush-Himalayas: productivity concerns and pollination problems. ICIMOD, Kathmandu

Pattamayutanon P, Angeli S, Thakeow P, Abraham J, Disayathanoowat T, Chantawannakul P (2015) Biomedical activity and related volatile compounds of Thai honeys from 3 different honeybee species. J Food Sci 80:M2228–M2240. https://doi.org/10.1111/1750-3841.12993

Pettis JS, Rose R, Lichtenberg EM, Chantawannakul P, Buawangpong N, Somana W, Sukumalanand P, Vanengelsdorp D (2013) A rapid survey technique for *Tropilaelaps* mite (Mesostigmata: Laelapidae) detection. J Econ Entomol 106:1535–1544

Potts SG, Biesmeijer JC, Kremen C, Neumann P, Schweiger O, Kunin WE (2010) Global pollinator declines: trends, impacts and drivers. Trends Ecol Evol 25:345–353. https://doi.org/10.1016/j.tree.2010.01.007

Pramual C (1994) DNA analysis of genetic diversity of *Apis cerana* Fabricius in Thailand. Dissertation, Chulalongkorn University, Bangkok

Rattanawannee A, Chanchao C, Wongsiri S (2007) Morphometric and genetic variation of small dwarf honeybees *Apis andreniformis* Smith, 1858 in Thailand. Insect Sci 14:451–460. https://doi.org/10.1111/j.1744-7917.2007.00173.x

Rattanawannee A, Chanchao C, Lim J, Wongsiri S, Oldroyd BP (2013) Genetic structure of a giant honey bee (*Apis dorsata*) population in northern Thailand: implications for conservation. Insect Conserv Divers 6:38–44. https://doi.org/10.1111/j.1752-4598.2012.00193.x

Ricketts TH, Regetz J, Steffan-Dewenter I et al (2008) Landscape effects on crop pollination services: are there general patterns? Ecol Lett 11:499–515. https://doi.org/10.1111/j.1461-0248.2008.01157.x

Rinderer TE, Oldroyd BP, De Guzman LI, Wattanachaiyingchareon W, Wongsiri S (2002) Spatial distribution of the dwarf honey bees in an agroecosystem in southeastern Thailand. Apidologie 33(6):539–543

Sanpa S, Chantawannakul P (2009) Survey of six bee viruses using RT-PCR in Northern Thailand. J Invertebr Pathol 100:116–119. https://doi.org/10.1016/j.jip.2008.11.010

Sanpa S, Popova M, Bankova V, Tunkasiri T, Eitssayeam S, Chantawannakul P (2015) Antibacterial compounds from propolis of *Tetragonula laeviceps* and *Tetrigona melanoleuca* (Hymenoptera: Apidae) from Thailand. PLoS One 10:e0126886. https://doi.org/10.1371/journal.pone.0126886

Sanpa S, Popova M, Tunkasiri T, Eitssayeam S, Bankova V, Chantawannakul P (2017) Chemical profiles and antimicrobial activities of Thai propolis collected from *Apis mellifera*. Chiang Mai J Sci 44:438–448

Saraithong P, Burgett M, Khongphinitbunjong K, Chantawannakul P (2012) *Apis dorsata* F: diurnal foraging patterns of worker bees in northern Thailand. J Apic Res 51:362–364. https://doi.org/10.3896/IBRA.1.51.4.12

Saraithong P, Li Y, Saenphet K, Chen Z, Chantawannakul P (2014) Bacterial community structure in *Apis florea* larvae analyzed by denaturing gradient gel electrophoresis and 16S rRNA gene sequencing. Insect Sci 22:606–618. https://doi.org/10.1111/1744-7917.12155

Seeley TD, Seeley RH, Akratanakul P (1982) Colony defense strategies of the honeybees in Thailand. Ecol Monogr 52:43–63. https://doi.org/10.2307/2937344

Sinpoo C, Williams GR, Chantawannakul P (2017) Dynamics of fungal communities in corbicular pollen and bee bread. Chiang Mai J Sci 44(4):1244–1256

Suntiparapop K, Prapaipong P, Chantawannakul P (2012) Chemical and biological properties of honey from Thai stingless bee (*Tetragonula leaviceps*). J Apic Res 51(1):45–45

Suwannapong G, Maksong S, Yemor T, Junsuri N, Benbow ME (2013) Three species of native Thai honey bees exploit overlapping pollen resources: identification of bee flora from pollen loads and midguts from *Apis cerana, A. dorsata* and *A. florea*. J Apic Res 52:196–201. https://doi.org/10.3896/IBRA.1.52.5.05

Sylvester HA, Limbipichai K, Wongsiri S, Rinderer TE, Mardan M (1998) Morphometric studies of *Apis cerana* in Thailand and the Malaysian peninsula. J Apic Res 37(3):137–145. https://doi.org/10.1080/00218839.1998.11100965

Tangjingjai W, Verakalasa P, Sittipraneed S, Klinbunga S, Lekprayoon C (2003) Genetic differences between *Tropilaelaps clareae* and *Tropilaelaps koenigerum* in Thailand based on ITS and RAPD analyses. Apidologie 34:513–524. https://doi.org/10.1051/apido:2003042

Tentcheva D, Gauthier L, Zappulla N, Dainat B, Cousserans F, Colin ME, Bergoin M (2004) Prevalence and seasonal variations of six bee viruses in *Apis mellifera* L. and Varroa destructor mite populations in France. Appl Environ Microbiol 70:7185–7191. https://doi.org/10.1128/AEM.70.12.7185-7191.2004

Thapa R, Wongsiri S (1997) Eupatorium odoratum: a honey plant for beekeepers in Thailand. Bee World 78:175–178. https://doi.org/10.1080/0005772X.1997.11099362

Theantana T, Chantawannakul P (2008) Protease and β-N-acetylglucosaminidase of honey bee chalkbrood pathogen *Ascosphaera apis*. J Apic Res 47:68–76. https://doi.org/10.1080/00218839.2008.11101426

Wanjai C, Sringarm K, Santasup C, Pak-Uthai S, Chantawannakul P (2012) Physicochemical and microbiological properties of longan, bitter bush, sunflower and litchi honeys produced by *Apis mellifera* in Northern Thailand. J Apic Res 51:36–44. https://doi.org/10.3896/IBRA.1.51.1.05

Warrit N, Smith DR, Lekprayoon C (2006) Genetic subpopulations of *Varroa* mites and their *Apis cerana* hosts in Thailand. Apidologie 37(1):19–30

Wattanachaiyingcharoen W, Oldroyd BP, Good G, Halling L, Ratnieks FLW, Wongsiri S (2002) Lack of worker reproduction in the giant honey bee *Apis dorsata* Fabricius. Insect Soc 49:80–85. https://doi.org/10.1007/s00040-002-8283-1

Wongsiri S (1989) The effects of the import of *Apis mellifera* L. to Thailand. In: Proceedings of the 4th international conference apicultural tropical climates, Cairo, 1989. International Bee Research Association, Cairo, London, pp 162–167

Wongsiri S, Tangkanasing P, Sylvester HA (1987) Mites, pests and beekeeping with *Apis cerana* and *Apis melifera* in Thailand. Am Bee J 127:500–503

Wongsiri S, Chen PP, Thapa R (1995) Other uses for bee products in Thailand. Br Bee J 123:144–149

Wongsiri S, Thapa T, Oldroyd B, Burgett DM (1996) A magic bee tree: home to *Apis dorsata*. Am Bee J 136:796–799

Wongsiri S, Chanchao C, Deowanish S, Aemprapa S, Chaiyawong T, Petersen S, Leepitakrat S (2000) Honey bee diversity and beekeeping in Thailand. Bee World 81:20–29. https://doi.org/10.1080/0005772X.2000.11099464

Wongsiri S, Thapa R, Sharma D, Bali K (2014) Bee-birds: ravagers of beekeepers, but saver of farmers. In: Gupta RK, Reybroeck W, van Veen JW, Gupta A (eds) Beekeeping for poverty alleviation and livelihood security. Springer, Netherlands, pp 355–378. https://doi.org/10.1007/978-94-017-9199-1_13

Yañez O, Gauthier L, Chantawannakul P, Neumann P (2016) Endosymbiotic bacteria in honey bees: *Arsenophonus* spp. are not transmitted transovarially. FEMS Microbiol Lett 363:1–7. https://doi.org/10.1093/femsle/fnw147

Chapter 13
Social Bees and the Current Status of Beekeeping in Indonesia

Sih Kahono, Panuwan Chantawannakul, and Michael S. Engel

Abstract Highly eusocial bees in Indonesia are diverse and encompass three groups: i.e., honey bees (Apini), stingless bees (Meliponini), and bumble bees (Bombini); each of which stores resources such as honey in their nests. The native peoples of Indonesia have used honey for a long time, hunting for honey from both wild honey bees and stingless bees. Although regional beekeeping has been developing from traditional to modern hives and methods, traditional beekeeping remains in practice using local honey bees and stingless bees. While data for national honey production is not recorded properly, scattered evidence demonstrates that honey production by hunting *Apis dorsata* has taken a large role in national honey production. Production from *A. mellifera* in Indonesia is tending to decrease owing to changing food sources, pests, and climate change. In the last decade, stingless bees that produce increasing amounts of medicinal honey, propolis, and their derivative products have gradually been developed. Indonesia needs to develop beekeeping by enhancing both existing natural ecosystems and artificial green environments as sources of food, and also promoting native stingless bees.

Keywords Social bees · Honey bees · Stingless bees · Bumble bees · Honey hunters · Traditional and modern beekeeping · Traditional knowledge · Medicinal honey · Propolis · Pests · Climate change

---

S. Kahono (✉)
Laboratory of Ecology, Zoological Division (Museum Zoologicum Bogoriense), Research Center for Biology, Indonesian Institute of Sciences (LIPI), Bogor, Indonesia
e-mail: sihk001@lipi.go.id

P. Chantawannakul
Department of Biology, Faculty of Science, Chiang Mai University, Chiang Mai, Thailand

M. S. Engel
Division of Entomology, Natural History Museum, University of Kansas, Lawrence, KS, USA

Department of Ecology and Evolutionary Biology, University of Kansas, Lawrence, KS, USA

Division of Invertebrate Zoology, American Museum of Natural History, New York, NY, USA

P. Chantawannakul et al. (eds.), *Asian Beekeeping in the 21st Century*,
https://doi.org/10.1007/978-981-10-8222-1_13

13.1 Introduction

Eusocial bees are those species that live in colonies with defined castes. The castes of eusocial species include a reproductive female or females (queens), numerous sterile females (workers), and males (drones). In highly eusocial species such as honey bees (Apini) and stingless bees (Meliponini), the castes are morphologically fixed, while in primitively eusocial groups, such as bumble bees (Bombini), there is an intergradation between minor worker through to queens, and major workers can assume the role of queen should the original gyne be killed. The most familiar eusocial bees are the three tribes of corbiculate Apinae: Apini, Meliponini, and Bombini; but social species exist in a variety of bee lineages (e.g., among some Halictinae and Xylocopinae) (Michener 1974, 2007), although these do not store honey. Workers are responsible for collecting resources, such as pollen and nectar, along with materials to build and protect the nest (Michener 1974, 2007). Within their nests, most eusocial corbiculate bees deposit honey, pollen, and propolis, and all of these are exploited by humans. The high species richness and uniqueness of the distribution of eusocial bees in Indonesia may relate to the country's broad geographic expanse, varied topological and environmental landscape, and complex geological history (Hill 2009).

The diversity of eusocial bees in Indonesia is quite high, consisting of five species of native honey bees (*Apis*) (Hadisoesilo 2001; Engel 2012), at least 46 species of stingless bees (Meliponini) (Table 13.1), and two species of bumble bees (*Bombus*). The distribution of the stingless bee *Tetragonula laeviceps* is quite large, covering the whole of Indonesia. However, the honey bees *Apis koschevnikovi* and *A. andreniformis*, and stingless bees *Heterotrigona itama*, *Tetrigona apicalis*, and *Lophotrigona canifrons* are distributed largely in the western part of Indomalaya, while the stingless bee *Wallacetrigona incisa* and honey bees *A. nigrocincta* and *A. dorsata binghami* are distributed in the eastern part of Indomalaya and within Wallacea (the island of Sulawesi) (Tim Ekspedisi 2016) (Table 13.1). The two species of *Bombus* are found only in the highlands of Sumatera and Java (Sakagami et al. 1990; Kato et al. 1992).

Eusocial bees have been exploited for their honey in Indonesia for a long time, with traditional hunting activities of both wild honey bees and stingless bees as well as traditional beekeeping practices. Local peoples in the various regions of Indonesia have developed different methods to their hunting. Almost all species of honey bees and stingless bees are targets for honey hunting. The most hunted honey bee is the native wild species, *A. dorsata*—the giant honey bee. *Apis dorsata* is believed to be the largest national supplier of honey (Hadisoesilo and Kahono 2011). The most sought after stingless bee is the most commonly distributed species that is well adapted to human environments, *Tetragonula laeviceps*. Honey hunting continues to dominate because it is easier and quicker than maintaining domestic stocks. Sustainable harvest systems are currently applied with attention to the safety of the bees, taking only a portion of the hive, and using clean harvesting and packaging equipment.

Table 13.1 Tabulation of species of stingless bees (Apinae: Meliponini) across Indonesian islands

Genus	Subgenus	Species	Sumatera	Java	Timor	Borneo	Sulawesi	Ambon	Maluku	Irian Jaya
Austroplebeia Moure		*cincta* (Mocsáry)								+
Geniotrigona Moure		*lacteifasciata* (Cameron)				+				
		thoracica (Smith)	+			+				
Heterotrigona Schwarz	*Heterotrigona* s.str.	*erythrogastra* (Cameron)				+				
		itama (Cockerell)				+				
	Platytrigona Moure	*flaviventris* (Friese)								
		hobbyi (Schwarz)				+				
		keyensis (Friese)								+
		lamingtonia (Cockerell)								+
		planifrons (Smith)								+
	Sahulotrigona Engel and Rasmussen	*atricornis* (Smith)								+
	Sundatrigona Inoue and Sakagami	*lieftincki* (Sakagami and Inoue)	+							
		moorei (Schwarz)	+	*		+				
Homotrigona Moure	*Homotrigona* s.str.	*aliceae* (Cockerell)	+							
		anamitica (Friese)				+				
		fimbriata (Smith)	+			+				
	Lophotrigona Moure	*canifrons* (Smith)	+			+				
	Odontotrigona Moure	*haematoptera* (Cockerell)				+				
	Tetrigona Moure	*apicalis* (Smith)	+	+		+				
		binghami (Schwarz)				+				
		vidua (Lepeletier)	+	+	+					
Lepidotrigona Schwarz		*javanica* (Gribodo)		+						
		latebalteata (Cameron)				+				
		nitidiventris (Smith)	+			+				

(continued)

Table 13.1 (continued)

Genus	Subgenus	Species	Sumatera	Java	Timor	Borneo	Sulawesi	Ambon	Maluku	Irian Jaya
		terminata (Smith)	+	+		+				
		trochanterica (Cockerell)	+			+				
		ventralis (Smith)	+	+		+				
Lisotrigona Moure		*cacciae* (Nurse)	+			+				
Papuatrigona Michener and Sakagami		*genalis* (Friese)								+
Pariotrigona Moure		*pendleburyi* (Schwarz)	+							
Tetragonula Moure	*Tetragonilla* Moure	*atripes* (Smith)	+			+				
		collina (Smith)	+							
		fuscibasis (Cockerell)	+			+				
	Tetragonula s.str.	*biroi* (Friese)								+
		clypearis (Friese)						+		+
		drescheri (Schwarz)	+	+		+				
		fuscobalteata (Cameron)	+			+	+			
		geissleri (Cockerell)				+				
		laeviceps (Smith)	+	+		+	+	+		
		melanocephala (Gribodo)				+				
		melina (Gribodo)	+			+				
		minangkabau (Sakagami and Inoue)	+							
		reepeni (Friese)	+			+				
		sapiens (Cockerell)							+	+
		sarawakensis (Schwarz)				+				
Wallacetrigona Engel and Rasmussen		*incisa* (Sakagami and Inoue)					+			
Totals			23	7	1	29	3	2	1	9

The higher classification follows that of Rasmussen et al. (2017), Engel and Rasmussen (2017), and Engel (unpubl. data). The asterisk indicates an unpublished locality for *Heterotrigona* (*Sundatrigona*) *moorei* (Schwarz)

Traditional beekeeping practices for honey bees and stingless bees usually do not remove the nests, retaining them in cavities of trees or hard objects. Honey bee farming practices for *A. cerana* have been partially augmented to modern beekeeping within wooden boxes and standard-sized frames. In general, the condition of modern beekeeping of *A. mellifera* in Indonesia is relatively stable in the number of beekeepers and hives; however, the production of honey tends to decline. Beekeeping has developed from traditional to modern methods, although traditional beekeeping practices are still conducted up to present and throughout the country.

In Java, which has suitable environments for mobile beekeeping systems for *A. mellifera*, beekeeping conditions are fluctuating and are tending to worsen due to the reduction of food resources, increase in pests, and the impact of climate change.

Managed beekeeping of stingless bees in Indonesia is definitely the ideal choice because it can be done in all types of natural environments. It can be conducted by ordinary people safely, cheaply, easily, and with less maintenance (Kahono 2015). Recently, beekeeping of local species of stingless bees has been growing significantly in every region across their distribution, progressively increasing the national production of medicinal honey, raw propolis, and derivative products.

13.2 Diversity of Social Bees

13.2.1 Honey Bees (Apini)

Honey bees comprise seven exclusively eusocial species, all classified in the genus *Apis* (Gould and Gould 1988; Ruttner 1988; Engel 1999; Michener 2007). The genus is the sole group in the corbiculate bee tribe Apini, and species live in perennial colonies, mostly within cavities although the more primitive species build nests in the open. Indonesia harbors the greatest diversity of honey bees, with at least five species native to the country (Engel 2012), namely *Apis* (*Micrapis*) *andreniformis*, *A.* (*Megapis*) *dorsata*, *A.* (*Apis*) *cerana*, *A.* (*A.*) *koschevnikovi*, and *A.* (*A.*) *nigrocincta,* the last endemic to Sulawesi. *Apis andreniformis* is one of the two dwarf species of honey bees; *A. cerana*, *A. koschevnikovi*, and *A. nigrocincta* are medium-sized species; while *A. dorsata* is the largest.

Two extralimital forms have either been recently recorded from Indonesia or are expected to occur within the country. *Apis* (*Micrapis*) *florea* has been reported in East Kalimantan (Kahono et al. 2012), and their recent occurrence has been recorded from Makassar and Maros (South Sulawesi), Pontianak (West Kalimantan), and Semarang (Central Java). The reason for the presence of *A. florea* in Indonesia is the matter currently being investigated and the species is a presumed adventive as no historical records are available for the species. A recently described species, *A.* (*A.*) *nuluensis*, is distributed on a plateau in Sabah (Tingek et al. 1996; Radloff et al. 2011), but is nothing more than a subspecific form of *A. cerana* (Engel 1999, 2012). Kalimantan has several highlands that are similar to Sabah, so it is likely that this subspecific form can be found in Kalimantan (Hadisoesilo 2001; Engel 2012).

13.2.2 Open-Nesting Honey Bees

13.2.2.1 *Apis* (*Micrapis*) *andreniformis* F. Smith, 1858

The dwarf honey bee *A. andreniformis*, also known as "Tawon Laler" (Javanese), "Lebah Kerdil" (Indonesian), or "Sireupeun" (Sundanese), is found at elevations of 0–500 m above sea level (Salmah et al. 1990; Otis 1991, 1996; Radloff et al. 2011). The abdomen of *A. andreniformis* is largely black to dark reddish brown, contrasting with the otherwise lighter bands of setae (Engel 2012). The species builds a single comb that is suspended from the branch of a shrub or small tree, and is usually protected by foliage. Nests are usually in easily accessible positions, and since the sting of *A. andreniformis* is not painful the bees are prone to hunting, which often kills the colony. The size of the nest frequently only as wide as an adult's palm, and the bees cannot be farmed. Although *A. andreniformis* only produces small quantities of honey, it is locally considered as a panacea for children.

Recently, the regional distribution of the bee is narrowing, and the species has recently become rare, and even suspected to be extinct in Java. In actuality, *A. andreniformis* is still present in Java as a rare component of the fauna in Banten and East Java. This species has been proposed for protection by the government.

13.2.2.2 *Apis* (*Megapis*) *dorsata* Fabricius, 1793

Apis dorsata, known locally as "Tawon Gung" (Javanese) or "Odeng" (Sundanese), has the largest body size of any species in the genus *Apis* (Ruttner 1988; Engel 1999; Michener 2007). The bees construct a large single comb, hanging under high tree branches (Dyer and Seeley 1994; Kahono et al. 1999), rock overhangs (Koeniger and Koeniger 1980), and parts of buildings (Reddy 1980; Reddy and Reddy 1993). This species can be found on almost all islands in Indonesia, except Maluku and Irian Jaya (Ruttner 1988). *Apis dorsata* performs seasonal migrations, stays at certain periods and places, and the bees return to the same nest site (Dyer and Seeley 1994; Koeniger and Koeniger 1980; Reddy and Reddy 1993).

Two of the three subspecies of *A. dorsata* are distributed in Indonesia, namely *A. d. dorsata* and *A. d. binghami* (Sakagami et al. 1980). The subspecies *A. d. dorsata* is the most widespread throughout India and Southeast Asia, including Palawan, Borneo, Sumatera, Java, Bali, Lombok to Timor, and the Kai Islands (Otis 1991), and southern parts of Maluku close to Timor such as Wetar, Sermata, Romang, and Moa islands. Since 2016, the islands of Ambon and Seram have been reported as newly harboring populations of *A. d. dorsata* (Lamerkabel, pers. comm.).

Apis dorsata binghami is found only on the island of Sulawesi and surrounding islands (Otis 1996). Its abdomen is more predominantly black with white pubescent bands, while the abdomen of *A. d. dorsata* is largely brownish with orange bands. In contrast to *A. d. dorsata* where many colonies may hang from a single tree at a given locality, *A. d. binghami* typically build only one or two colonies at a given locality (Hadisoesilo 2001; Nagir et al. 2016), with at most ten colonies (Hadisoesilo 2001).

13.2.3 Cavity-Nesting Honey Bees

13.2.3.1 *Apis* (*Apis*) *cerana* Fabricius, 1793

Apis cerana is known locally as "Lebah Madu" (Indonesian), "Tawon Gula" (Javanese), "Nyiruan" (Sundanese), or "Dondora" (Mamasa), and has a smaller body size compared to the other two species of domesticated honey bees, *A. nigrocincta* and *A. koschevnikovi*. Currently, *A. cerana* is widespread on almost all islands in Indonesia (Ruttner 1988). Indonesian populations of *A. cerana* are usually recognized as the subspecies *A. cerana indica*, although the distinctiveness and utility of subspecies in *A. cerana* has been challenged, with many Indonesian populations clustering more with *A. cerana javana* (Radloff et al. 2010). *Apis cerana* is quite variable across Indonesia and in association with latitude and elevation, reflective of the genetic intermingling of diverse populations within the country (Hadisoesilo 2001), further vitiating the utility of subspecies which would only intergrade among one or more traits (Radloff et al. 2010). Thus, subspecies should not be recognized in *A. cerana* as they do not correspond to natural, genetic units.

13.2.3.2 *Apis* (*Apis*) *koschevnikovi* Enderlein, 1906

The honey bee *A. koschevnikovi* is known regionally as "Lebah Merah" because the color of its body is slightly reddish. This species is distributed in the western part of Wallacea (Ruttner et al. 1989; Otis 1996; Hadisoesilo et al. 1999) and is slightly larger than *A. cerana* (Hadisoesilo 2001). Drone flight times and the structure of the male endophallus differ from those of *A. cerana* and the latter in particular serves to support the specific status of *A. koschevnikovi* (Tingek et al. 1996; Hadisoesilo and Otis 1996; Engel 1999).

13.2.3.3 *Apis* (*Apis*) *nigrocincta* F. Smith, 1861

Until recently, *A. nigrocincta* was found only on Sulawesi and smaller surrounding islands, for example Sangihe (Otis 1996; Damus and Otis 1997). *Apis nigrocincta* is slightly larger than *A. cerana* (Hadisoesilo and Otis 1996; Hadisoesilo 1997), with the body coloration, clypeus, and metafemora more yellow (Hadisoesilo 2001). While the structure of the male endophallus does not differ appreciably from that of *A. cerana* (Hadisoesilo 1997), the lack of a pore in the drone cell cap is a distinctive trait excluding *A. nigrocincta* from a clade comprising *A. cerana* and *A. koschevnikovi* (Engel 1999). In addition, the drone flight time is different from that of *A. cerana* (Hadisoesilo and Otis 1996), at least where these species co-occur.

13.2.4 Stingless Bees (Meliponini)

In Indonesia, stingless bees have different names among indigenous peoples, such as "Lebah Getah" (Indonesian), "Klanceng" (Javanese), "Teuweul" (Sundanese), "Galo-galo" (Minang, West Sumatera), "Merang" or "Katappe" (Mamasa, West Sulawesi), and "Tannese" (Kaili, Central Sulawesi). At least 46 species in ten genera of stingless bees have been recorded from Indonesia (Sakagami et al. 1990; Rasmussen 2008; Rasmussen et al. 2017) (Table 13.1). Specimens in research collections (e.g., Museum Zoologicum Bogoriense, University of Kansas Natural History Museum), unpublished reports, and student university theses have revealed several candidates for new records of known species as well as the potential for new species (e.g., Tim Ekspedisi 2016; Suriawanto et al. 2017). Unlike honey bees, almost all species of stingless bees can be moved into hives and the bees are unlikely to depart or escape. Meliponiculture is well developed in many countries (Cortopassi-Laurino et al. 2006), and there is considerable potential to more fully exploit these species in a sustainable manner throughout Indonesia.

13.3 Historical Records and Recent Developments in Beekeeping

13.3.1 Captive Beekeeping: A. cerana *and Related Species*

Traditional beekeeping is usually conducted as a side activity relative to the principal form of employment. Local communities in Java, Sumatera, Kalimantan, Sulawesi, Bali, and Lombok conduct traditional beekeeping of *A. cerana* in simple wooden boxes without frames, or in the cavity of palms, ferns, and other tree trunks. Similar systems are also practiced with *A. koschevnikovi* in Kalimantan and *A. nigrocincta* in Sulawesi. Local peoples in Kalimantan trap wild colonies of *A. cerana* and *A. koschevnikovi* by keeping empty wooden boxes on the edge of the forest. After the boxes are occupied by new colonies escaping predation from monkeys and honey bears, the boxes are moved to the villages (Fig. 13.1). These traditional beekeeping methods are still being used, although in recent times they are diminishing.

Fig. 13.1 Traditional hive models of *Apis* (*Apis*) *koschevnikovi* in East Kalimantan

The modern style of beekeeping used for *A. mellifera*, that is, framed hives, has been replicated on a smaller scale with *A. cerana*. Development of mixed farming systems of bees with plantations of fruit trees and industrial plants is currently growing in several places in Indonesia (Saepudin 2011). Traditional harvesters extract honey by squeezing or putting cut honey cells on filters, but modern harvesters use small-sized honey extractors.

People in some areas modify the models of hives and beekeeping activities to adapt them to their own environmental conditions. Hives with frames of different sizes are used by beekeepers in the highlands of Bogor and Sukabumi (Fig. 13.2). Hives with double entrances are used in Halmahera, Ambon, and the Mollucas (Lamerkabel, pers. comm.). In the area of Kuningan, beekeepers of *A. cerana* place their hives in natural environments, such as the edge of steep lands. People in the village of Pager Ageung, Tasikmalaya (West Java) and Bawean Island, Gresik (East Java) have their own beekeeping practices, letting the colonies escape from the hives during the lean seasons and then recapturing them in the flowering seasons.

Currently fixed (i.e., non-moving) beekeeping of *A. cerana* is on the rise, especially in natural environments rich in foods owned by local governments, beekeeping associations, plantation companies, and individuals. However, in many parts of Indonesia people still use traditional systems and do not rely on more modern methods. Recently, beekeeping of *A. cerana* has gradually developed on the islands of Ternate, Halmahera, Seram, and Maluku, using local hive modifications and relying on existing plants, such as coconut, nutmeg ("Pala"), Eucalyptus, cloves, and other local food plants.

Pests, habitat alterations, and climate change are the primary causes of diminishing food resources that impact colonies and decrease honey production. Considerable time, investment, and both pure and applied research are necessary in order to address all of these matters as they relate to Indonesian apiculture.

Fig. 13.2 An innovative hive of *Apis* (*Apis*) *cerana* at Sukabumi, West Java

13.3.2 Mobile System of Beekeeping: Apis mellifera

Modern beekeeping of *A. mellifera* requires capital investment for materials such as hives and beekeeping equipment, livestock mobility, and payment for manpower. Since the western honey bee was introduced to Indonesia, it is likely that the production of honey has not shown significant progress. Several problems have arisen due to lack of regulation of bee hive placement that sometimes causes miscommunication and rejections. Many potential food plants are part of gardens owned by individuals or institutions. The rejection of hive placement comes not only from local people but sometimes also from local governments. Landowners, ordinary people, stakeholders, and official government representatives have to be educated of the benefits of the bees—all bees, not just *A. mellifera*—to pollinate plants. Indeed, significant educational efforts need to be made that emphasize the use of Indonesia's native social and solitary bees as effective pollinators, particularly those that can be managed through apiculture (Apini), meliponiculture (Meliponini), bombiculture (Bombini), or other practices (e.g., Alkali bees).

Pests and climate change are the most serious problems for beekeeping in Indonesia. Pests decrease bee populations while climate change leads to uncertain phenology, altered floral and bee distributions, and reduction in flowering. Although the use of *A. mellifera* in beekeeping was initiated in many parts of Java, Sumatera, Kalimantan, Sulawesi, and West Papua, it has only become well established in Java. The landscape of Java has quite extensive areas of monocultures of industrial and agricultural lands that provide sufficient forage resources required by *A. mellifera*. However, there are no recent data for honey production to be able to evaluate the development of this practice of beekeeping in Indonesia. Although the number of beekeepers and bee hives is relatively stable, the above problems have not been resolved, which might indicate that beekeeping of *A. mellifera* has not led to improvements, and honey production may be gradually declining. Indeed, reliance on a non-native species, such as *A. mellifera*, likely leads to decreased yields since the bees are not adapted to the climate of Indonesia. In this regard, greater development of beekeeping with native populations of *A. cerana* would likely ameliorate many issues in both natural and agricultural landscapes.

13.3.3 Open Nests: **Apis dorsata**

Apis dorsata cannot be captive-bred because of its migratory and aggressive behavior. The bees seasonally occupy certain nesting patches that are used for honey harvesting. Many areas of the Indonesian archipelago have the potential to be used in honey production. Well-known areas of *A. dorsata* honey producers are Sumbawa, Lake Sentarum, and Ujung Kulon National Park. Their contributions to honey production are higher than any other species throughout Indonesia (Hadisoesilo and Kahono 2011).

Local wisdom plays a role in collecting honey of *A. dorsata*. Through local knowledge, people know when bees are coming and when to harvest. Honey hunting in Java and Sumatera is conducted during the day and hunters wear a thick long-sleeved shirt, but usually no mask. They use a simple smoker and some of them pray or read mantras before climbing. In Kalimantan, honey hunting is mostly done at night and some hunters do local traditional rituals. Following local wisdom, communities surrounding Lake Sentarum install "tikung" and those in Bengkulu install "sunggau" in order to attract migrant colonies of *A. dorsata* (Hadisoesilo and Kuntadi 2007). In empty former nesting patches, the hunters in Bogor, West Java bend down tree trunks to horizontal positions in an effort to attract migrant colonies (Fig. 13.3).

Forest fires, both natural and intentional, cause nesting trees and food resources to be diminished. Cultural festivals also impose problems for migrating bees owing to the cutting of dozens of large trees from the forests, among them nesting sites of *A. dorsata*. To protect nesting trees for *A. dorsata*, local communities of Riau, Kalimantan, Sumbawa, and the Mollucas protect the trees by local regulatory prohibitions.

Cooperative training in sustainable honey-harvest practices for *A. dorsata* by local governments and NGOs such as JMHI, JMHS, and their related collaborators is necessary to increase honey production, guarantee hygienic production, and insure the safety of the bee colonies.

Fig. 13.3 Harvesting colonies of *Apis* (*Megapis*) *dorsata* by "tikung" at Lake Sentarum

13.3.4 Common Stingless Bees

Apiculture (beekeeping of honey bees) is often limited in Indonesia owing to availability of resources, the presence of hive pests, and changes in flowering season caused by climate change. These reasons have fostered the development of stingless beekeeping (meliponiculture). The advantages of stingless bees include safety (owing to the lack of stings), cost-effectiveness, simplicity in handling, high ability to adapt to limited environments, and production of valuable materials such as medicinal honey and propolis (Kahono 2015).

Local communities in Java, Bali, Lombok, Sumbawa, Sumatera, Sulawesi, Ambon, Ternate, and Halmahera are familiar with traditional stingless beekeeping methods. The colonies nest in the cavities of cut tree trunks, bamboo, simple wooden boxes, and/or hardened materials. The beekeepers usually place nests around houses to make them easy to care for and harvest. *Tetragonula laeviceps* has been farmed in human settlements of many places on the main islands of Indonesia, while other small-sized bees such as *T. minor*, *T. fuscobalteata*, and *T. clypearis* have also been farmed in their local areas where they occur. Creative ideas for making wooden hives, with a variety of models and sizes, have been produced at several places in Java (Fig. 13.4).

Stingless beekeeping of bigger forest species such as *Tetrigona apicalis*, *Lepidotrigona terminata*, *Geniotrigona thoracica*, and *Lophotrigona canifrons*, which are collected from Java, Sumatera, and Kalimantan, is conducted in semi-intensive systems using a section of a large tree containing a colony, with surrounding box for the honey and propolis (Fig. 13.5). However, these practices are less developed than those for *Tetragonula* with the exception of *Heterotrigona itama*.

Fig. 13.4 Model of innovative nests of *Tetragonula* spp. in Java (**a**) Bogor, (**b**) Tasikmalaya, and (**c**) Yogyakarta

Traditional beekeeping of *Wallacetrigona incisa*, a genus of stingless bee endemic to Sulawesi, is still practiced by the local people of the highlands of Mamasa-West Sulawesi, by putting cut logs near houses. A new model of stingless beekeeping using wooden, vertical, terraced hives was first applied in the highland forests of Luwu Utara (Fig. 13.6). Nowadays, Sulawesi is the largest producer in Indonesia of medicinal honey and raw propolis, which are mostly sold to other islands and are partially exported to neighboring countries. Recently, stingless beekeeping of *T. fuscobalteata* and *T. clypearis* using bamboo nests under bamboo houses has been developed on the islands of Halmahera, Ternate, and Ambon (Fig. 13.7) (Lamerkabel, pers. comm.).

13.4 Bee Forage Resources

Forests, plantations, and other green areas are rich in food for eusocial bees. In evergreen country, flowers and gummy plants are available throughout the year (Backer and Bakhuizen van den Brink 1963), but these have not yet been exploited as forage resources in beekeeping. The availability of these plants in large numbers throughout the year is a guarantee of success for beekeeping efforts should they be developed.

Fig. 13.5 A traditional hive of *Heterotrigona (Heterotrigona) itama* at Kalimantan Timur

Compared to the other islands, Java has many plantations and wild plants that flower in different places and times that are suitable for the moving beekeeping system of *A. mellifera* (Mashudi and Suwanda 1988; Sihombing 1997). Many plants flower throughout the year, such as coconut and Calliandra. Java formerly had a flowering calendar that gave guidance for the movement of *A. mellifera*, but it is no longer accurate because environmental conditions are deteriorating, a process that is unfortunately likely to continue. Climate change impacts the flowering period and has led to declining floral populations. There are some reliable species of plants that support the moving system of beekeeping in Java such as "Aren" (*Arenga pinnata*), "Kayu Putih" (*Eucalyptus* spp.), "Albisia" (*Paraserianthes falcataria*), "Longan" (*Euphorbia longan*), "Rambutan" (*Nephelium lappaceum*), "Durian" (*Durio zibethinus*), "Mangga" (*Mangifera indica*), rose apple (*Eugenia* spp.), "Alpokad" (*Persea americana*), "Jeruk" (*Citrus* spp.), "Randu" (*Ceiba pentandra*), cashew (*Anacardium occidentale*), Calliandra (*Calliandra calothyrsus*), "Sonobrit" (*Dalbergia sissoo*), rubber (*Hevea brasiliensis*), "Petai" (*Leucaena glauca*), "Akasia" (*Acacia mangium*), coconut (*Cocos nucifera*), and palms (Mashudi and Suwanda 1988).

Relative to honey bees, stingless bees have smaller body sizes and fewer numbers of individuals within the colony. They can utilize diverse flowers of small to large size, and from everything ranging from shrubs to large trees. A colony of

Fig. 13.6 A new innovative vertical, wooden hive of *Wallacetrigona incisa* (the photograph was taken by Paimin Ponijan)

stingless bees does not need as many food plants, so this may be partially why stingless bees thrive better than honey bees when food is limited.

13.5 Pests and Natural Enemies

Pests in beekeeping are very specific and control measures are already well known. The common pests in captive beekeeping are *Varroa destructor*, caterpillars of the Galleria moth, wasps of the genus *Vespa*, and ants. Every location has its own local pests, such as monkeys and honey bears in Sumatera and Kalimantan. Beekeeping of *A. mellifera* usually follows the standard management practices for colonies, but in *A. cerana* there is less attention so many pests attack the colonies. Greater attention and investment needs to make toward addressing pests of hives of *A. cerana*, as well as addressing any potential pests of *A. koschevnikovi* and *A. nigrocincta*.

In nature, *Pernis* sp., an eagle bee-eater, specializes on *A. dorsata*. Its migration route follows those of the bees (Prawiradilaga, pers. comm). In the locations with large populations of the eagle, the bees usually nest in protected positions. Further research needs to be done on the behavior of the eagles, as well as the predator-prey interactions of the bees in the presence of eagles.

Fig. 13.7 New innovations for a bamboo house of stingless bees in Ambon, Ternate, and Halmahera (the photograph was taken by Jakobus Lamerkabel)

Feral nests of stingless bees are well protected by propolis (Wille and Michener 1973; Michener 2007, 2013). Only strong animals such as honey bears, monkeys, or mice are capable of occasionally damaging nests. However, traditional hives that are frequently opened can be more vulnerable to such pests.

13.6 Other Eusocial Bees in Indonesia

Two species of bumble bees are found in the highlands and mountains: *Bombus* (*Melanobombus*) *rufipes* which is scattered on the islands of Java and Sumatera (Kahono 2000, 2009; Kato et al. 1992), and *B.* (*Megabombus*) *senex* on Sumatera (Salmah et al. 1990) with unpublished occurrences on Java (pers. obs.). Currently, the vertical distribution as well as the number and extent of habitats has shifted up in accordance with changes in habitat conditions (Kahono 2000). *Bombus* serve as primary pollinators of highland plants and agricultural crops (Kahono 2001). The decline and disappearance of these plants has impacted bumble bee populations (Kahono 2009). Elsewhere in the world, bombiculture (bumble beekeeping) has been developed as an effective method for agricultural pollination of certain crop species, particularly within glasshouses. The use of bumble bees can significantly increase crop yields for certain floral species. Such practices have never been explored within Indonesia.

Within the nest, the bees keep small amounts of honey and stored pollen. Indigenous people sometimes find nests in the wild and consume the honey and pollen. The population of *B. rufipes* has been gradually declining due to narrowing habitats caused by human pressure, mainly due to deforestation of the mountains and by the intensive application of pesticides (Kahono 2009). Unless efforts are made to conserve these habitats, Indonesia may lose its populations of these bumble bees.

Aside from the highly eusocial corbiculate bees discussed in this work, additional lineages of bees in Indonesia also exhibit some degree of social behavior, including primitively eusocial societies. Small carpenter bees (Ceratinini) and allodapine bees (Allodapini) are known to have diverse social biologies (e.g., Michener 1974, 2007; Rehan et al. 2009), and the same is true for halictines (Michener 1974, 2007). These groups are diverse throughout Indonesia and while none produce honey, they are important pollinators in native habitats and represent critical social lineages for understanding the evolutionary development of insect societies. The nesting biology of these groups remains understudied throughout the region. Natural habitats need to be conserved in order to promote these bees, as these smaller social lineages as well as their diverse solitary relatives are just as critical to the survival of the native flora as are the more conspicuous and large societies of honey bees and stingless bees.

13.7 Future Perspectives

Natural and artificial environments as food resources for bees are available throughout Indonesia. There are social bees adapted to live in each environment, and utilizing native, local social bees in these habitats should be a priority in the development of national beekeeping efforts. Displacing endemic bees and moving the bees to different geographic locales and ecological situations is damaging, disrupting their native distributions and causing negative ecological stresses not only on the translocated species but also to those local bees (social and solitary) native to the area. In fact, translocating bees can also impact local vertebrate populations. For example, the introduction of *A. mellifera* into areas in Australia dramatically impacted native nectar-feeding birds (Paton 1993). The cutting of nesting trees of *A. dorsata* needs to be prevented by the introduction of new regulatory laws. Except for research, cutting nesting trees of stingless bees and their removal from the forest is not recommended due to the potential of upsetting the balance of the forest ecosystems as well as the colonies themselves. Instead, traditional meliponiculture developed in a sustainable manner within local forests should be encouraged as this will help natural forests as well as the bordering agricultural areas. Forest fires and their resulting smoke in Sumatera and Kalimantan must be brought under control and burning of forests ceased as these are damaging to beekeeping in Indonesia.

Each region of Indonesia produces propolis with different active compounds that affect its benefits. However, Indonesia has not yet maximized utilizing local propolis rather than importing propolis of non-native species at high prices. Instead, there should be an Indonesian industry of native propolis which can benefit local areas as well as be a potential export.

Indonesia harbors a remarkably rich and unique diversity of bees, including the greatest generic diversity of Asian stingless bees and number of species of true honey bees (*Apis*). It is therefore imperative to develop further sustainable practices for managing and protecting our native bees, as well as to continue surveys and exploration into the systematics and biology of Indonesia's pollinating bees, both solitary and social.

Acknowledgements The authors are grateful to Mr. Paimin Ponijan, Syaiful Alaydrus, M.Si, Dr. Dewi Prawiradilaga, NGO APMP SAMPAN, and Ir. Jacobus Lamerkabel, M.Si. for sharing information and photographs, and to Dr. Djunijanti Peggie for reading an initial draft of the manuscript. This is a contribution of the Division of Entomology, University of Kansas Natural History Museum.

References

Backer CA, Brink Jr RCBVD (1963) Flora of Java, vol 1 (Spermatophytes only). NVP Noordhof, Groningen, The Netherlands

Cortopassi-Laurino M, Imperatriz-Fonseca VL, Roubik DW et al (2006) Global meliponiculture: challenges and opportunities. Apidologie 37(2):275–292

Damus MS, Otis GW (1997) A morphometric analysis of *Apis cerana* F and *Apis nigrocincta* Smith populations from Southeast Asia. Apidologie 28:309–323

Dyer FC, Seeley TD (1994) Colony migration in the tropical honey bee *Apis dorsata* F (Hymenoptera: Apidae). Insectes Soc 41:129–140

Engel MS (1999) The taxonomy of recent and fossil honey bees (Hymenoptera: Apidae; *Apis*). J Hymenopt Res 8(2):165–196

Engel MS (2012) The honey bees of Indonesia (Hymenoptera: Apidae). Treubia 39:41–49

Engel MS, Rasmussen C (2017) A new subgenus of *Heterotrigona* from New Guinea (Hymenoptera: Apidae). J Melittol (73):1–16

Gould JL, Gould CG (1988) The honey bee. WH Freeman, New York

Hadisoesilo S (1997) A comparative studies of two species of cavity-nesting honey bees of Sulawesi. Dissertation, University of Guelph, Guelph

Hadisoesilo S (2001) The diversity of indigenous honey bee species of Indonesia. Biodiversitas 2(1):23–128 (translated)

Hadisoesilo S, Kahono S (2011) The potential of forest honey bee *Apis dorsata* in Sumbawa Besar. Expedition Report of Indonesian Forest Honey Association (JMHI) (translated)

Hadisoesilo S, Kuntadi S (2007) Local wisdom in "beekeeping" of a forest honeybee (*Apis dorsata*). Forest Department, Research and Development of Forestry, Center Research and Development for Forestry and Conservation (translated)

Hadisoesilo S, Otis GW (1996) Drone flight times confirm the species status of *Apis nigrocincta* Smith, 1861 to be a species distinct from *Apis cerana* F in Sulawesi, Indonesia. Apidologie 27:361–369

Hadisoesilo S, Meixner M, Rutter F (1999) Geographic variation within *Apis koschevnikovi* Buttel-Reepen, 1906 in Borneo. Treubia 31:305–311

Hill R (2009) Southeast Asia's changing palaeogeography. Blumea 54:148–161

Kahono S (2000) Bees and wasps as pollinators at tropical rain forest of Gunung Halimun National Park and their distributions in Indonesia. National seminar on utilization of natural bioresources and its management. Faculty of Biology-Chrystian University of Satyawacana, Salatiga (translated)

Kahono S (2001) Roles and problems of insect pollinators in Indonesia. Fauna Indonesia 5(2):9–16 (translated)

Kahono S (2009) Initial studies of the function of Mount Ciremai as an important habitat for bumblebee, *Bombus rufipes* Lepeletier, 1836 (Hymenoptera: Apidae). J Biologi Indonesia 5(3):387–390 (translated)

Kahono S (2015) Development of LIPI beekeeping models for education, ecotourism, and production that can be implemented to the community. Technical Report of LIPI Features Activities 2015 (translated)

Kahono S, Amir M, Nakamura K (1999) Seasonal migration and colony behavior of the tropical honeybee *Apis dorsata* F (Hymenoptera: Apidae). Treubia 31(3):283–297

Kahono S, Lupiyaningdyah P, Erniwati H et al (2012) Potency and utilization of insect pollinators to increase palm oil production in Api-Api village, district of Sumur, Penajam Paser Utara, Kalimantan Timur. Zoo Indonesia 21(2):23–34 (translated)

Kato M, Salmah M, Nagamitsu T (1992) Colony cycle and foraging activity of a tropical-montane bumblebee, *Bombus rufipes* (Hymenoptyera, Apidae) in South Asia. In: Nakamura K, Katakura H (eds) Evolutionary biology and population dynamics of herbivorous ladybeetles in Indonesia. Nakamura K and Katakura H, Sapporo, pp 81–90

Koeniger N, Koeniger G (1980) Observations and experiments on migration and dance communication of *Apis dorsata* in Sri Lanka. J Apic Res 19:21–34

Mashudi, Patra K, Suwanda O (1988) Honey bees, bees honey in Indonesia Year 2000. PAP, Jakarta, p 268 (translated)

Michener CD (1974) The social behavior of the bees: a comparative study. Harvard University Press, Cambridge

Michener CD (2007) The bees of the world, 2nd edn. The Johns Hopkins University Press, Baltimore

Michener CD (2013) The Meliponini. In: Vit P, Pedro SRM, Roubik DW (eds) Pot-honey: a legacy of stingless bees. Springer, Berlin, pp 3–17

Nagir MT, Atmowidi T, Kahono S (2016) The distribution and nest-site preference of *Apis dorsata binghami* at Maros forest, South Sulawesi. Indonesia J Insect Biodivers 4(23):1–14

Otis GW (1991) A review of the diversity of species within *Apis*. In: Smith DR (ed) Diversity in the genus *Apis*. Westview Press, Boulder, pp 29–49

Otis GW (1996) Distributions of recently recognized species of honey bees (*Apis* spp.) in Asia. J Kansas Entomol Soc 69(4):311–333

Paton DC (1993) Honeybees in the Australian environment. Bioscience 43:95–103

Radloff SE, Hepburn C, Hepburn HR et al (2010) Population structure and classification of *Apis cerana*. Apidologie 41(6):589–601

Radloff SE, Hepburn HR, Engel MS (2011) The Asian species of *Apis*. In: Hepburn HR, Radloff SE (eds) Honeybees of Asia. Springer, Berlin, pp 1–22

Rasmussen C (2008) Catalog of the Indo-Malayan/Australasian stingless bees (Hymenoptera: Apidae: Meliponini). Zootaxa 1935:1–80

Rasmussen C, Thomas JC, Engel MS (2017) A new genus of Eastern Hemisphere stingless bees (Hymenoptera: Apidae), with a key to the supraspecific groups of Indomalayan and Australasian Meliponini. Am Mus Novitates 3888(3888):1–33

Reddy C (1980) Studies on the nesting behavior of *Apis dorsata* F. Int Conf Apic Trop Climate 2:391–397

Reddy SM, Reddy CC (1993) Studies on the distribution of nests of giant honey bee (*Apis dorsata* F). Indian Bee J 55:36–39

Rehan SM, Richards MH, Schwarz MP (2009) Evidence of social nesting in the *Ceratina* of Borneo (Hymenoptera: Apidae). J Kansas Entomol Soc 82(2):194–209

Ruttner F (1988) Biogeography and taxonomy of honeybees. Springer, Berlin

Ruttner F, Kauhausen D, Koeniger N (1989) Position of the red honey bee, *Apis koschevnikovi* (Buttel-Reepen, 1906), within the genus *Apis*. Apidologie 20:395–404

Saepudin R (2011) Productivity of a honey bee (*Apis cerana*) in integrated agricultural system of coffee plantation. Dissertation. Graduate School of IPB (translated)

Sakagami SF, Matsumara T, Ito K (1980) *Apis laboriosa* in Himalaya, the little known world largest honeybee (Hymenoptera, Apidae). Insect Matsumurana 19:47–77

Sakagami SF, Inoue T, Salmah S (1990) Stingless bees of central Sumatra. In: Sakagami SF, Ohgushi R, Roubik DW (eds) Natural history of eusocial wasps and bees in Equatorial Sumatra. Hokkaido University Press, Sapporo, pp 125–137

Salmah S, Inoue T, Sakagami SF (1990) An analysis of apid bee richness (Apidae) in Central Sumatra. In: Sakagami SF, Ogushi R, Roubik DW (eds) Natural history of eusocial wasps and bees in Equatorial Sumatra. Hokkaido University Press, Sapporo, pp 139–174

Sihombing DTH (1997) The science of honey beekeeping. Gadjah Mada University Press, p 243 (translated)

Suriawanto N, Atmowidi T, Kahono S (2017) Nesting sites characteristics of stingless bees (Hymenoptera: Apidae) in Central Sulawesi, Indonesia. J Insect Biodivers 5(10):1–9

Tim Ekspedisi (2016) Mount Gandang Dewata, Mamasa, Mamuju: Diversity of flora, fauna, micobes, and bioresources utilization in kabupaten Mamasa, Sulawesi Barat. Ekspedition Report of Deputy of Life Sciences – LIPI (translated)

Tingek S, Koeniger G, Koeniger N (1996) Description of a new cavity dwelling species of *Apis* (*Apis nuluensis*) from Sabah, Borneo with note on its occurrence and reproductive biology (Hymenoptera, Apoidea, Apini). Senckenberg Biol 76:115–119

Wille A, Michener CD (1973) The nest architecture of stingless bees with special reference to those of Costa Rica (Hymenoptera: Apidae). Rev Biol Trop 21(Suppl 1):1–278

Chapter 14
Management and Conservation of Philippine Bees

Cleofas R. Cervancia

Abstract This chapter highlights the status and trends in bee research and development in the Philippines. The diversity of bee species, bee plants, and management of stingless bees for pollination, production of bee products, and threats to beekeeping were discussed. This focuses on studying the biology of native honey bee species including stingless bees to promote sustainability of beekeeping and pollination services. In order to support to the beekeeping industry, the Philippine National Standards for Code for Best Beekeeping Practices and Honey were established by the Bureau of Agriculture and Fishery Standard.

Keywords *Apis cerana* · *Apis dorsata* · *Apis breviligula* · *Apis andreniformis* · *Apis nigrocincta* · Stingless bees · Solitary bees · *Aethina tumida* · *Varroa* spp. · *Tropilaelaps* spp.

14.1 Introduction

The Philippines is home to diverse species of social and solitary bees. Of the nine species of honey bees in the world, five are native to the Philippines, namely: *Apis cerana, Apis dorsata, Apis breviligula, Apis andreniformis,* and *Apis nigrocincta* (Cervancia 2003a). Two species, *A. andreniformis* and *A. dorsata* are found only on the island of Palawan, while *A. nigrocincta* was observed in Mindanao by Damus and Otis (1997). There are at least 260 identified pollinator species that belong to non-*Apis* groups (Baltazar 1966). *Apis mellifera* is used in commercial beekeeping; however, being an exotic species, it is more susceptible to microbial diseases (American foulbrood, European foulbrood, and Chalkbrood), parasitic mites (*Varroa* spp. and *Tropilaelaps* spp.), bird predation, and small hive beetle infestation. In 2014, the small hive beetle entered the Southern Philippines (Mindanao) and wiped out 80% of the *A. mellifera* colonies (Cervancia et al. 2016).

C. R. Cervancia
Institute of Biological Sciences, University of the Philippines Los Banos, College, Laguna, Philippines
e-mail: crcervancia@up.edu.ph

P. Chantawannakul et al. (eds.), *Asian Beekeeping in the 21st Century*,
https://doi.org/10.1007/978-981-10-8222-1_14

Honey bee research in the Philippines is interdisciplinary, covering bee biology, genetics, pathology, chemistry, apitherapy, and biomathematics. The University of the Philippines Los Banos (UPLB) is the only institution in the Philippines doing relevant research on bees. Its bee program focuses on native bee species. The technologies developed for the propagation of stingless bees for pollination and production of bee products are milestones. Based on the benefit-cost ratio, propagation of stingless bees is commercially viable (Locsin et al. 2017). The Philippine National Standards for Honey and Code of Best Beekeeping Practices are now being implemented in country. This chapter discusses the diversity of Philippine bee species and advances in bee research development and extension and strategies for conservation of native bee species.

14.1.1 Diversity

14.1.1.1 *Apis cerana* Fabr

Based on morphometric analysis, Tilde et al. (2000) reported that there are at least four populations of *A. cerana* in the Philippines (Fig. 14.1). The bees from Palawan were unequivocally distinct and separate from other Philippine samples. The bees of Luzon were different from those of Visayas and Mindanao and within Luzon, the bees from the highlands differed clearly from the lowlands. Based on neighbor-joining and parsimony analyses, Smith et al. (2000) found five major groups of haplotypes: an Asian mainland group, a Sundaland group, a Palawan group, Luzon Mindanao group, and *A. nigrocincta*; however, De la Rua et al. (2000) found only four haplotypes that discriminate among the bee populations from the different islands.

14.1.1.2 *Apis andreniformis* Smith

De Guzman et al. (1992) confirmed the presence of *A. andreniformis* in the Philippines. The species is found only on the island of Palawan. Its morphology was compared with *A. florea* collected from Thailand. Worker bees of *A. andreniformis* are generally black in color with black hairs on the hind tibia. In contrast, *A. florea* is red-brown with white tibial hairs. The wing venation has a much larger cubital index than *A. florea*. Unlike *A. florea*, *A. andreniformis* is highly defensive of its nest and performs shimmering waves when disturbed.

Apis nigrocincta Smith was first reported by Damus and Otis (1997) to be present on Mindanao and Sangihe, a small island between Mindanao and Sulawesi. The species was differentiated from *A. cerana* through morphometric analysis. In areas where *A. nigrocincta* and *A. cerana* live together, they can be differentiated by their color and size. *A. nigrocincta* is larger and yellowish in color, while *A. cerana* is smaller and darker in color.

Fig. 14.1 Map of *A. cerana* sampling locations. Number of samples is given in parentheses. Palawan—1: Brooke's Point (7); 2 Quezon (8); 3: Aborlan (7); 4: Puerto Princesa (11); 5: Roxas (6); 6: Tatay (3); 7: El Nido (4). Luzon northern lowland—8: Batac (5); 9: San Vincente (1). Luzon highland—10: Baguio City and region (8). Luzon southern lowland—11: San Jose City (1); 12: Laguna City and region (7). Visayas—13: San Joaquin (5); 14: Valencia (4); 15: Negros Occidental (1); 16: Argao (2); 17: Isabel (2). Mindanao—18: Ozamiz City (1); 19: Carmen City (2); 20: Bunawan (1); 21: Polomolok (1); 22: Panabo (3); 23: Kidapawan (3); 24: Tacurong (1); 25: Davao del Sur region (6); 26: General Santos City (1). (Adapted from Tilde et al. 2000)

14.1.1.3 *Apis breviligula* Maa

The giant honey bee, *A. breviligula* is found on the islands of Luzon (except Palawan), Visayas, and Mindanao. Based on DNA analysis, Lo et al. (2010) support Maa's classification of species status for *A. breviligula* separate from *A. dorsata*. There is a geographical separation of the two species. *A. dorsata* is confined to Palawan while *A. breviligula* is found in all other areas in the Philippines. The color of *A. breviligula* is completely black with grey stripes while *A. dorsata* has distinct yellow stripes on the abdomen. The mating flight is between 18:00 and 18:15 in the evening. The honey produced by both species ranges from 40 to 60 kg per colony during the honey flow season (March–May). Colting and Cervancia (2004) observed that the dimension of giant bee nests ranged from 30 to 200 m long and 30–150 m wide that were inclined from 10 to 50°.

14.1.1.4 Species of Wild Pollinators

Baltazar (1966) published a Catalogue of Philippine Hymenoptera to which major species of pollinators belong. In the list, there are Colletidae (genus *Hylaeus*, 15 species), Halictidae (genus *Halictus* 44 species; genus *Sphecodes* 7 species; genus *Nomiodes* 7 species; genus *Nomia* 25 species); Melittidae (Ctenoplectra 1 species, *C. vagans*), Megachilidae (genus *Lithuge* 1 species *L. scabrosus*; genus *Dianthidium* 1 species, *D. minutissimum*; genus *Stelis* 1 species, *Stelis* sp.; genus *Parevaspis* 3 species; genus *Heriades* 2 species; genus *Megachile* 29 species; genus *Coelioxys* 10 species), Anthoporidae (genus *Nomada* 11 species; genus *Anthophora* 9 species; genus *Thyreus* 9 species), subfamily Xylocopinae (genus *Allodape* 8 species; genus *Ceratina* 15 species; genus *Xylocopa*, 49 species), and family Apidae (genus *Bombus* 6 species; genus *Trigona* 7 species). In addition, there are at least seven species of stingless bees, but the only domesticated species are *Tetragonula biroi*, *T. laeviceps*, and *T. iridipennis*. In large-scale crop pollination, *T. biroi* is used because it can easily be managed (Cervancia et al. 2012).

14.2 Historical Records and Development of Beekeeping

Filipinos have long been keeping native bees, *A. cerana*, in earthen pots and hollow logs. Harvesting honey from wild colonies is practiced by indigenous peoples where the hunters perform rituals before harvesting the combs, in order to be protected from bee stings. Larvae and pupae are commonly cooked and eaten by members of hunter's family. Honey was used in traditional medicine to cure coughs, colds, and wounds. It is still a common belief that when wild bees nest on your house, it will bring good luck to the family.

Modern beekeeping using *A. mellifera* began in 1913 when colonies were introduced by an American named Schultz (Otanes 1926; Gabriel 1986). A breakthrough in bee health research was the identification of bee mites, *Varroa jacobsoni* (now known as *Varroa destructor*) by Delfinado (1963; Delfinado and Baker 1974). The mites associated with *A. cerana* and *A. mellifera* in the Southern Tagalog region were identified by Malabanan and Corpuz-Raros (1998).

Bee research started in 1968 when Dr. Roger Morse from Cornell University, USA joined the Department of Entomology, University of the Philippines Los Banos (UPLB) in collaboration with Dr. Francisco Laigo, a Professor at UPLB. They published an Extension Bulletin on Beekeeping (Morse and Laigo 1968a), and investigations of predatory birds (Morse and Laigo 1968b) and bee mites (Laigo and Morse 1968) were carried out. Morse and Laigo (1969a, b) published a monograph on *Apis dorsata,* (which is now identified as *A. breviligula*). They mapped the distribution pattern of *A. breviligula* in Mt. Makiling. In 2004, Manila-Fajardo et al. surveyed the same area in Mt. Makiling and observed that the number of nests had decreased and occupied the lower altitudes. Some aspects of the defensive behavior of *A. breviligula* were observed by Woyke et al. (2006) on a nest in Cavite City, Luzon island. Flipping of the abdomen was correlated with the temperature of the nest curtain. This behavior was compared with *A. laboriosa* and *A. dorsata* in India (Woyke et al. 2007). During mating flights, it was established that the proportion of the workers accompanying the drones is higher than the number of drones. This is a strategy to protect the drones from predators (Woyke et al. 2005). In 1980s, a noted botanist, Dr. Pacifico Payawal started melissopalynological studies. This led to the identification of pollen sources for *A. mellifera* (Payawal 1984; Payawal et al. 1986, 1991a, b; Tilde and Payawal 1987, 1992a, b), *A. cerana* (Tilde et al. 2003), and *A. dorsata* (Manila-Fajardo and Gonzales 2010). Pollination studies progressed, with consideration of pollinator–plant interactions. These include pollination of mango (Fajardo et al. 2008), bitter gourd (Deyto and Cervancia (2009), cucumber (Bergonia and Cervancia 1992; Cervancia and Forbes 1993a, b), passion fruit (Rodriguez and Cervancia 1999), Brassica (Rubin-Reyes and Cervancia 1999), calamondin (Manila-Fajardo et al. 2003), coffee (Manila-Fajardo 2011), reforestation species (Escobin and Cervancia 1999; Escobin et al. 2004), and mangrove species (Almazol and Cervancia 2013, 2014; Cervancia and Almazol 2014; Almazol et al. 2014). There was a variation in the foraging patterns among bee species (Cervancia et al. 1994; Barile and Cervancia 1995; Forbes and Cervancia 1993). The colony establishment is also dependent on the type of ecosystem (Manila-Fajardo and Cervancia 2003).

Morphometric diversity studies followed in which Tilde et al. (2000) identified various populations of *A. cerana* in the Philippines. This study was complemented with mtDNA analysis (Smith et al. 2000; De la Rua et al. (2000). Other studies on diversity using isozymes were those of Tenorio et al. (2002) and Tendero et al. (2007).

Proliferation of adulterated honey was a setback to the industry, thus studies on characterization of honey were done. Honey sold in known supermarkets was not all authentic (Tilde and Payawal 1992a; See et al. 2011a). To establish standards for honey, honeys from different species of bees were compared (See et al. 2011b) and used in the establishment of Philippine National Standards for Honey (PNS/BAFS 2016a). Another hive product of prime importance in medicine is propolis, and its properties were elucidated by Matienzo-Macale et al. (2011), Alvarez et al. (2013) and Belina-Aldemita et al. (2013). Apitherapy research progressed starting with the testing of honey and propolis alginate against microorganisms and stroke in mice (Collantes et al. 2016; De Guzman et al. 2016; Desamero et al. 2016; Ang et al. 2016). Because of the medicinal properties of bee products, the protocol for production of medical grade honey has been proposed (de Guzman 2017). The protocol harmonizes with the Philippine National Standard for Code for Best Beekeeping Practices (PNS/BAFS 2016b).

With the problems associated with *A. mellifera* in the Philippines, especially from mites (Beaurepaire et al. 2015; Anderson et al. 2004; Cervancia and Fajardo 2002), American foulbrood disease (Cervancia et al. 2013; Montecillo et al. 2014) and most recently, the small hive beetle (Cervancia et al. 2016), the country focused more on conservation and harnessing the potential of native bees for pollination (http://teca.fao.org/technology). Mitigating measures against diseases and pests include the shook swarm method (Fajardo and Cervancia 2011), use of probiotics (Montecillo et al. 2014), and irradiation to eradicate microbial pathogens (de Guzman et al. 2011). With the observed resistance against commercial miticide, formic acid combined with sepium was found effective under low temperature (Sison 2016).

To facilitate decision-making in utilizing bees for pollination, mathematical models were devised for foraging behavior and determining the optimal location of bee hives in a farm requiring pollination services (Rabajante et al. 2009; Esteves et al. 2010; Tambaoan et al. 2011; Ciar et al. 2013; Gavina et al. 2014; Jatulan et al. 2015; Real et al. 2016).

14.3 Bee Forage Resources

The bee plants of the Philippines were listed in Table 14.1. The most important pollen and nectar sources are *Cocos nucifera, Mimosa pudica, Mimosa diplotricha, Leucaena leucocephala, Melaleuca citrina, Zea mays, Pterocarpus indicus,* and *Muntingia calabura.* Actual observation of foraging behavior of bees is complemented with pollen analysis in honey and bee bread gathered from combs. Except for *P. indicus,* which flowers only from February to May, all the mentioned plants bloom year-round.

Table 14.1 Common bee plants in the Philippines (based on pollen grains identified in honey and bee bread). Blooming period may vary with locality

Scientific name	Common name	Blooming period
Thunbergia grandiflora (Rox.ex Rottler) Roxb.	Skyflower	All year round
Alternanthera ficoidea (L.) Sm.	Joyweed	All year round
Amaranthus spinosus L.	Spiny amaranth	All year round
Amaranthus viridis L.	Green amaranth	All year round
Celosia argentea L.	Cock's comb	All year round
Anacardium occidentale L.	Cashew	December–February
Mangifera indica L.	Mango	November–May
Centella asiatica (L.)	Centella	All year round
Cascabela thevetia (L.)	Yellow oleander	December–February
Ilex sp.	Holly	March–May
Actinophleus macarthuri (Wendl.) Becc.	Mac Arthur palm	All year round
Adonidia merrillii Becc. (Manila palm)	Manila palm	All year round
Aegiceras floridum Roem. & Schult.	Black mangrove	February–June
Areca catechu L.	Betel nut palm	All year round
Elaeis guineensis (Jacq.)	African oil palm	All year round
Dypsis lutescens (Wendl.) Beentje and Dranst	Golden cane palm	All year round
Cocos nucifera L.	Coconut	All year round
Heterospathe philippinensis (Becc.)	sagisi palm	All year round
Martinezia caryotaefolia Kunth	Fishtail palm	All year round
Nypa fruticans Wurmb	Nipa palm	All year round
Asclepias curassavica L.	Milkweed	All year round
Ageratum conyzoides L.	Goat weed	All year round
Bidens pilosa	Beggarticks	October–November

Table 14.1 (continued)

Scientific name	Common name	Blooming period
Cosmos caudatus Kunth	Cosmos	All year round
Emilia sonchifolia (L.) DC	Lilac tassel flower	October–April
Helianthus annuus L.	Sunflower	October–May
Sonchus oleraceus L.	Sow thistle	All year round
Tagetes erecta	Marigold	All year round
Tridax procumbens	Wild daisy	All year round
Tithonia diversifolia	Wild sunflower	October–March
Vernonia cinerea	Bitter leaf	All year round
Wedelia biflora (L.)DC	Creeping oxeye	All year round
Impatiens sp.	Touch-me-not	All year round
Begonia sp.	Begonia	All year round
Alnus japonica (Thunb.) Steud.	Alder	April–November
Trichodesma sp.	Borage	All year round
Canarium ovatum Engl.	Pili nut	All year round
Campanula sp.	Bell flower	All year round
Trema amboinensis (Willd.) Blume	Charcoal tree	All year round
Lonicera sp.	Honeysuckle	April–July
Carica papaya L.	Papaya	All year round
Casuarina equisetifolia L.	Agoho	All year round
Terminalia catappa L.	Beach almond	All year round
Commelina diffusa (Linn.)	Alibangon, kulasi	All year round
Ipomoea aquatic Forssk.	Kangkong	All year round
Ipomoea batatas (L.) Lam.	Sweet potato	All year round
Ipomoea nil (L.) Roth	Ivy, morning glory	All year round
Ipomoea triloba L.	Morning glory	All year round
Merremia sp.	Bulak-bulakan	All year round
Brassica juncea (L.) Czem	Mustard	All year round
Raphanus sativus (L.)	Radish	All year round
Cucumis melo L.	Watermelon	March–May
Cucumis sativus L.	Cucumber	All year round
Cucurbita maxima Lam.	Squash	All year round
Lagenaria siceraria (Molina) Standl.	Bottle gourd	All year round
Luffa cylindrica Roem.	Sponge gourd	All year round
Momordica charantia L.	Bitter gourd	All year round
Sechium edule (Jacq.)	Chayote	All year round
Trichosanthes sp.	Snake gourd	All year round
Dillenia philipinensis Rolfe	Catmon	August to December
Codiaeum variegatum (L.) Juss.	Croton	All year round
Excoecaria agallocha L.	Blinding tree	March–August
Hevea brasiliensis Müll. Arg.	Rubber	All year round

(continued)

Table 14.1 (continued)

Scientific name	Common name	Blooming period
Macaranga tanarius (L.) Müll. Arg.	Rubber tree	January to April
Manihot esculenta Crantz	Cassava	September–October
Aeschynomene sp.	Joint vetch	All year round
Calopogonium mucunoides Desv.	Wild ground nut	December–March
Cassia siamea Lam.	Yellow cassia	June–December
Centrosema pubescens Benth.	Butterfly peas	December–March
Crotalaria sp.	Rattlebox	December–March
Erythrina sp.	Coral tree, dapdap	Whole year
Gliricidia sepium (Jacq.)	Kakawate, madre de cacao	December–April
Glycine max (L.) Merr.	Soybean	All year round
Intsia bijuga (Colebr.) Kuntze	Borneo teak	All year round
Leucaena leucocephala (Lam.) de Wit	Ipil-ipil	All year round
Mimosa diplotricha C. Wright	Sensitive plant, makahiyang lalaki	All year round
Mimosa pudica L.	Sensitive plant, makahiyang babae	All year round
Pithecellobium dulce (Roxb.) Benth.	Manila tamarind	October to April
Prosopis sp.	Mesquite	September–March
Pterocarpus indicus Willd.	Narra	April–May
Tamarindus indicus L.	Tamarind, sampaloc	April–October
Vitex negundo L.	Lagundi	All year round
Vitex parviflora Juss.	Molave	May–October
Persea americana Mill.	Avocado	January–March
Lagerstroemia speciosa (L.) Pers.	Banaba	May–August
Xylocarpus granatum Koen.	Cedar mangrove	January–November
Lansium domesticum Corr.	Lansones	March–May
Moringa oleifera Lamk.	Moringa	January to May
Muntingia calabura L.	Aratiles	All year round
Musa sapientum L.	Banana	All year round
Aegiceras corniculatum (L.) Blanco	Black mangrove	October to March
Callistemon citrinus (Curtis) Skeels	Bottle brush	All year round
Eucalyptus sp.	Eucalyptus	January–March
Psidium guajava L.	Guava	All year round
Syzygium sp.	Bintang	February–March
Syzygium cumini (L.) Skeels	Duhat	February–March

Table 14.1 (continued)

Scientific name	Common name	Blooming period
Syzygium samarangense (Blume) Merr. & L.M. Perry	Macopa	March–April
Averrhoa bilimbi L.	Camias, bilimbi	All year round
Averrhoa carambola L.	Carambola, star fruit	All year round
Passiflora edulis Sims	Passionfruit	All year round
Antidesma bunius (L.) Spreng.	Bignay	September–October
Zea mays L.	Corn	All year round
Imperata cylindrica (L.) Raeusch.	Cogon grass	All year round
Antigonon leptopus Hook. & Arn.	Cadena de amor	All year round
Rhizophora mucronata Lam.	Red mangrove	August–December
Coffea spp.	Coffee	January–June
Scyphiphora hydrophyllacea C.F.Gaertn.	Nilad	February–June
Citrofortunella microcarpa (Bunge) Wijnands	Calamansi	All year round
Citrus grandis (L.) Osbeck	Pomelo	All year round
Mimusops elengi L.	Cherry	All year round
Pouteria campechiana (Kunth) Baehni	Tiesa	October–February
Nephelium lappaceum L.	Rambutan	March–May
Brugmansia suaveolens (Humb. & Bonpl. ex Willd.) Bercht. & C. Presl	Angel's trumpet	All year round
Capsicum annuum L.	Pepper	All year round
Nicotiana tabacum L.	Tobacco	All year round
Solanum wendlandi Hook.f.	Blue potato creeper	All year round
Phyla nodiflora (L.) Greene	Cape vine	All year round

14.4 Bee Pests

14.4.1 Parasitic Mites

Varroa jacobsoni Oudemans and *Varroa destructor* Anderson and Trueman are also found in the Philippines. The commonly used miticides are fluvalinate and coumaphos. Due to observed resistance of the bees to commercial miticides, formic acid is used for mite control. However, at higher temperature, broods die and the queen stops laying in treated colonies. Application is therefore carried out only during cooler months (December to February). *Tropilaelaps* was first reported in the Philippines by Delfinado and Baker (1961) on *A. mellifera* and also found to infest *A. breviligula* (Laigo and Morse 1968; Anderson and Morgan 2007). Concurrent infestation by *Tropilaelaps* mites *and V. destructor* was observed on *A. mellifera* (Cervancia 2003b). The miticides for *V. destructor* are also used to control *T. clareae.*

14.4.2 Small Hive Beetles

In June 2014, the small hive beetle (SHB; *Aethina tumida* Murray) was first detected in *A. mellifera* colonies in Lupon, Davao Oriental, Philippines. It is not known how the beetles were introduced. Traps, good hygienic practices, and mite management are recommended to control SHB.

14.4.3 Predatory Birds

The species that are major threats to beekeeping are *Merops philippinus, Merops viridis*, *Chaeturia gigantean dubia*, and *Hirundapus celebensis*. The birds forage on the apiaries from September to May. Losses due to bird predation are estimated to be around 40%. Beekeepers disperse the hives in groups of 10–20 in order to make the apiary less attractive to the birds (Cervancia et al. 1999).

14.4.4 Wax Moth

The lesser wax moth (*Achroia grisella* F) and greater wax moth (*Galleria mellonella*) are also present in apiaries. The wax moths infest the stickies during storage. To kill the immature stages of moths, stickies are exposed to sunlight for 30 min, and later wrapped in newspaper before storage.

14.4.5 Wasps (Vespa spp.)

High wasp infestations cause abandonment of colonies. Mechanical destruction of wasps in the apiaries is practiced. Reducing hive entrances is one way of preventing the entry of the wasp in the hive. However, the wasps are seen flying near the hive and catching bees entering the hives.

14.4.6 Frogs

Three species of frogs are observed preying on bees at night: *Bufo marinus, Rana catesbeiana,* and *Rana erythraea.* To prevent predation, the height of hive stand is elevated to 60 cm above the ground.

14.5 Bee Diseases

The most common bee diseases such as American foulbrood (AFB), European foulbrood (EFB), and Chalkbrood disease are observed in almost all *A. mellifera* colonies. Infection of AFB has been significantly reduced due to proper treatment and management. There are two reported virus diseases in *A. mellifera* and *A. cerana* in some areas of Luzon. These are the Deformed Wing Virus (DWV) and Aphid Lethal Paralysis Virus (ALPV). Interestingly, the viruses were not observed in areas without *A. mellifera* colonies (Beaurepaire and Moritz 2014). The microbial diseases and their control measures are as follows:

AFB is one of the most threatening epidemic brood diseases of *A. mellifera* in the Philippines. Infected colonies are burned, and the use of antibiotics is not allowed. In case of mild infection, the "shook swarm" method is used. European foulbrood (EFB) can be found in all areas where *A. mellifera* hives are kept. Like in AFB, control is through the "shook swarm" method. Chalkbrood disease is prevalent during cold months and occurs in weak colonies. Management of the disease includes removing of the mummies from the bottom board, strengthening of the colony by providing adequate food, and re-queening if the queen is failing.

14.6 Future Prospects of Bee Research, Development, and Extension

Studies on different species of bees, especially the solitary bees, are still lacking. Our ongoing research on genetic diversity and population dynamics of these species will contribute to the global assessment of bee and other pollinator populations. Conservation strategies can be crafted based on the knowledge of bee biology. Stewardship programs designed for an agroecosystem will warrant pollinator protection. The current government program of bee management training aims to disseminate science-based beekeeping technologies.

References

Almazol AE, Cervancia CR (2013) Floral biology and pollination of three mangrove species (*Aegiceras floridum* Roem. & Schults., *Scyphiphora hydrophyllacea* Gaertn. f, and *Xylocarpus granatum* Koen.) in Pagbilao mangrove forest, Quezon province, Philippines. J Nat Stud 12(2):39–47

Almazol AE, Cervancia CR (2014) Pollen sources of bees in Pagbilao mangrove ecosystem. Philipp Ent 28(2):155–170

Almazol AE, Cervancia CR, Buot IE et al (2014) Floristic composition and physiognomy of Pagbilao mangrove, Quezon province, Philippines. IAMURE Int J Ecol Conserv 7(1). https://doi.org/10.7718/ijec.v7i1.725
Alvarez PL, Micor JR, Hizon-Fajardo AB et al (2013) Identification of flavonoids and phenolic compounds in propolis from stingless bee (*Tetragonula biroi* Friese). Philipp Ent 27(1):91–99
Anderson DL, Cervancia CR, Sito A et al (2004) Varroa mites and their host relationships in the Philippines. In: Camaya EN, Cervancia CR (eds) Bees for new Asia. Proceedings of the 7th Asian apicultural association conference and 10th bee net symposium and technofora, pp 177–178
Anderson DL, Morgan MJ (2007) Genetic and morphological variation of bee parasitic *Tropilaelaps mites* (Acari:Laelapidae): new and re-defined species. Exp Appl Acarol 43(1):1–24
Ang MJC, Virtucio RCP, Collantes TMA et al (2016) Healing activity of gamma-irradiated propolis-and honey-alginate dressing from the Philippine stingless bees (*Tetragonula biroi* Friese) on mouse model of incisional wound. In: Paper presented at the apimondia Philippines symposium and workshop on indigenous bees, Tagaytay City, Philippines, 1–4 Feb 2016
Baltazar CR (1966) A catalogue of Philippine hymenoptera. Pacific Insects Monogr 8:1–488
Barile GE, Cervancia CR (1995) Foraging and nesting behavior of *Trigona biroi* Friese (Hymenoptera:Apidae). Philipp Ent 9(6):555–556
Beaurepaire AL, Moritz RFA (2014) Honeybees health in Asia. Legato meeting W.P. Biocontrol and pollination (www.legato-project.net)
Beaurepaire AL, Truong TA, Fajardo AC et al (2015) Host specificity in the honeybee parasitic mite, *Varroa* spp. in *Apis mellifera* and *Apis cerana*. PLoS One 10(8):e0135103. https://doi.org/10.1371/journal.pone.0135103
Belina-Aldemita MD, Lechuga JA, Mateo JMC et al (2013) Detection of phenolic compounds in stingless bee (*Tetragonula Biroi Friese*) propolis and five tree sources using tandem liquid chromatography-mass spectrometry. J Nat Stud 12(2):30–38
Bergonia EA, Cervancia CR (1992) Comparison of hand and honey bee pollination of cucumber (*Cucumis sativus* L). Philipp J Sci 121(3):255–262
Cervancia CR (2003a) Philippines: haven for bees. Honeybee Sci 24(3):129–134
Cervancia CR (2003b) Population dynamics of beemites *Varroa destructor* Anderson and Trueman and *Tropilaelaps clareae* Delfinado and Baker. Metro Manila Commission Professorial Chair Lecture, 22 January 2003. University of the Philippines Los Banos
Cervancia CR, Almazol AE (2014) Foraging behavior of floral visitors of three mangrove species (*Aegiceras floridum* Roem. & Schults., *Scyphiphora hyrophyllacea* Gaertn. f., and *Xylocarpus granatum* Koen.) in Pagbilao Mangrove forest, Quezon province. Philip Ent 90–98
Cervancia CR, Fajardo AC (2002) Competitive exclusion between *Varroa destructor* Anderson and Trueman and *Tropilaelaps clareae* Delfinado and Baker. In: Proceedings of the 6th Asian apicultural association conference, Bangalore, India, 24 Feb–2 March 2002
Cervancia CR, Forbes MF (1993a) Pollination of cucumber (*Cucumis sativus* L): density of bees needed for effective pollination. Philipp J Sci 122(1):129–132
Cervancia CR, Forbes MF (1993b) Pollination of pechay (*Brassica pekinensis* Rupr) and radish (*Raphanus sativus* L.) Philipp J Sci 122(1):129–132
Cervancia CR, Apacionado FF, Tilde AC (1994) Foraging behavior of three *Xylocopa* species in UP Los Banos vicinity. Philipp J Sci 123(2):155–160
Cervancia CR, Forbes MF, Alvaro RL (1999) Honey bee predation of *Hirundapis celebensis* (Sclater) and *Merops viridis americanus* (P.L.S.Muller). Philipp J Sci 128(1):15–19
Cervancia CR, Fajardo AC, Manila-Fajardo AC et al (2012) Management of Philippine bees. UP Los Banos, Laguna. ISBN 978-971-547-272-2
Cervancia CR, Fajardo AC, Sabino NG et al (2013) Prevalence of American foul brood (AFB) disease of honey bee, *Apis mellifera* (L) in the Philippines and its pathogenicity to *Apis cerana* Fabricius. Philipp Ent 27(1):75–90
Cervancia CR, De Guzman LI, Polintan EA et al (2016) Current status of small hive beetle infestation in the Philippines. J Apic Res. https://doi.org/10.1080/00218839.2016.1194053

Ciar RR, Bonto LS, Hale PM et al (2013) Foraging behavior of stingless bees, *Tetragonula biroi* Friese: distance, direction and height of preferred food. https://www.researchgate.net/publication/257811529

Collantes TMA, Bariuan JV, Desamero MJM et al (2016) Healing activity of gamma-radiated propolis-and honey-alginate dressing from the Philippine stingless bees (*Tetragonula biroi* Friese). Paper presented at apimondia Philippines symposium and workshop on indigenous bees. 1–4 Feb 2016. Taal Vista Hotel. Tagaytay City, Philippines

Colting LM, Cervancia CR (2004) Nesting behavior of giant honey bee (*Apis dorsata* Fabricius) in the Philippines. In: Camaya EN, Cervancia CR (eds) Bees for new Asia. Proceedings of the 7th Asian apicultural association conference and 10th Bee net symposium and technofora, pp 177–178

Damus M, Otis G (1997) A morphometric analysis of *Apis cerana* F. and *Apis nigrocincta* Smith populations from Southeast Asia. Apidologie 28(5):309–323. https://doi.org/10.1051/apido:19970507

De Guzman ZM (2017) Promotion and utilization of organic production system and irradiation technology in the production of safe and quality bee products. Paper presented at the 26th DA-BAR Technology Forum, Manila, Philippines, 9 August 2017

De Guzman LI, Forbes MF, Cervancia CR et al (1992) *Apis andreniformis* Smith in the Philippines. (Notes). J Apic Res 31(2):119

De Guzman ZM, Cervancia CR, Dimasuay KGB et al (2011) Radiation inactivation of *Paenibacillus larvae* and sterilization of American foul brood (AFB) infected hives using Co-60 gamma rays. Appl Radiat Isot 69:1374–1379

De Guzman ZM, Baldos DT, Asa AD et al (2016) Development of radiation-sterilized honey alginate wound dressing for exudating wounds. In: Paper presented at the apimondia Philippines symposium and workshop on indigenous bees. 1–4 Feb 2016 Taal Tagaytay City, Philippines, 1–4 Feb 2016

De la Rua P, Tilde AC, Simon U et al (2000) Mitochondrial DNA analysis of *Apis cerana* F. In the Philippines. Heredity 84:124–130

Delfinado MD (1963) Mites of the honey bee in Southeast Asia. J Apic Res 2(2):113–114

Delfinado MD, Baker EW (1961) *Tropilaelaps*, a new genus of mite from the Philippines (Laelapidae(s.lat.): Acarina). Fieldiana Zool 44(7):53–56

Delfinado MD, Baker EW (1974) Varroidae: a new family of mites of honey bees (Mesostigmata: Acarina). J Wash Acad Sci 64(1):4–10

Desamero MJM, Villablanca MA, Bariuan JV et al (2016) Honey and propolis abrogate neurologic deficit and neuronal damage in the hippocampus and cerebral cortex of ischemic stroke rats. Int J Appl Res Nat Prod 10(1):7–15

Deyto RC, Cervancia CR (2009) Floral biology and pollination of ampalaya, *Momordica charantia* L. Philipp Agric Sci 92(1):8–18

Escobin RP, Cervancia CR (1999) Insect floral visitors of four reforestation tree species in Mt. Makiling, Luzon, Philippines. Philipp J Sci 127(3):169–179

Escobin RP, Payawal PC, Cervancia CR (2004) Pollination syndrome of four reforestation tree species in Mt. Makiling, Luzon, Philippines. Philipp Agric Sci 87(2):182–190

Esteves RJP, Villadelrey MC, Rabajante JF (2010) Determining the optimal distribution using mixed integer programming. J Nat Stud 9(1):79–82

Fajardo AC, Cervancia CR (2011) Mitigating disease outbreaks in your apiary. The Shook Swarm Method. UPLB Bee Program Publication. 4p

Fajardo AC, Medina JR, Opina OS et al (2008) Insect pollinators and floral visitors of mango (*Mangifera indica* L. cv Carabao). Philipp Agric Scientist 91(4):372–382

Forbes MF, Cervancia CR (1993) Foraging behavior of *Apis cerana F.* and *Apis mellifera L.* in Majayjay, Laguna. Phil J Sci 122(1):129–132

Gabriel BP (1986) Prospects for industrial entomology in the Philippines. Philipp Ent 4(6):525–534

Gavina MZ, Rabajante JF, Cervancia CR (2014) Mathematical programming models for determining the optimal location of beehives. Bull Math Biol. https://doi.org/10.1007/s11538-014-9943-9

Jatulan EO, Rabajante JF, Banaay CGB et al (2015) A mathematical model of intra-colony spread of American foul brood in European honey bee (*Apis mellifera* L.) PLoS One 10(12):e0143805. https://doi.org/10.1371/journal.pone.0143805
Laigo FM, Morse RA (1968) The mites *Tropilaelaps clareae* in *Apis dorsata* colonies in the Philippines. Bee World 49:116–118
Lo N, Gloag RS, Anderson DL et al (2010) A molecular phylogeny of the genus *Apis* suggests that the giant honey bee of the Philippines, A. *breviligula* Maa, and the plains honey bee of southern India, *A. indica* Fabricius, are valid species. Syst Entomol 35:226–233
Locsin AA, Cuevas AC, Polintan EA et al (2017) Economic analysis of stingless bee production in the Philippines. Paper presented at the international meliponine scientific conference, Convention and Carnival, Malaysia 16–18 May 2017
Malabanan JM, Corpuz-Raros LA (1998) Mites associated with *Apis mellifera* Linnaeus and *Apis cerana* Fabricius in southern Tagalog region, Philippines. Philipp Ent 1(2):123–135
Manila-Fajardo AC (2011) Reproductive ecology of *Coffea liberica* w. Bull ex Hiern var. *liberica* in Lipa, Batangas, Philippines. Dissertation, Botany University of the Philippines Los Baños
Manila-Fajardo AM, Cervancia CR (2003) Performance of honey bees *Apis cerana* Fabricius in three ecosystems in Laguna. Philipp Agric Sci 86(2):146–157
Manila-Fajardo AC, Gonzales AKBM (2010) Pollen sources of *Apis dorsata* Fabr in Mainland, Palawan, Philippines. Philipp Ent 24(2):137–149
Manila-Fajardo AC, Pitargue FC, Cervancia CR (2003) Pollinators and floral characteristics of calamondin (x *Citrofortunella microcarpa* (Bunge) Wijjnands). Philip Agric Scientist 86(2):131–133
Matienzo-Macale A, Lamorena LM, Cervancia CR (2011) Characterization of propolis collected by stingless bee, *Trigona biroi* Friese. J Nat Stud 9(2)/10(1):73–76
Montecillo AD, Sabino NG, Fajardo AC et al (2014) Screening of lactic acid bacteria from *Apis mellifera* L. and *Trigona* spp. against *Paenibacillus larvae* (White) causing American foulbrood disease of honey bees. Philipp Ent 28(1):32–42
Morse RA, Laigo FM (1968a) Beekeeping in the Philippines. Farm Bull. 27. UP College of agriculture. p 56
Morse RA, Laigo FM (1968b) The Philippines spine-tailed swift, *Chaetura dubia* Mc Gregor. Philipp Ent 1(2):138–143
Morse RA, Laigo FM (1969a) *Apis dorsata* in the Philippines (including annotated bibiliography) Monograph No.1. Philipp Ent, p 96
Morse RA, Laigo FM (1969b) Control of beemites *Varroa jacobsoni* Oudemans and *Tropilaelaps clareae* Delfinado-Baker with chlorobenzilate. Philipp Ent 1(2):144–148
Otanes FQ (1926) Honey bee and how to raise them. Philipp Agric Res 9(2):149–173
Payawal PC (1984) Pollination of crop plants and the beekeeping option. Philipp Agric 75(3–4):131–144
Payawal PC, Tilde AC, Manimtim AL (1986) Year-round pollen sources of Italian honey bees (*Apis mellifera* L) in the Philippines I Los Banos area. Philipp Agric 69:217–225
Payawal PC, Tilde AC, Manimtim AL (1991a) Year-round pollen sources of Italian honey bee (*Apis mellifera* L) in the Philippines II Metro Manila. Philipp Ent 7:593–604
Payawal PC, Tilde AC, Manimtim AL (1991b) Year-round pollen sources of Italian honey bee (*Apis mellifera* L) in the Philippines III selected areas. Philipp Agric 74(4):503–509
PNS/BAFS (2016a) Philippine National Standard for code of good beekeeping practices, p 186
PNS/BAFS (2016b) Philippine National Standard for honey, p 185
Rabajante JF, Jacildo AJ, Figuera RB (2009) Modeling the area strict searching strategy of stingless bees (*Trigona biroi* Friese) as quasi random walk process. J Nat Stud 8(2):15–31
Real RCR, Jose EC, Rabajante JF et al (2016) A Mathematical model of emergent food site selection behavior of stingless bees (*Tetragonula biroi* Friese). Paper presented at the apimondia Philippines symposium and workshop on indigenous bees, Tagaytay City, Philippines, 1–4 Feb 2016

Rodriguez GA, Cervancia CR (1999) Insect pollination of passion fruit, *Passiflora edulis* sims var *flavicarpa* Degener. Philipp J Sci 128(2):120–124

Rubin-Reyes M, Cervancia CR (1999) Floral visitors and pollination of Chinese mustard, *Brassica campestris* L. Philipp J Sci 128(1):32–37

See MAD, Dorado SA, Manila-Fajardo AC et al (2011a) Physico-chemical characteristics of commercial honey in select supermarkets in Metro Manila, Philippines. Philipp Ent 25(2):187–198

See MAD, Manila-Fajardo AC, Fajardo AC, Cervacia C (2011b) Physico-chemical properties of Philippine honey and their implication in the establishment of standard for tropical honeys. Korean J Apic 26(1):45–48

Sison MLJ (2016) Efficacy of formic acid-gel-*Gliricidia sepium* formulation and fluvalinate against beemites, *Varroa destructor*. Paper presented at the apimondia Philippines symposium and workshop on indigenous bees. Tagaytay City, Philippines, 1–4 Feb 2016

Smith DR, Villafuerte LS, Otis GW et al (2000) Biogeography of *Apis cerana* F. and *A. nigrocincta* Smith: insights from mtDNA studies. Apidologie 31(2):265–280

Tambaoan RS, Rabajante JF, Esteves RJP et al (2011) Prediction of migration path of a colony of bounded-rational species foraging on patchily distributed resources. Adv Stud Biol 3(7):333–345

Tendero B, Gaza HL, Tandang RN (2007) Isozyme analysis of the genetic variation in populations of European honey bee, *Apis mellifera carnica* P from selected apiaries. Philipp Ent 21(2):159–172

Tenorio MLE, Laude RP, Tandang RN (2002) Esterase and glucose-6-$P0_4$ dehydrogenase isozymes at different developmental stages of the giant honey bee, *Apis dorsata* F. Philipp Agric Sci 85(3):295–298

Tilde AC, Payawal PC (1987) Note: a qualitative pollen analysis of honey samples. Philipp Agric 70:67–70

Tilde AC, Payawal PC (1992a) Commercial honey in the Philippines 1. Pollen grain analysis. Philipp Agric 75(1):81–87

Tilde AC, Payawal PC (1992b) Commercial honey in the Philippines 2. Physical and chemical analysis. Philipp Agric 75:89–91

Tilde AC, Fuchs S, Koeniger N et al (2000) Morphometric diversity of *Apis cerana* Fabricius with in the Philippines. Apidologie 31:249–263

Tilde AC, Payawal PC, Cervancia CR (2003) Pollen spectra among natural populations of *Apis cerana* Fabricus. Philipp Agric Sci 86(1):27–37

Woyke J, Wilde J, Wilde M et al (2005) Workers often predominate in 'dusk drone flights' of giant honey bee, *Apis dorsata*. J Apic Res 44(3):130–131

Woyke J, Wilde J, Wilde M et al (2006) Abdomen flipping of *Apis dorsata brevigulla* worker bees correlated with temperature of nest curtain surface. Apidologie 37:501–505

Woyke J, Wilde J, Wilde M et al (2007) Comparison of defense body movements of *Apis laboriosa*, *Apis dorsata dorsata* and *Apis dorsata breviligula* honey bees. J Insect Behav 21(6):481–494

Chapter 15
Conclusions and Future Perspectives

Panuwan Chantawannakul, Samuel Ramsey, Geoffrey Williams, and Peter Neumann

Abstract In this chapter, we briefly provide future directions for study that we believe will broadly benefit the bees and people of Asia and abroad. Asia maintains the highest biodiversity of native honey bee species on the planet, in addition to the introduced European honey bee. This intermingling of species has promoted a number of virus-host shifts that potentially threaten honey bees because of a lack of coevolution. This places Asia in a unique position—as a potential source of novel parasites and diseases to other parts of the globe, as well as a venue to study those parasites and diseases that have spread to the reaches of our planet in their native Asian honey bee hosts. Our knowledge of the coevolutionary arms race between Asian honey bees and their native parasites is sparse. Future efforts should be directed towards filling important gaps in our fundamental knowledge of honey bee defenses so that novel, sustainable management strategies can be developed.

Keywords *Apis* spp. · Honey bee · Parasite · Novel treatment strategies · *Neocypholaelaps* · *Paenibacillus larvae* · Small hive beetle · *Tropilaelaps* spp. · *Varroa* spp. · Virus

P. Chantawannakul (✉)
Department of Biology, Faculty of Science, Chiang Mai University, Chiang Mai, Thailand

S. Ramsey
Department of Entomology, University of Maryland, College Park, MD, USA

G. Williams
Department of Entomology and Plant Pathology, Auburn University, Auburn, AL, USA

P. Neumann
Institute of Bee Health, Vetsuisse Faculty, University of Bern, Bern, Switzerland

Agroscope, Swiss Bee Research Centre, Bern, Switzerland

P. Chantawannakul et al. (eds.), *Asian Beekeeping in the 21st Century*,
https://doi.org/10.1007/978-981-10-8222-1_15

The unparalleled honey bee biodiversity of Asia holds tremendous benefits for the region, but also poses a number of unique challenges. Uniting the genus *Apis* via the introduction of the European honey bee into Asia has created a context where inter-species disease transmission is common. Further research is paramount to maintain the health of this dynamic ecosystem in the face of constant ecological perturbation. Our understanding of the biology of honey bees (both endemic and introduced) is still lacking in key areas, as is our grasp of the unique challenges posed by these emerging host shifts. Throughout this book, we have discussed areas of study that are urgently needed. In doing so, we hope to promote greater efficiency in the scientific process and to encourage greater respect for an animal closely associated to humans—the honey bee. We summarize these priority areas here.

1. *Bee Parasites*

   Several honey bee parasites have only been studied in *A. mellifera*, despite their impact on other *Apis* species. Further attention should be given to the life-cycle of these parasites and to details of their biology that facilitate host shifts. It is also important to focus more attention on the less studied parasites of Asian honey bees that are currently affecting *A. mellifera*, or will most likely do so in the near future. Specifically, knowledge about the biology of *Varroa* spp., *Tropilaelaps* spp., and *Nosema ceranae*, as well as resistance mechanisms of Asian honey bees against these parasites, would benefit tremendously from more detailed study of their original virus-host relationships. Attention should also be given to the effects of lesser known parasites, such as *Neocypholaelaps*, as well as indirect impacts of insect pests such as wax moths and small hive beetles. The latter has just been introduced to Asia.
2. *Bee viruses*

   Details of the impact of honey bee viruses on bees in Asia are sparse, despite the ever-expanding distribution of these pathogens. Even less is known about their interaction with potential vectors, such as parasitic mites. Because most bee viruses are RNA viruses, their mutational rate is especially high, which could facilitate shifts among different host species. Genomic study on viral strains and variants in Asia is still in its infancy, but would shed light on virus-host interactions. It could also deepen our understanding of viral impacts within the genus *Apis*, and non-*Apis* genera too (e.g., *Bombus*). Furthermore, synergistic effects should be explored, particularly if honey bees are infected with more than one virus.
3. *Novel treatment strategies*

   Pesticide exposure is known to have profound lethal and sublethal impacts on pollinators. However, some of the most significant pesticide exposure to honey bees comes from our management of in-hive pests. The development of novel treatment strategies for both crop pests and honey bee diseases is integral to achieve sustainable populations of healthy honey bees. Pathogens desperately in need of effective non-chemical control measures include *Varroa* spp., *Tropilaelaps* spp., *Paenibacillus larvae,* and small hive beetle. Novel treatment may be developed from the diversity of experiences and traditional knowledge of local beekeepers with a rich history of beekeeping in the region.

In conclusion, the honey bees of Asia should be considered a valuable resource. Not only are they critical to regional biodiversity and food security, but they also provide opportunities to understand the coevolution of important introduced parasites of the European honey bee in their native hosts. The marvelous diversity of Asian honey bees, as well as our intriguing relationship with them over millennia, is something truly fascinating and warrants our warm embrace.

Kolibri 360 GmbH
Ferdinand-Jühlke-Straße 7
99095 Erfurt, Deutschland
produktsicherheit@kolibri360.de